Planning and Managing the Safety System

Planning and Managing the Safety System

Edited by
Ted S. Ferry and Mark A. Friend

Lanham • Boulder • New York • London

Published by Bernan Press
An imprint of The Rowman & Littlefield Publishing Group, Inc.
4501 Forbes Boulevard, Suite 200, Lanham, Maryland 20706
www.rowman.com
800-865-3457; info@bernan.com

Unit A, Whitacre Mews, 26-34 Stannary Street, London SE11 4AB

Copyright © 2017 by Bernan Press

All rights reserved. No part of this book may be reproduced in any form or by any electronic or mechanical means, including information storage and retrieval systems, without written permission from the publisher, except by a reviewer who may quote passages in a review.

British Library Cataloguing in Publication Information Available

Library of Congress Cataloging-in-Publication Data
Names: Ferry, Theodore S., editor. | Friend, Mark A., editor.
Title: Planning and managing the safety system / [edited by] Theodore S. Ferry and Mark A. Friend.
Description: Lanham : Bernan Press, [2017] | Includes bibliographical references and index.
Identifiers: LCCN 2016045136 (print) | LCCN 2016046091 (ebook) |
 ISBN 9781598887747 (cloth : alk. paper) | ISBN 9781598887754 (Electronic)
Subjects: | MESH: Safety Management
Classification: LCC R859.7.S43 (print) | LCC R859.7.S43 (ebook) | NLM WA 485 | DDC 610.28/9—dc23
LC record available at https://lccn.loc.gov/2016045136

∞™ The paper used in this publication meets the minimum requirements of American National Standard for Information Sciences—Permanence of Paper for Printed Library Materials, ANSI/NISO Z39.48-1992.

Printed in the United States of America

Contents

Preface		vii
Acknowledgments		ix
1	Planning and Managing Safety: A History *Mark A. Friend, Tracy L. Zontek, and Burton R. Ogle*	1
2	The Safety Function *Mark A. Friend*	17
3	Planning and Organizing the Safety System *Mark A. Friend*	35
4	Safety Management Framework *Kathy S. Friend and Mark A. Friend*	47
5	Directing the Safety System *Mark A. Friend*	67
6	Controlling the Safety Program *Mark A. Friend*	79
7	Staffing and Job Processes—A Safety Perspective *Mark A. Friend*	95
8	Safety and Health Training and Education *Mark A. Friend*	109
9	Ergonomics *Mark D. Hansen*	123

10	Process Safety Management *Mark D. Hansen*	165
11	Systems Safety Engineering *Henry A. Walters*	185
12	Accident Investigation *Mark A. Friend*	209
13	Financial Aspects for Safety and Health Professionals *Mark D. Hansen*	233
14	The Recordkeeping Function—A Legal Perspective *Michael O'Toole*	271
15	Industrial Hygiene Within the SMS *Tracy L. Zontek and Burton R. Ogle*	295
16	Radiation Safety *Tracy L. Zontek and Burton R. Ogle*	305
17	Employee Health and Wellness *Tracy L. Zontek and Burton R. Ogle*	319
18	Emergency Response Planning *Kimberlee K. Hall*	345
19	Hazardous Materials and Waste Management *Tracy L. Zontek and Kimberlee K. Hall*	379
Index		403
About the Authors		419

Preface

Working on this text has been a privilege. Theodore (Ted) Ferry was instrumental in helping me obtain my doctorate from West Virginia University (WVU). As a fellow West Virginian from nearby Clarksburg, Ted was directed to me by the chair of the Safety Studies Department at WVU, Dr. C. Everett Marcum. At the time Ted was working at the University of Southern California in their Aerospace Institute of Safety and Management. Dr. Marcum also helped me enlist some other premier safety professionals, including Fred Manuele, whom I have come to highly respect professionally and personally; Mr. William Tarrants, whom I became acquainted through a series of site visits with NIOSH; Frank Bird, Jr., whom I knew through interactions at various professional conferences; William Pope, another well-respected author of safety material; Dr. Julius Morris, professor at the University of Hawaii; and Jim Findlay, a Canadian engineer and safety expert. Discussing philosophies and beliefs with these safety giants greatly impacted my beliefs regarding safety and the overall approach I took. Hopefully, it has also had an impact on the many students I've taught at WVU, Murray State University, East Carolina University, and Embry-Riddle Aeronautical University.

I began using his text at WVU and again at Murray State. After reviewing as many texts as I could find on the topic of safety management, I realized that Ferry's text was, by far, superior to the rest in the field. I adopted the book and required it of all students in the MS program at Murray. After his passing, I was very disappointed at his absence and the fact that he was no longer available to revise the text. In the late 1990s Dr. Jim Kohn, a great friend and colleague, updated the book with some of the best minds in the field. The text he did is still relevant and helpful to all who use it. My personal approach was a little different. Ferry's original text emphasized safety management systems, but when the book was originally written, this concept was simply referred to as *system safety*. In this current rewrite, my colleagues and I have taken Ferry's approach and updated it to reflect the latest research and materials available in safety management systems (SMS). We tie much of the book to SMS and provide more of a management and technical safety background than is usually found in books or articles related to SMS.

The book initially provides a background on how the overall field of safety began abroad and eventually developed in the United States. Following that description is an overview of the field of safety, as it exists today. From there, we addressed the basic management functions, including planning, organizing, directing, controlling, and staffing—as they relate to SMS. All of this is placed within the safety management or SMS framework, as is addressed in various regulations and standards. Following the discussion of management applications, the book explores key safety topics from an SMS perspective, including accident investigation, recordkeeping, industrial hygiene, wellness, emergency response, hazardous materials, and radiation. These are areas often dealt with by the safety professional and members of the safety team. We really attempted to provide a document useful to current safety professionals and soon-to-be professionals in colleges and universities. Thank you Ted Ferry.

Acknowledgments

Many thanks to the authors who contributed to this book—in both its original and revised editions. Thanks also to my former professors and colleagues in the former Department of Safety Studies at West Virginia University and faculty at Murray State University and Embry-Riddle Aeronautical University. Special thanks also extend to Dr. James (Jim) P. Kohn, a former long-time friend and colleague at East Carolina University. Of course, I owe a debt of gratitude to Ted Ferry, who provided input and insight into my doctoral dissertation and wrote the early editions of this text. Sincere thanks and appreciation to my wife Kathy, who provides inspiration and support in all our professional endeavors.

Chapter 1

Planning and Managing Safety

A History

Mark A. Friend, Tracy L. Zontek, and Burton R. Ogle

CASE

Mike Williams investigated a fatality in his company where a worker was assigned to clean filters in a process involving relatively high temperatures. There were six filters accessed by a walkway on the first floor and overhead walkways on the second and third. The rest area and water fountains were on the first floor. The second two floors had no place to sit or take a break. The area of work was often over 100° F, so employees were instructed to take frequent breaks, and drink lots of fluids. It was near the end of the work shift on a hot summer day, and Jim Hunt, an older worker with a history of heart disease, was hurrying to complete his task. He completed the cleaning process for all eighteen filters and punched the time clock to leave at 3:30 p.m., right on time. When he approached the gate, he stopped his vehicle and slumped over the steering wheel. The guard was unable to revive him, so he called 911. Jim was pronounced dead on arrival at the local hospital. The cause of death was a heart attack. In spite of OSHA requirements, the accident was not reported, since the death occurred after work hours. Following the investigation, Mike Williams reported the fatality was after hours and completely the fault of the employee, since he did not follow instructions in terms of taking appropriate breaks and drinking lots of fluids.

INTRODUCTION

Safety and health programs are valuable and efficient business tools; however, the senior manager must understand what makes a safety and health program successful and how the program should be adjusted to fit the organization and situation. The background for that understanding is a vital link in creating a sound, intelligent executive approach to safety and health.

THE SAFETY AND HEALTH MOVEMENT

How concerned a country is about safety and health is closely linked to its level of industrialization, its type of government, and its social outlook. In countries where great value is placed on individual life, there is naturally a greater degree of concern. The safety movement in the United States illustrates this point very clearly. In many aspects of safety and health, there are countries which have long been the leaders—and in some cases, are thought to be ahead of the United States. Conversely, in some countries considered industrially well advanced, little has been done from the safety and health perspective. Among emerging nations there is everything from complete commitment to disdain for the very idea.

In several respects, Canada and certain European countries have many advanced approaches to safety and health and have implemented them more rapidly than the United States. Many countries have been so innovative and forward-looking that some of their concepts deserve universal acceptance. These differing levels of safety and health awareness evolve gradually, as the country itself evolves. The growth of the safety and health movement in the United States illustrates the process.

When considering the history and the evolution of occupational health and safety, it is important to consider the past and how it influences the present. This section presents the chronological evolution of the movement worldwide until the development of modern day legislation. In the United States and other developed countries, there are extensive occupational health and safety, legally enforceable regulations; however, in developing and transition countries, the evolution of occupational health and safety regulations is decades or centuries behind. Some companies choose to move production to developing or transition countries, because of the low cost of labor and lack of governmental oversight. These decisions must be made in terms of economic, quality, and social consciousness factors. Current regulations in developed countries came about due to changing education and social consciousness. As the world becomes more flat and globalization a way of life, consideration of safety as a system becomes even more critical. Impacts can be easily felt at home from decisions made oceans away.

The Beginning

The evolution of the field began as early as 300 BC, Hippocrates recognized certain dangerous practices in specific trades. He is also responsible for making the first record of lead poisoning, and for changing medicine from a superstition to a scientific practice. About 500 years later, Pliny the Elder recorded the dangers inherent when working with zinc and sulfur, and even recommended that workers in dusty trades wear animal bladders over their faces as respirators, to prevent the inhalation of dust. In the second century, Galen developed several theories on anatomy and pathology. He was the first to record the dangers of acid mists to copper miners, but gave no solutions, writing simply that certain occupations seemed to be a menace to health.

The Middle Ages

During the Middle Ages, little was done to improve work standards. The only advancement during this time period was the provision of assistance to sick guild members and their families. The universities of the twelfth and thirteenth centuries underwrote considerable experimentation. However, the study of occupational hazards was ignored. There were few advancements in the field of safety and industrial hygiene until 1473, when Ulrich Ellenbog's pamphlet on occupational diseases, including his occupational hygiene instructions, was published. During the fifteenth and sixteenth centuries, the hazards of mining were widely considered. Two important authors during this period were Phillipus Paracelsus and Georgias Agricola. Paracelsus wrote a treatise on occupational diseases in mining, and described such diseases as disturbances of the lungs, stomach, and intestines that resulted from the digging and washing of gold, silver, salt, alum, sulfur, lead, copper, zinc, iron, and mercury. He theorized that these miners' diseases were caused by vapors, and that contact with these metals should be avoided as these diseases were incurable. Paracelsus was also opposed to certain theories and practices then in use, and recommended specific remedies instead of bleeding. Agricola wrote *De Re Metallica*, which told of health dangers to miners. He described the miners' diseases, as did Paracelsus, attributing them to improper ventilation, and went as far as recommending protective masks for the miners.

About 100 years later, Bernardo Ramazzini wrote the first complete book dealing with occupational diseases, entitled *De Morbis Artfactum Diatriba*. Ramazzini believed in the welfare of the worker. He visited workshops and factories of all trades, noting the diseases common to each trade, and observed the workers' habits. The book discussed several diseases resulting from the inhalation of dust and fumes, and recommended reforms to alleviate many hazards, but he was ignored for more than a century. Ramazzini was the first to advise physicians to ask a sick man, "Of what trade are you?" This provided significant insight into the role of occupational health and overall health.

BEGINNINGS OF LABOR LEGISLATION—
THE EUROPEAN MOVEMENT

Labor legislation has played an important part in the progress and development of industrial hygiene and safety. Many of England's working customs date back to the fourteenth century. The Statute of Labourers of 1351 required that every able-bodied man or woman not owning land or already engaged in other work must accept any job offered. This law was in effect until the turn of the nineteenth century. The laws and customs of England were inherited by the colonies and practiced even after they were removed from the books.

One objective of charities in the seventeenth and eighteenth centuries was to found houses of industry in which small children, some less than five years old, were trained for apprenticeships with employers. The apprentice system gave rise to the labor legislation that swept through the nineteenth century. One development of this legislation was recognition of the need to control work hazards. Control of work hazards thus became a basic principle for organized safety programs. A movement to determine the causes of injuries, and to eliminate those causes, had finally begun.

Powered Machinery and Child Labor

The rapid development of power-driven machinery and its effects on manufacturing led to the increased use of children in English textile factories. In 1784, an outbreak of fever through the mills in Manchester led to the first governmental action for safety. By 1795, Manchester had formed a city board of health, which recommended legislation to reduce work hours and to alleviate improper working conditions in the factories. The first legislative step toward the prevention of injury and protection of labor in English factories occurred in 1802, with the passage of the Health and Moral Apprentices Act. This act was aimed directly at the ills of the apprentice system, which then involved a large number of pauper children who worked long hours under the worst possible conditions. Inspectors were appointed to enforce the act, and to direct the adoption of such sanitary regulations as they might think proper (Macdonald, 1868). Therefore, the conditions of children workers in England were a major factor in prompting safety legislation.

Women in the Early Work Force

Possibly owing to the regulation of child labor, a labor force of women emerged in the textile mills in the nineteenth century. In 1844, Parliament passed legislation reducing the maximum allowable working time for women to twelve hours a day. And for the first time ever, this legislation contained specific provisions for worker health and safety. Compensation for preventable injuries caused by unguarded mining machinery was also initiated, with the passage of the Mines Act of 1842. The Mines Act prohibited women, girls, and boys under the age of ten from working underground; the act also called for mine inspectors. By 1843, the first mine inspectors made their report; two years later, however, women were still working underground. It became obvious that mere legislation was not the answer; a method of enforcement was necessary. The amount of safety and labor legislation continued to rise, and so did the accident rate.

Mine Safety Inspections

Because of the increasing accident rate in the mining industry, the British government organized a mine safety inspection program in 1850. By 1855, a new mine safety act listed specific areas of safety regulation for the safety inspectors to investigate: ventilation, guarding of unused shafts, the proper means for signaling, correct gauges and valves for steam boilers, and indicators and brakes for powered lifting equipment. After several disastrous mining accidents and explosions, the mine act was extended to the Mines Act of 1860. However, investigations revealed evidence of incompetent management, and also revealed a total disregard of the existing safety laws. Because of these investigations, a new regulation was passed, the Coal Mines Act of 1872. That same year, health and safety laws for the metal mines were also codified in England. This was the beginning of a complete industrial code, regulating specific dangers to health, life, and limb. This act of 1872 extended the safety regulations of previous acts, and included requirements for the employment of competent and certified managers and an increased number of inspections. Other new regulations included the mandatory use of safety lamps, and requirements for securing of roofs and sides. This act also established that willful neglect of safety provisions by employers, as well as by miners, would be punishable by imprisonment at hard labor.

Workshop Safety in the 1800s

During the early part of the nineteenth century, factory and workshop legislation continued to develop. In 1833, Parliament passed the Factory Act, establishing mandatory factory inspections and limiting workers' hours. Lord Ashley's "Great Factory Act" of 1844 required the fencing of gears and shafts. This was the first law requiring the guarding of hazardous machinery. Several additional trades were brought under control with the Factory Act of 1864. This act was the first to respond to results of experiments of medical and sanitary observers, by requiring ventilation for the removal of gases, dusts, and fumes originating from machinery. The Workshop Regulation Act of 1867 made the factory acts inclusive to all places in Great Britain which made or finished articles or parts for sale.

Labor Legislation: The U.S. Movement

In 1776, Independence Year, the United States was a nation of planters and farmers. There were few skilled workers, and business was primarily "cottage industry," with occasional roving jacks-of-all-trades to repair wagons, pots, guns and the like. At this time, there was little machinery in the thirteen colonies, but this was soon to change. Since world markets were unable to provide goods during the Revolutionary War, it became important to develop local industries to meet these needs. Textile mills were the first to develop. Working conditions were generally poor. The seven-day work week with fourteen-hour days was standard. With a plentiful labor supply and a tough, profit-minded management, improvement in the work environment was slow. Labor unions in those years strove to obtain benefits for widows and orphans, rather than promote worker safety. For the most part safety and health were largely ignored during this period and into the mid-1800s. The country grew from small farming communities into large agricultural areas. A period of great social change began, along with rapid industrial development, bringing with it certain safety and health issues.

The period from 1850 to 1900 was a traumatic one for the United States, but one that laid much groundwork for the safety and health movement. The Civil War in the 1860s produced an unexpected surge of industrial development, as the country sought to meet the demands of a war economy. Railroads developed, and together with waterways, became great highways for moving materials and encouraging even more rapid growth. With these developments, political tension and social upsets were commonplace. The working man bore the brunt of industrial problems in hazardous, unsafe working environments. There is no record of the total number killed and maimed in work-related accidents, and the existence of occupational diseases was not even recognized. If a factory worker was injured or killed, no one paid the survivors or family anything. Usually they became burdens on society with no provisions for their welfare. These intolerable conditions brought forth reformers who lamented the human toll of the workplace and began to agitate for improvements. Workers' compensation had already been an established as a national standard in Germany under Bismarck, and American reformers now urged that it be adopted in this country.

During this period, coal mining in the United States flourished, but with it came a terrible toll of accidents, injuries, and deaths. Anthracite mining in Pennsylvania was particularly dangerous. Among railroad workers there were few who had not lost a finger or a limb. Labor unions became increasingly active; however, their militant tactics in opposing management often delayed safety and health progress. A few states did experiment with safety laws for the workplace, and there were occasional signs of progress. Surveys at the time showed a terrible toll of accidents and ever-increasing health problems in the workplace. The country was moving rapidly toward full industrialization, and by the 1880s could truly be considered an industrial nation. About this time there was movement for an eight-hour work day and other improvements in working conditions, but again there was great strife based on opposition to union activity and hopes for working reforms were dashed once more.

In the United States, factory inspections were initiated in Massachusetts by 1867. Two years later, Massachusetts established the first state Bureau of Labor Statistics, which was charged with researching the causes of accidents. As a result of the Bureau's research on deaths and injuries from industrial accidents, Massachusetts passed the first state law compelling that dangerous moving machinery be guarded. Research also revealed the part fatigue plays in causing accidents. This aided in the fight for shortening the work day to ten hours.

Employee Benefit Association

Along with these formalized safety programs, several companies had begun paying compensation to employees disabled in work accidents. In 1908, the International Harvester Company initiated such a program, called the Employees Benefit Association. Membership into the association was voluntary; however, a high percentage of the employees participated in the program. This program received mixed reactions from those outside the company. It was viewed as both visionary and liberal, or as a revolutionary attempt to head off all legislation designed to benefit the working man.

Common Law Doctrine and Worker Injury Suits

At this time, most suits brought by injured employees were based on common law doctrines. Common law doctrines divided industrial accidents into two groups: those accidents where no one was at fault, and those accidents where either the injured employee, a fellow employee, or the employer were identifiably at fault. Also, the employer was permitted to use these common law defenses in refuting liability:

1. *Assumption of the Risk*—the employee knew about the risk at the time of employment and he or she accepted this risk;

2. *Contributory Negligence*—the employee in some way contributed to the accident, relieving the employer from liability; and
3. *Fellow Servant Rule*—the employer could not be held liable if one or more fellow employees was the cause of the accident.

A court would rarely hold the employer liable, and the possibility of an employee gaining any form of compensation from the employer was very small. Although court decisions involving common law defenses were proper in principle, they were lacking in social justice.

Workers Compensation Introduced

While regulations were beginning to evolve in the United States, the great legislative reform that initiated the organized safety movement in the world would not have progressed very far without the passage of workers compensation laws. Bismark of Germany, in 1885, led the way, and although his law covered only sickness, it was the first compulsory compensation act for workers. Great Britain had earlier required compensation for death or injuries from boiler explosions, but this legislation was designed to protect the passengers of steamboats, not the workers. Workers compensation laws, more than anything else, made employers appreciate the unrecoverable losses which were suffered when the services of their finest workers are lost in accidents on the job. Great Britain, Austria, Hungary, France, Italy, and Russia soon followed Germany's example in requiring some form of workers compensation.

Workers Compensation in the United States

Recognition of inequalities led to the passage of workers compensation legislation. The first compensation law in the United States to be declared constitutional was passed by the federal government in 1908. Two early state compensation laws, Maryland's in 1902 and New York's in 1910, were declared unconstitutional because of conflicts with the "due process of law" provisions in the Fourteenth Amendment. In 1911, Wisconsin and Washington passed the first state compensation laws that were declared constitutional. Soon after, several states followed, and by 1928, forty-three states plus Hawaii, Puerto Rico, and Alaska had passed workers compensation laws.

The workers compensation laws eliminated any consideration of fault of the accident. Under these laws, the employer is held liable if an injury arises "out of and in the course of" employment. This last statement is open to interpretation. Owing to liberal court interpretations of that statement, plus rising medical costs and increases in workers compensation benefits, insurance costs for workers compensation coverage have risen dramatically. The total cost of workers compensation insurance premiums can be substantial depending on the state and risk classification. Risk is determined by two factors: the frequency of on-the-job injury and the severity of injury. These continuous increase in costs have had two main effects. Management is becoming more concerned with the effectiveness of occupational health safety programs, and health and safety professionals have greater responsibility for proper accident investigation and reporting. The effectiveness of the industrial hygiene program can have a direct impact on the workers compensation costs.

It is often said that the passage of workers compensation laws in the United States encouraged early efforts in accident prevention. The first compensation law was passed around the beginning of the twentieth century. Obviously, these laws awakened employers to the fact that accidents are costly and so should be prevented, if only to reduce losses. But as pointed out earlier, the idea that little or no safety effort existed before 1900 is false. With the passage of compensation laws, and the high price of medical care, most companies had developed well-organized safety programs prior to the passage of the compensation laws. According to David Stewart Beyer's *Industrial Accident Prevention,* as cited in Russell DeReamer's *Modern Safety and Health Technology,* even before 1914, machine guards and safety devices were marketed by safety-product manufacturers. Such products as hard hats, safety goggles, sweep guards, circular saw guards, and safety cans for flammable liquids were also already available.

The first half of the twentieth century saw great growth in the safety and health movement. Under President Theodore Roosevelt, many workplace studies were made to determine working conditions. There was new—and generally enforced—legislation for the workplace, including mines, boats, and railroads. A few managers began to recognize that workplace conditions affecting safety and health were directly and indirectly tied to profits. Lessening accidents and injuries was seen as lowering expenses. In general though, U.S. industry still had a long way to go. In Europe many good safety and health ideas and concepts had already been developed, but the United States was slow to adopt them, calling them foreign and unsuitable.

It took some dramatic and sometimes tragic events to wake the U.S. public up to the need for more protection in the workplace and more stringent safety regulation. A fire at the Triangle Shirtwaist Factory in New York City in 1911 produced just such a response. A shocked public learned the deaths of over 100 workers were directly caused by unsafe conditions. More alarming was the discovery that locked exit doors had trapped workers in the burning building. Public and legislative pressure caused management to take a closer look at the need for worker protection and safety.

World War I gave yet another boost to the expanding industrial structure, and in some unexpected ways, aided the safety and health movement. Oddly enough, the war really started progress in the use of personal safety equipment. A safety helmet (based on the soldier's trench helmet) was developed, and a respirator (based on the poison gas mask) came into industrial use. Even the first-aid kit was a wartime development. By the 1920s, companies were starting to compete for safety awards, and many new safety laws were put into effect. By the late 1930s, the American National Standards Institute (ANSI) had developed nearly 400 safety standards for industrial use. To illustrate how effective this activity was, in 1930 the accident frequency rate was 17.5. This rate refers to the number of injuries resulting in lost work time per a given number of hours worked. With the increased emphasis on safety, by 1940 this had dropped dramatically to 11.04. Many of the advances during this time could be traced to the activities of America's first female Secretary of Labor Frances Perkins. Although the first U.S. safety book appeared in the late 1920s, it was not until 1931 that H. W. Heinrich's *Industrial Accident Prevention* was published, and came to be generally accepted as a standard reference for a number of years afterwards.

World War II provided another impetus for increased safety and health activity. Without such an effort, the United States could easily have lost the war due to accidental losses at home and abroad during that period of scarce resources and greatly increased industrial activity. In fact, the accident frequency rate climbed back to a wartime high of 15.39 at one point. Safety and health action was obviously needed. In response to this, educational programs for training safety personnel were developed, and safety became a required way of doing business. Unfortunately, after the war most of this activity died off; schools dropped safety courses, and the great emphasis on safety—so essential during the war years—gradually diminished.

Uncertain Times for Safety

Despite decreased emphasis on safety and health activity during the 1950s, the accident rate was brought down by 1955. Obviously, safety programs developed during the war years had their effect. The reduced rates enabled safety personnel to show management how safety paid. Reducing lost work time meant higher productivity and greater profits. By the mid-1960s, the United States went through a period of soul-searching and some experimentation. Ralph Nader burst on the scene to put industry and management on trial for their allegedly poor safety practices. The public itself became increasingly aware of workplace health and safety hazards. Congress, in response to this growing social concern, came along with many pieces of legislation including the Occupational Safety and Health (OSH) Act on December 29, 1970, and took effect April 28, 1971. The OSH Act created three agencies: Occupational Safety and Health Administration (OSHA), National Institute for Occupational Safety and Health (NIOSH), and the Occupational Health and Safety Review Commission (OSHRC).

OSHA was established within the Department of Labor. The mission of OSHA is to assure safe and healthful working conditions for working men and women by setting and enforcing standards and by providing training, outreach, education, and assistance. Under OSHA standards, employers must provide a safe and healthful work environment free of serious hazards. Ignorance is not a valid defense for failure to comply with OSHA regulations or provide a safe workplace. Under the general duty clause, OSHA compliance officers can cite employers for hazardous workplace conditions even if no specific standard exists. OSHA standards are laws and must be approved by Congress and the President. As a result, it is difficult for OSHA to promulgate laws on emerging issues in a timely fashion. While not enforceable, NIOSH strives to develop guidelines for emerging issues.

NIOSH was established within the Department of Human and Health Services, Centers for Disease Control and Prevention, under the provisions of the Occupational Safety and Health Act. NIOSH is engaged in research, education, and training related to occupational safety and health. NIOSH has three strategic goals: conduct research to reduce work-related illnesses and injuries, promote safe and healthy workplaces through interventions, recommendations, and capacity building, and enhance international workplace safety and health through global collaborations. These goals are achieved through the work of a diverse set of professionals, including epidemiology, medicine, nursing, industrial hygiene, safety, psychology, chemistry, statistics, economics, and many branches of engineering. NIOSH also provides on-site evaluations of potentially toxic substances, technical information concerning health and/or safety conditions, technical assistance for controlling on-the-job injuries, technical assistance in the areas of engineering and industrial hygiene, and assistance in solving occupational medical and nursing problems at the request of the employer.

OSHRC is an independent federal agency that decides on contests of citations or penalties resulting from OSHA compliance visits. OSHRC functions as a two-tiered administrative court that conducts hearings and administrative law judges rule on evidence, and then a commission provides discretionary review of administrative law judge's rulings. This provides a method of due process if OSHA compliance actions cannot be mediated or resolved.

Initially, the new safety programs consisted mostly of inspections. The answer to all safety problems was to inspect the facilities and operations, look for equipment faults and hazards, and try to correct them. This was far from adequate, as it soon became obvious that *people*, as well as equipment, were involved in most accidents. A more comprehensive approach involved emphasizing the correction of *both* unsafe conditions and behaviors.

It soon became apparent that occupational disease was also an important factor and had reached proportions where it had to be controlled. From emphasizing safety only, the country entered an era of emphasizing health aspects, as well. Items such as noise and toxic substances in the workplace became prime targets. OSHA standards now include requirements for employers to prevent exposure to some infectious diseases, ensure the safety of workers who enter confined spaces, prevent exposure to harmful chemicals, provide respirators or other safety equipment, and provide training for certain dangerous jobs. Incidents resulting in deaths must be reported within eight hours after learning of it (Occupational Safety and Health Administration [OSHA], 2014).

It also became more acceptable to talk about defining safety and health in terms of management roles and responsibilities, as well as terms of hardware and facilities. This trend continues today. Prime movers in this trend are the American National Standards Institute (ANSI) and the American Industrial Hygiene Association (AIHA) with their creation and promotion of the ANSI Z10, Occupational Health and Safety Management Systems Standard (OHSMS), that devotes a complete section to management leadership. "This standard provides critical management system requirements and guidelines for improvement of occupational safety and health," said Gary Lopez, CSP, ASSE Standards Development Committee Chair (Smith, 2012). The concept of developing a Safety Management System (SMS) to provide management leadership, employee participation, and continual improvement is spreading through a number of industries. The author expects SMS to become the standard for management leadership and continual improvement in safety, as it creates a platform for a strong safety culture.

Legislation has made the United States one of the leaders in the progress of industrial hygiene and safety. The three major pieces legislation to date are the Metal and Nonmetallic Mine Safety Act of 1966, the Federal Coal Mine Health and Safety Act of 1969, and the Occupational Safety and Health Act of 1970. Furthermore, the Environmental Protection Agency (EPA) has a multitude of legislation that affects industrial hygienists, such as air and water permits, radon, hazardous waste and others. The Metal and Nonmetallic Mine Safety Act details mine health and safety standards. This act created the Federal Metal and Nonmetallic Safety Board of Review, and also provided for mandatory reporting of all mine accidents, injuries and occupational diseases. The federal Coal Mine Health and Safety Act was an attempt to establish the highest possible standards of health protection for the coal miner. This act gave the federal government the power to withdraw miners from any mine deemed unsafe by the government. The act also established requirements intended to assure working conditions free of dust and fumes while the miners are underground. Among these are standards regarding dust control, use of respiratory equipment, proper ventilation, and medical examinations for the miners. During the period when legislation was evolving, private organizations also began developing to promote safety and health among U.S. and international employees.

Safety Organizations—Their Beginnings

In 1912, 200 men interested in accident prevention met in Milwaukee. This meeting was titled the First Cooperative Safety Congress and was sponsored by the Association of Iron and Steel Electrical Engineers. The following year a committee was formed, and the National Council of Industrial Safety was organized. The goals of the council were broadened over the years, and the name was changed to the National Safety Council (NSC). Today, the NSC is the leading safety organization in the United States. Its interests include traffic, home, farm, school, and industrial safety.

American Society of Safety Engineers

In 1911, paralleling these other developments, three insurance company representatives met in New York City, forming the United Association of Casualty Inspectors. This was the beginning of the organization that today is known as the American Society of Safety Engineers (ASSE). The ASSE has become an important influence in the careers of safety engineers and has advanced fundamental safety knowledge. ASSE provides a voice by promoting advocacy, education, and standards development through the expertise, leadership, and commitment of its members. Occupational health and safety advances are presented in the peer-reviewed journal *Professional Safety*.

American National Standards Institute

The American National Standards Institute (ANSI) was founded in 1918 and provides the research for the development and publication of standards, including safety and health standards. The mission of ANSI is to increase the global competitiveness and quality of life of U.S. business and citizens by promoting and facilitating voluntary consensus standards and conformity assessment systems, and safeguarding their integrity. ANSI's health standards are advisory; only after adoption by a federal, state, or local health agency do they become mandatory. Many regulatory agencies have adopted their standards, which must be purchased. Health and safety professionals have free access to all OSHA and EPA laws; however, when ANSI standards are incorporated, the actual ANSI standard must be purchased.

American Conference of Governmental Industrial Hygienists

The American Conference of Governmental Industrial Hygienists (ACGIH) began in 1938 to provide a forum for the discussion of difficulties and experiences among industrial hygienists. ACGIH was formed to encourage the interchange of experience among industrial hygienists, as well as to collect and make accessible such information and data. ACGIH publishes *Threshold Limit Values and Biological Exposure Indices*—occupational health standards for workers that are updated annually by leading scientists and field industrial hygienists. The TLVs (free to members) are considered to be the most conservative and up-to-date; although, they are not legally enforceable. Since OSHA standards can only be changed through an act of Congress after much debate and influence by lobbyists, updates and progress has been slow and limited. Many of the original OSHA permissible exposure limits that were promulgated in 1970 were based on the 1968 ACGIH TLVs and have not been updated.

American Industrial Hygiene Association

The American Industrial Hygiene Association (AIHA), organized in 1939, was established to encourage the exchange of information of the different sciences, such as chemistry and toxicology and their application to industrial hygiene. This organization is the nucleus of industrial hygiene development and is responsible for the recognition of industrial hygiene as a science. AIHA and its members are the front line in worker health and safety—from research to implementation in the field. Members range from students, industry, private business, labor, government, and academia. The premier industrial hygiene peer-reviewed journal, *Journal of Occupational and Environmental Hygiene (JOEH)*, is a joint publication of ACGIH and AIHA. In addition, AIHA publishes *The Synergist*, a monthly magazine with in-depth news and information about the occupational and environmental health and safety fields.

These organizations demonstrate how professional groups develop in response to a need and then serve to lead the way with emerging issues. There are many other safety organizations and professional societies in the United States and throughout the world that actively provide a forum for an exchange of ideas through which professionals are able to learn from one another. These organizations administer professional development conferences and continuing education for health and safety professionals and are a vital part of information sharing and solutions to pressing industrial hygiene issues. There are no secrets in safety.

Where Does the United States Stand Today?

While the regulations reflect greater social pressure for safety of all employees, to some extent they also represent a response to management's earlier lack of action in providing a safer and healthier work environment. Government has been slow in responding to worker protection. According to a report by the AFL-CIO (2015), it would take OSHA 140 years to inspect all facilities under its jurisdiction just once. That is up from once every eighty four years in 1992. In some states, add an additional ten years, because the state OSHA program is even further understaffed (American Federation of Labor and Congress of Industrial Organizations [AFL-CIO], 2015).

Safety—On and off the Job

In 2013, the rate of fatal work injury for U.S. workers was 3.2 per 100,000 full-time equivalent (FTE) workers compared to a final rate of 3.4 per 100,000 in 2012 (Bureau of Labor Statistics [BLS], 2014). Not all of the safety and health problems are in the workplace. According to the National Safety Council (2013), approximately nine out of ten deaths and about seven out of ten medically consulted injuries occurred off-the-job. Production time lost due to off-the-job injuries accounted for approximately 235 million days in 2011 as compared to only 60 million for on-the-job injuries. The results of both types of injuries are felt in the workplace. Outside injury or illness may not appear in the direct costs of workers' compensation or damaged equipment, but the failure of trained employees to appear for critical jobs has many hidden and not-so-hidden costs, such as delayed production, uncompleted deliveries, training expenses for new employees and more. These costs emphasize the need to also consider and include off-the-job safety as a target.

CONCLUSION

The evolution of safety legislation throughout Europe, the United States, and the rest of the world has formed a foundation upon which a strong safety and health program can be built. It also provides the framework for an effective SMS to be built into the overall organization. When combined with existing standards, the legislation has become the basis and the impetus for an effective approach to the overall safety system.

REFERENCES

American Board of Industrial Hygiene. (2014). *About ABIH*. Retrieved from: http://abih.org/.

American Conference of Governmental Industrial Hygienists. (2016). *Defining the Science of Occupational and Environmental Health*. Retrieved from: http://www.acgih.org/home.

American Federation of Labor and Congress of Industrial Organizations (AFL_CIO). (2011). *Death on the Job the Toll of Neglect*. Retrieved from: http://www.aflcio.org/content/download/6485/69821/version/1/file/dotj_2011.pdf.

American Industrial Hygiene Association. (2016). *AIHA Protecting Worker Health*. Retrieved from: https://www.aiha.org/about-aiha/Press/Pages/default.aspx.

American National Standards Institute. (2015). *ANSI*. Retrieved from: http://www.ansi.org/.

Bureau of Labor Statistics (BLS). (September 11, 2014). *Census of Fatal Occupational Injuries Summary, 2013*. Retrieved from: http://www.bls.gov/news.release/cfoi.nr0.htm.

Friend, M.A. and Kohn, J.P. (2014). *Fundamentals of Occupational Safety and Health* (6th Ed.). London, UK: Bernan Press.

Macdonald, A. (1868). The Law Relative to Masters, Workmen, Servants, and Apprentices. London, UK: William Mackenzie.

National Institute for Occupational Safety and Health (NIOSH). (2015). *Providing National and World Leadership to Prevent Workplace Illnesses and Injuries*. Retrieved from: http://www.cdc.gov/niosh/.

National Safety Council. (1983). *Accident Prevention Manual for Industrial Operations*, National Safety Council: Itasca, IL.

National Safety Council. (2013). *Injury Facts*. Retrieved from: http://www.mhi.org/downloads/industrygroups/ease/technicalpapers/2013-National-Safety-Council-Injury-Facts.pdf.

Occupational Safety and Health Administration (OSHA). (2014). *All about OSHA*. Retrieved from: https://www.osha.gov/Publications/all_about_OSHA.pdf.

Smith, S. (September 27, 2012). ANSI Z10-2012 Standard Provides the Blueprint to Create an EHS Management System. *EHS Today*. Retrieved from: http://ehstoday.com/consensus/ansi-z10-2012-standard-provides-blueprint-create-ehs-management-system-0.

Chapter 2

The Safety Function

Mark A. Friend

CASE

Bob Renee was recently hired as the safety director of Hawkins Supply, a household products distributor with distribution centers on the East Coast. He was on the job for a very short time when he began noticing employees not following some basic safety rules that both orientation and annual refresher training clearly addressed. Bob approached one employee who failed to wear his safety glasses and hard hat during his routine work and told the young man he needed to do so. The employee gave Bob a dirty look and said, "You're just the safety director. You don't tell me what to do. I report to somebody else. Get lost." Bob was immediately taken aback and repeated his directive. The young man reluctantly donned his sear, but Bob believes the next time he observes the young worker, PPE will again be missing. Bob is now thinking about how he will handle this the next time.

IS SAFETY REALLY FIRST?

Safety is not and cannot be first. It is profits, mission, product, service, and the strength and growth of the business that are first. Safety is important because it can either support or damage these things; it is not something separate, a function by itself. Safety and health expertise is needed to support the organization's objectives, so it should be fully integrated into plans and operations. Safety is a concern of all company functions and should, therefore, not merely be imposed as an afterthought. The integration of safety into an operation is almost completely and directly related to management's perception of the function. If it is seen as an operator or technician problem, then it will never be fully integrated into or support the company. If it is realized as a management responsibility with staff support, then it can succeed.

Companies want efficient production, sales, and service to maximize profits and to ensure the company continues. Efficient production requires facilities, equipment, materials, and human resources, and management is responsible for supplying them. While doing so, support services, such as engineering, purchasing, product inspection, maintenance, and research, are also needed. By incorporating safety into support services, it is merged at the functional level and becomes an integral part of the operation. The human resources function is a good example. Personnel selection, placement, training, employee relations, and motivation all involve safety and health considerations. Building safety and health into each of these human resource functions and continually auditing them helps assure the functions are safe and support the overall organizational mission. This concept applies to any organizational function.

SAFETY: AN INTERDISCIPLINARY FUNCTION

Research on safety and health problems is often conducted by experts from a single discipline. For example, human factors (or ergonomics) experts frequently work on special safety/health problems, such as mishap causation or back injuries. Since the researcher may approach the problem from one perspective, this may mean it is only seen as a human factors problem and studied within that context for solutions. When two or more experts from different fields study the problem, expecting to coordinate their findings and complement each other's work, there may be a more balanced approach, as when a design engineer and a human factors specialist work on the back problem and coordinate their complementary findings.

Bringing several disciplines together to work among and between themselves for an integrated solution that considers a problem in light of all the disciplines is considered an interdisciplinary approach. It is only when they work together from the start, seeking a whole solution, do the disciplines become interdisciplinary. Thus, the human factors specialist, the design engineer, the marketing expert, the safety professional, and others who work as a team produce an integrated approach.

SAFETY PROBLEMS: SYMPTOMS OF MANAGEMENT FAILURE

The business organization is a system composed of many subsystems. There are many systems managed in the organization, including financial systems, personnel systems, production systems, maintenance systems, and so forth. *Any accident, injury, unsafe act, workplace health problem, or unsafe condition indicates a failure in one of the systems and, therefore, a failure in the management system.* The safety and health professional acts as a system evaluator, often in collaboration with experts from other disciplines, advising management on how functional parts of the system should operate in regard to safety and health. Safety becomes a resource for getting the job done, not a regulatory burden that hampers the system. Again, **safety errors** made within the organization are, in fact, **errors in the management system**.

Accident/injury investigation subscribes to the multiple causation theory; that is, there are always several causes present that allow or even encourage accidents to happen. Accident investigations attempt to determine causes and all causes are then traced to *their* underlying causes, largely by asking, "Why" for each cause. When this is done in depth, it is *always* determined that management actions or inactions are contributing factors to the event.

Take the simple example of a ladder that breaks in a stockroom when a clerk climbs to a high storage bin. It could be the ladder had not been properly and regularly inspected (a lower management error). Asking why this happened, the investigator might find a procedure for regular inspection of ladders had not been established (a management failure), the regulations regarding inspection were not known (a management failure), an inspection system for equipment in general was not in effect (a management failure), such ladders were of an inferior quality and should not have been purchased in the first place (a staff failure), and funds to purchase proper equipment were not made available (staff and management failures). A closer review of the overall operation might show these management and staff failures were also causing similar problems throughout the organization, including costly delays and production difficulties, worker unrest, and labor organization concern over the use of unsafe equipment. From a simple broken rung, several cause factors (multiple causation) and many examples of poor management and staff work may be detected. Once such conditions are recognized and corrected, the entire operation will improve. Thus, the safety function can help improve the entire system.

SAFETY AS A MANAGEMENT FUNCTION

There is little difference between the safety function and any other organizational function. It starts with senior management building a safety policy compatible with other policies in the company. The achievable goals and objectives are set, coordinated, and carried out the same way as with other functions. The safety and health function is parallel to and interacts with the other *management* functions, such as the control of quality, cost, innovation, and production. It is not enough for management to merely support the functions; it must lead the way. The organization needs management's active leadership, not just passive support. *Safety competes with every other functional part of the organization for corporate resources.* Management decides what functions receive the corporate resources, and these decisions are typically based on return on investment. The parts of the organization showing the highest returns on investment receive the funds necessary to perform their functions. The safety and health professional must build the financial case to receive appropriate funding to carry out the function.

Management directs the safety and health effort by integrating the safety function into the overall corporate mission; setting realistic goals, planning, organizing, and establishing controls for safety, and ultimately directing the safety effort. While top management assigns responsibility and authority for safety and health to the line managers and hold them accountable for the results, top management retains the overall accountability and responsibility for the safety function as it is carried out throughout the company by individuals reporting directly or indirectly to top management.

Merely because safety functions are well developed, starting with policy and continuing through line and staff accountabilities does not mean safety is valued by senior management or considered on the same level as other functions. Other functions, such as production, marketing, and cost control, may hold more interest and concern.

Line and staff managers can and should be accountable for those safety and health functions **they control**. Using the broken ladder illustration, purchasing cannot be held accountable for the stockroom clerk's actions in climbing a defective ladder. Purchasing can be accountable for buying an inferior and possibly unapproved product—if they have the resources and directions needed to purchase proper equipment. The safety professional is not responsible for an accident in the storeroom where there was no direct control but may be at least partially responsible and accountable if he or she did not give purchasing good guidance. It is also a safety responsibility to make equipment inspections (or see that it is done) and advise management on the need for certain guidelines and procedures. Safety cannot be held responsible for line functions but is responsible for warning and advising line and staff. The responsibility of the safety professional is to monitor and advise management regarding safety. The responsibility of management is to assure that safety is built and operating properly in the system. Typically, where there is a conflict between line-and-staff goals, staff must make a persuasive case, because the line part of the organization has final authority. If an employee perceives any conflict between directives coming from an individual staff member and his or her own line manager, the employee normally considers the line manager to be "the boss" and tends to respond to the line management directive.

Evaluating the Safety Function

A company's level of safety can be assessed by asking each the following:

- Is either the chief executive officer, the chief operating officer, or the equivalent serving as the accountable executive?
- Is a professional manager, qualified in terms of education and experience and directly reporting to the accountable executive, managing, monitoring, and coordinating the system-safety process?
- Has the organization developed, documented, and fully implemented a comprehensive system-safety plan accessible to all employees?
- Are safety policies, procedures, and processes based on the safety management plan clearly defined, documented, communicated, and implemented throughout the organization?
- Is the system-safety plan clearly tied to the organizational mission, aligned with corporate objectives, and integrated throughout the organization?
- Have procedures been clearly developed and implemented for response to accidents, incidents, and operational emergencies?
- Are expectations, accountabilities, responsibilities and authorities for safety-related policies, processes, and procedures clearly defined, documented, and communicated throughout the organization?
- Are system-safety competencies for all positions identified and documented providing evidence that all competency requirements are met for all positions?
- Are there clear standards for acceptable operational behavior as it relates to system safety for all employees?
- Can appropriate responsibilities and authorities be clearly traced from any level or position to the accountable executive in the organization?
- Are resources in terms of personnel and budgets appropriate for implementation and maintenance of system safety throughout the organization?
- Are processes clearly defined, documented, communicated, and implemented to ensure all hazards likely to cause death, physical harm, or equipment or property damage identified and documented at all stages of the system, process, and product life cycles?
- Have safety-risk-analysis processes for all identified hazards been clearly defined, documented, communicated, and implemented throughout the organization?
- Have risk-control and mitigation processes above an appropriate, acceptable, and clearly defined level of risk been implemented and applied to all hazards?
- Is a closed-loop system for the reporting of safety issues by any and all employees and relevant constituents without fear of reprisal strongly

encouraged and clearly communicated? Is anonymity available when preferred?
- Does the organization have a clear and effective system in place to monitor, measure, evaluate, and document the performance and effectiveness of all risk controls?
- Does the organization continuously monitor operational data, including products and services received from contractors, safety reports, and employee safety feedback to determine and document conformity to established risk controls and evaluate system safety performance?
- Are there clear and relevant system-safety outputs generated regularly, thoroughly reviewed, and incorporated into policies, procedures, and processes by top management?
- Are there timely and appropriate periodic reviews of all system-safety procedures and processes to ensure relevance and appropriateness?
- Are comprehensive audits of system safety performed at least annually on all safety-related functions of operational processes, including those performed by contractors, to verify safety performance and evaluate the effectiveness of safety risk controls?
- Do audits demonstrate all system-safety functions are in conformance with the safety management plan and are utilized for continuous improvement of system-safety processes and performance?
- Do auditors possess appropriate professional qualifications and are they independent of any processes or work evaluated.
- Are procedures clearly defined, documented, communicated, and implemented to collect data and investigate incidents, accidents, and instances of regulatory noncompliance for identification of potential new hazards or risk-control failures?
- Does top management demonstrate the growth of a positive safety culture throughout the organization by publication and distribution to all employees of senior management's stated commitment to safety?
- Is there documented, frequent, and visible demonstration of management commitment to system safety through both verbal and written communications?
- Does the organization ensure all personnel are appropriately trained and competent to perform duties related to safety? Training scope is commensurate with required competencies and responsibilities in the safety system. Initial and periodic safety training for all employees is clearly outlined, scheduled, and performed.
- Are safety training requirements met for all new hires and transfers within the organization?
- Is appropriate safety refresher training for all employees updated and instituted no less than annually?

- Does top management document communication of safety outputs throughout the organization, rationale behind controls, and preventive or corrective actions?
- What is the purpose of the organization? If this is clear to management then **everything the safety function does should support this purpose.**

Safety and Health as Avoidance of Loss

A prominent management consultant said, "It is the first duty of business to survive. The guiding principle is not the maximization of profits, but the avoidance of loss." While profits are vital to survivability, the avoidance of loss truly maximizes profit. Avoidance of loss is directly and indirectly related to safety at all levels and in all operations, and it characterizes the safety function very well. Operating errors produce accidents, injuries, and economic losses. These operating errors and unplanned events occur mostly when human and physical resources are poorly handled.

The causes of operating errors and unplanned events may have little relationship to their end results. The end results (accidents and injuries) may be considered fortuitous, and by themselves should not dictate the amount of effort expended to bring about changes in operating procedures. A reflex movement, a broken machine part, or even chance may all play contributing roles in the end results. There may be hundreds of mistakes without a single recordable accident, injury, or property-damage loss. Chance alone may make a difference. The end results of unplanned events is are unpredictable; however, the number of undesired, unplanned events is one way of measuring managerial performance—far better than the injury rate. Merely keeping track of actual injuries or accidents may not give the full picture of how many unplanned events ineffective management is allowing, but locating operational errors *that could potentially result in an accident* is a technique for loss avoidance.

James Reason's Accident Causation, or Swiss-Cheese Model, proposes that an accident occurs because several human errors happen in such a way as to make the accident unavoidable (Reason, 1990). The model focuses both on the organizational hierarchy and human error. For example, when a new solvent is purchased, no effort is made to determine its adverse effects in closed quarters, nor has appropriate training regarding the use of the hazardous material been provided (fallible decision). The product is shelved in the warehouse and becomes accessible by front-line managers. A front-line manager made a poor decision when obtaining the product from storage and insisting it be used prior to training in closed-quarters to remove stains from a product (line-management deficiencies). The worker was not properly trained on the use of respiratory equipment or on the information from the safety data sheet. In this unsafe culture he felt ongoing pressures to do his job quickly and proceed without proper training (preconditions). He continued to work in spite of a foul odor and light headedness (unsafe act) and unmistakable warnings of danger (inadequate defenses).

- **Active errors** (also called unsafe acts) are the proximate causes of the accident. The worker used the solvent, even though he had no training on the hazardous material or its adverse effects (Reason, 1990).
- **Latent errors** are the remaining elements in the organization contributing to the accident: senior managers purchasing decisions, line-management pressures, unsafe climate and culture coupled with lack of training, and pressure to produce. Without these errors, the accident could have been prevented (Reason, 1990).
- **Window of opportunity** refers to the opportunity for those active and latent errors to contribute to an accident. Had the worker been properly trained and not been pressured to proceed, he would have prevented the accident . . . this time. Had the material not been made available before the training took place, the accident could have been avoided. Yet, the latent errors remain unresolved, waiting for their opportunity (thus a "window of opportunity") to strike (Reason, 1990).
- **Causation chain** refers to the alignment of all necessary windows of opportunity at all levels in the organization, thus leading to the occurrence of a particular accident. That is, the causes of most accidents can be traced back to "windows of opportunity" opened at all levels in the organization (Reason, 1990).

Loss Control

The *loss control concept*, as some practitioners refer to safety and health management, holds that safety and health management should "locate and define operational errors involving incomplete decision-making, faulty judgment, administrative miscalculations, poor management, and individual practices leading to incidents which downgrade the system and provide sound advice to management as to how such mistakes can be avoided."

The circle is complete. Safety and health management should do exactly what management wants to do in any other functions; that is, it should contribute to the survival of the organization. Its aims and goals are the same as senior management's overall objectives. It is up to the safety professional to have management see that safety clearly supports the organization's objectives. It is up to management to assure that safety objectives clearly align with the organizational mission. That alignment helps management accomplish its mission.

ALIGNMENT OF SAFETY AND HEALTH OBJECTIVES

Again, the ideal safety and health program will have the same objectives as that of any other corporate function. Peter Drucker and Warren Bennis discuss "multiple objectives of business," and some of the safety implications:

Profitability. This objective refers to the manager's desire to maximize profits (or avoid losses). Reducing production, personnel, and equipment costs are part of making operations more profitable, as is taking up organizational slack. Accidents and injuries cost money, thus reducing profits. Safety, therefore, has an impact on profitability. Both management and safety should recognize that increased use of new and improved technologies, processes, and products provides new and different hazards. As changes are introduced, new hazards associated with those changes can cause injury, property damage, and losses of profitability.

Market Standing. Safety and health issues can have an adverse effect on product or service marketing, actually lowering a company's market standing. Liability suits, rumors of an unsafe product, and newsworthy accidents that occur within the firm can endanger this standing.

Productivity. This is directly influenced by time-consuming accidents and incidents that shut down a production line, delay deliveries, incapacitate skilled operators, or destroy essential equipment. Less easily measured, but still costly and just as detrimental to production, are the losses of efficiency and morale that come and remains after a serious accident or incident. With just-in-time operations, such delays can, in a few hours, throw a well-organized system with minimum storage and timely delivery and shipping into a costly domino reaction with effects reaching far beyond the directly affected production line.

State of Resources. Whenever accidental losses of resources occur, money that could well be used for more profitable pursuits may be diverted to solve needless safety and health losses. Losses could include replacing lost workers and equipment, designing new tasks, and incurring legal and insurance costs.

Service. A market position and profitability may rest heavily on the quality and speed of the service supplied. When safety and health problems occur and must be explained or defined, the ability to give quality service to customers is reduced. An unsafe product can result in enormous service costs if repair or replacement is necessary.

Innovation. The development of new products and services is often essential to company growth. The need to incorporate safety and health considerations radically affects innovation from initial concept through design and production. The cost of anticipating hazards prior to production must be taken into account if the company is to innovate and operate *profitably.*

Social Contribution. In terms of public responsibility, safety and health considerations have far-reaching effects. Manufacturing processes that, knowingly or unknowingly, pollute the environment or result in the distribution of products which later prove to be unsafe, can result in needless costs and poor public opinion. Failures to accept social responsibility can lead to increased public attention on the company and members of the management team. There are increased costs to the company due to direct costs and the downgrading of its image.

Identification. This involves a management and staff commitment to the same organizational objectives. The safety and health issues concerning a process or product can seriously divide management and staff. What makes a product safer in design and use may make it harder to produce or sell. When management and staff both understand that safety is tied to the prime objectives of service, profit, and mission, organizational goals are more easily reached.

Social Influence. This refers to the distribution of power and influence within an organization to achieve expectations. Effective safety and health performance should become an expectation. To achieve this expectation, responsibility must be placed throughout the organization and carefully tied to accountability. This is usually done by building safety expectations into job descriptions and periodically reviewing those expectations against actual performance in regular job evaluations for each and every manager. Evaluations consider how well the job is done in terms of creating a safe environment for all direct reports.

Collaboration. Safety and health issues are involved at all levels and may involve conflicts or problems between staff and management, departmental interests, and more. Conflict is natural when various interests compete for limited funds in the budget. Effective management will focus on safety as a common goal that affects mutual interests and achieves needed cooperation at all levels. Such common goals are found between safety and every other function.

Revitalization. Even though things may be running smoothly, a new product, new production method—even a change in organizational structure—will always involve new safety and health problems, real or potential. These must be fully integrated into management's total safety and health programs. Change, itself, is constant and will force revitalization if an organization is to stay alive. With a market edge good for only a brief period, revitalization becomes a constant process.

Integration. There is the need to have overlapping objectives and responsibilities for smooth operations. This may be especially crucial for safety and health functions, which cannot be confined to a single department. They requires multi-level and inter-department management approaches to keep them from "falling through the cracks."

Perhaps some governmental organizations, public institutions, and non-profit organizations may consider themselves mission-oriented rather than profit-oriented; nevertheless, they run on a business-like basis to achieve their objectives. All the above roles apply to them.

Exactly How Is Management Involved?

Once the manager identifies and evaluates a safety and health problem he or she

1. Takes proper corrective actions.
2. Weighs and assumes the related risks.
3. Does nothing and becomes guilty of oversights and omissions. Oversights and omissions are traced to

   a. A less than adequate management system, and/or
   b. A lack of specific control factors.

How these factors connect to management functions and objectives is shown in figure 2.1. It outlines an adequate management system and specific control factors needed. The figure highlights senior management, middle management, and staff roles and shows where the assumption of risk, oversights, and omissions enter the system. Specific faults or corrective actions with management systems and specific controls always relate to each other. The curved line with arrows connecting the circled **"Specific Control Factors"** and **"Management System,"** illustrates that a fault related to one circle can be traced to the other. For example:

1. A problem with the block *technical information* under **"Specific Control Factors"** might relate to **"Management System"** rectangles of *policy and directives* or shortage of suitable *staff* to provide *technical information.*

2. Out of the **"Management Systems"** circle, failing to provide adequate *budgets* could easily result in *maintenance* and *supervision* problems under **"Specific Control Factors."**

Accidents and Hazards—Management Failure

In times when economy of operation is necessary, management seeks to reduce costs by streamlining operations, cutting back on equipment purchases, reducing stockpiles, tightening up on maintenance, increasing efficiency of production, and cutting down on overhead personnel. Safety and health activities are sometimes considered surplus. The senior manager who sees safety and health as a marginal activity may be overlooking the potential for large monetary savings and productivity improvements without costly

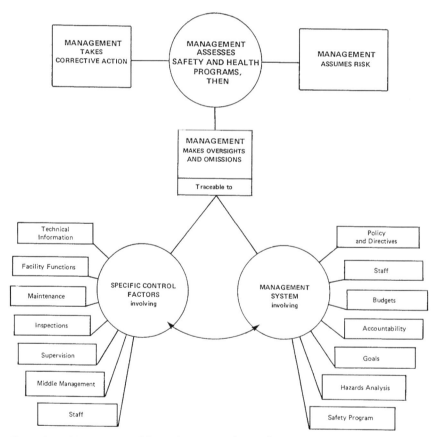

Figure 2.1 Management Either Takes Corrective Action or Assumes Risks and Makes Oversights and Omissions Related to Control Factors and the Management System

expenditures. Unwanted loss of resources is connected to management and makes its way through the chain of command to the points in the organization where the losses occur.

This systematic way to achieve savings stems from the simple assumption that **all accidents and hazards are indications of management failures and a failure of the management system.** It applies to all organizations. Management system failures may ultimately cause many production losses, accidents, and hazards. Those failures may go on for years, continually causing large, cumulative losses without anyone being aware of them. They may be simple inefficiencies, errors, or oversights. A system developed by William Fine routinely analyzes accidents, hazards, near misses, and plain poor operation as a means of detecting these inefficiencies, oversights, and errors on management's part.

Hypothesis

It is assumed that management will be responsible for the causes and existence of every hazard. The rationale for accepting this hypothesis is that in investigating every accident, one can always find some degree of management involvement or activity that might have prevented the event in some way. With this base, the investigation of each hazard, accident, and injury is directed at identifying the underlying management involvement. *Management* refers to all levels of above the front-line supervisor, including all staff and support functions such as personnel, supply, maintenance, security, and safety. The immediate supervisor is excluded because the key to prevention and correction lies above the supervisor. The supervisor is the front-line manager—one level above the nonmanagement employee. The supervisor generally has little time to be a creative manager. His or her duties are usually prescribed and so demanding that the supervisor must concentrate just on those responsibilities upper management emphasizes the most. The supervisor is the last in a line of managers and may often be the fall guy who, while not given the resources to do the job properly, is still held responsible for worker safety and health. Finally, while he or she may be the closest to the actual work tasks being carried out, the supervisor may have the least opportunity or say over the safety and health aspects of those tasks.

Applying the System

The method consists of investigating each event with management introspect. A single question is answered at every turn and in connection with every fact and circumstance uncovered. The question is: *"Where did management fail?"* It is what management did or did not do that permitted the event or situation. In some cases, the management failures are obvious. In others, they are remote and indistinct, but they are always present. Consider the facts. Management advertises, screens, trains, places, and supervises all employees. Management determines when employees come to work, when they go home, where they stand, their total working environment. They may even dictate when employees can eat and go to the restroom. Management "holds all the cards," and employees have to play the hand they are dealt. When accidents occur, it is the management system at fault.

On the other hand, there is no reason to feel defensive about a system that deliberately seeks failures or, more precisely, failures in the structure and systems. Prompt attention to management inefficiencies, uncovered by the technique, will indicate management's concern and interest to personnel and will "buy" appreciation and boost overall employee morale. The benefits of seeking management involvement is demonstrated by the following example borrowed from an actual industrial case and illustrating the technique in operation.

Case of the Improvised Scaffold

This situation involved an improvised scaffold constructed of stepladders, planks and beams, fastened with rope. There was no question that it was unsafe, with potential for serious injury or even a fatality. Simply correcting the unsafe situation by buying a new scaffold would be simple. Investigating the condition to determine the underlying management errors yielded some surprising results. It was found that a properly manufactured scaffold was not available for the crew because someone in a higher management position wanted to operate at the lowest possible cost, avoiding new equipment purchases. It was also found that the improvised scaffold had been assembled and disassembled for relocation at different job sites at least once a week throughout the previous year and probably for a much longer time.

Each time the improvised scaffold was erected or removed, it required three to four employees working about three hours each time or approximately 1,872 hours per year. In use it required frequent inspection and tightening, consuming more hours. Moreover, the crew working on it was never at ease.

What if they had purchased a new scaffold? Two employees could easily erect a scaffold this size in one-half hour (only 52 hours per year). Simple arithmetic shows that an investment of $3,000 for a new scaffold saves over tens of thousands of dollars in labor costs annually and greatly improves the safety status, efficiency, and morale of the work crew.

This case dramatically brought to top management's notice how much its "penny wise and pound foolish" policy was costing. They were saving a few dollars by using improvised equipment but incurring tremendous hidden costs. Reviewing and changing this penny-pinching policy throughout the organization brought about tremendous savings, while eliminating many potentially costly hazards.

A cursory investigation would not have yielded the resulting benefits. Safety investigation should not simply be profit motivated, but the moral is clear. In most cases, management's lack of involvement in safety and health hazards is merely symptomatic of overall inefficient operations. Investigating with the question, "Where did management fail?" can actually save the company money in the long run. It can pinpoint problems that are not confined to safety issues, but affecting overall purchasing, quality control, personnel training, and production policies. Of course, the investigating system is only half of the story; action is not complete until the proper level of management takes corrective action.

Exactly How Is the Board of Directors Involved?

Safety Responsibilities often not Understood

Boards of directors may be naïve about their safety responsibilities. Since senior executives do not often see the significance of public concern for safety and because they seldom have safety expertise on the board, they may not act on safety matters until forced to do so by unions or legislation. When boards do give attention to the subject of safety, it may only be to weigh the cost of complying with safety regulations against the cost of not doing so. This represents a submissive and defensive posture, as opposed to taking a safety and health initiative. Top management cannot be completely faulted, if they have not had education leading to their enlightenment or safety expertise at the top advisement level. If the board does not seem interested in safety and health as a corporate function, it may be because they do not understand the relationship to overall corporate health. The board just may not understand how the safety and health functions act in their best interest.

Changes in Safety Management Direction

Safety includes protecting of all corporate resources, handling product liability, dealing with hundreds of thousands of chemicals, using exotic metals, overseeing robotics, complying with national and international governmental regulations, intervening in workplace violence, and more. Are boards of directors concerned with these issues? With average costs of accidents, insurance rates, and the expenses associated with litigation skyrocketing, boards are naïve not to take an interest.

If board policy is aimed at simple legal compliance, only a thorough examination of a particular situation accompanied by a safety perspective will determine if minimum compliance is cost-effective. Would the board consider that the cost of premature equipment replacements due to damage or destruction by accident is far greater than the depreciation reserves started when the equipment was purchased? Do they know that safety programs built on defensive reactions are costly, and that defensive-type programs, particularly in simple response to legislation, become progressively more expensive and inefficient?

The potential cost of problems involving noise, radiation, temperatures, toxic dusts, fumes, gases, vapors, and smokes may well exceed the cost of physical injury and property damage from related accidents—particularly in the short term. The board of directors has ultimate responsibility for all aspects of the company, to include safety and health. Without any safety background among board members, they likely depend on guidance from accounting firms and corporate counsel. A solid board should confront safety and health issues as it would any other major financial management process—with strong subject-matter expertise. It should realize that short-term investments in issues not necessarily resulting in a quick return and result in long-term savings and ultimately higher profits.

Corporate Strategies and Objectives

Corporate safety strategies must be matched to corporate objectives, and strategic planning must align safety and health issues with the best interests of the entire corporation. If the board can pull safety and health into the planning and directing processes and make the safety function an integral part of nearly every operation in every location, it can have widespread, favorable results. It must at least be able to ask the right questions and insist on the right answers.

Is Compliance Enough?

Many companies tend to think of safety and health strictly in terms of compliance—providing the minimum required by law, on the theory that compliance alone is quite costly and exceeding requirements is more costly yet. In fact, the reverse may be true. In the long run, it is more profitable to operate safely, and minimum compliance yields minimum benefits.

REFERENCES

Drucker, Peter. (1954). *The Practice of Management.* New York: Harper & Row.
Fine, William T. (1976). *A Management Appraisal in Accident Investigation.* Silver Springs, MD: U.S. Naval Surface Weapons Laboratory.

Findlay, James V. and Kuhlman, Raymond L. (1980). *Leadership in Safety.* Loganville, GA: Institute Press.

Knox, N.W. and Eicher, R.W. (1976). *MORT User's Manual SSDC-4.* Washington: Government Printing Office.

Petersen, Dan. (1978). *Techniques of Safety Management*, 2nd ed. New York: McGraw-Hill.

Pope, William C. (1976–1980). *Systems Safety Management Series.* Alexandria, VA: Safety Management Information Systems.

Reason, James. (1990). *Human Error.* UK: Cambridge University Press.

Chapter 3

Planning and Organizing the Safety System

Mark A. Friend

CASE

Larry Hugh has been director of safety for nearly a decade, but he is unhappy with the overall safety program in his company. After careful consideration and discussions with numerous vendors, Larry purchases a comprehensive safety and health program from an online source. The vendor states that it will meet all OSHA compliance requirements and fit into any organization. After the first few months, Larry realizes the program is poorly aligned with the overall mission of his company. After discussions with the CEO and many of the workers, Larry decides to scrap the recently purchased program and build one based on the overall needs of the company and its workers. He carefully aligns the new one with the company mission.

INTRODUCTION

Long before any program is put into effect, senior management can do much to determine its ultimate success—or failure. In particular, this chapter will concentrate on the necessary advance steps to be taken in developing and executing a safety and health system—both by senior management and key positions within the organization.

THE PLANNING FUNCTION

The planning function consists of four major areas: (1) forecasting, (2) goal and objective setting, (3) policy making, and (4) budgeting. All fall within the responsibilities and authorities of various positions within the organization. Forecasting the market and outlining future direction of the organization falls to the highest levels of management and must precede the other functions. Setting goals and objectives then defines the organization's future direction more specifically and falls within the parameters established by forecasts and planned direction of the organization.

The safety system is designed and based on the organization where it is located. Once overall objectives regarding safety are established, realistic goals are incorporated into the organizational plan. First, goals may simply be targets of performance. For example, the corporate goal during the coming year could be a 15 percent reduction in OSHA recordables or first-aid cases. It might also be something like a maximum dollar payout for workers compensation benefits. Some organizations chart quantitative targets or trends in an attempt to lower their overall costs. Although these are often easily measured and compared to overall performance, they don't necessarily reflect the best approach to improving the overall safety system. Performance measures, in terms of how well managers and all employees are engaged in the overall safety system, are generally considered stronger indicators and, therefore, better company targets or goals.

As goals are established, the overall well-being of the organization must be considered. The goal of any safety program is to improve the overall management system, and, as part of that, the overall safety system. Merely establishing and even possibly accomplishing those goals without consideration of the systems themselves is nearly impossible. Designing of the safety system begins with the establishment of an overall, comprehensive plan. The various components of the organization must be considered relative to how safety will become integrated into the system. The overall goal is to improve the system by building safety into each and every part.

Policies provide guidance in carrying out these goals and objectives. Budgeting is the necessary last step. In planning for safety and health, the four areas should operate as interrelated activities, culminating in a successful system, outlined by a written program normally contained in one or a series of safety manuals.

In any safety and health system, the planning function must be consistent with the following (Federal Aviation Administration [FAA], 2014):

1. *Appropriateness.* Executive management must consider the future, the goals and objectives, of a company in order to develop an appropriate program to them.
2. *Accountability.* Senior management must assume overall responsibility and accountability for the program.

3. *Delegation.* Safety and health duties must be logically delegated throughout the management structure, though final responsibility and accountability are retained by senior management.
4. *Key Personnel.* The key persons responsible for carrying out the plan should be identified. This is usually a function of position.
5. *Definition.* The safety and health relationships between management, staff, and line should be clearly defined. This is usually done through job description.
6. *Measurement.* There should be a method of assuring that each member knows his or her exact safety and health responsibilities, and an effective means for measuring what has been done.
7. *Support.* All management and staff must provide adequate support to the safety and health function.
8. *Identification.* There should be detailed plans for identifying, controlling, and auditing workplace safety and health hazards.
9. *Monitoring.* There should be a system to monitor individual, divisional, and corporate performance in safety and health matters.
10. *Information.* Managers and staff should provide information and results from this monitoring to senior management for decisions.
11. *Training.* Safety and health training needs should be clearly identified and periodically reviewed.
12. *Resources.* Executive management should provide adequate resources for carrying out the safety and health program, through budgeting, allocation, and personnel.

SAFETY POLICY AND OBJECTIVES

Prior to the establishment of objectives, the organizational mission and primary goals should be reviewed. Any and all objectives, including those of the safety program, should fall within parameters set forth by the mission and its related goals. The safety objectives should contain (FAA, 2014):

1. A commitment of the organization to fulfill the organization's safety objectives.
2. A clear statement about the provision of the necessary resources for the implementation of the safety system.
3. A safety reporting policy that defines requirements for employee reporting of safety hazards or issues.
4. A policy that defines unacceptable behavior and conditions for disciplinary action.

The safety policy must be signed by the accountable executive and communicated throughout the organization. It also needs to be regularly reviewed by the accountable executive to ensure it remains relevant and appropriate to organization. Policy and objectives are not the same. Policy is general in nature, while program objectives and goals are very specific, spelling out the expected end results. Policy gives notice that the program *will* be carried out; objectives state the end results expected as policy *is* carried out. Policy may state, "It is company policy to prevent injury to employees," while the objective may state specifically, "The objective is to reduce the number of lost-time accidents to 25% of its present level." This is not yet spelled out how it is to be done; it is up to individual managers to see that it is done and to explain the process and goals to their subordinates. Their interpretation may not apply to the entire company, but only to their operation or department. The lower the level of implementation, in fact, the more specific it becomes, though in the end it all depends on policy. Thus, there is a need to assure that all policies are communicated throughout the organization. When management is convinced that a positive policy for safety and health is worthwhile, and where this policy has been communicated throughout the organization, success follows.

Safety and health policy is usually prepared by lower-level managers and staff, and signed by the senior executive. Effective policy comes into being only through the effort of the entire organization, from senior management down and certainly with the help of the safety professional. While policies influence the direction of the entire organization, they should not be "carved in stone," and should be reviewed and revised periodically. Since senior management generally has the clearest overview of present conditions, recent and proposed changes, and anticipated conditions, it should take an active role in these policy reviews and revisions.

Policy Problems

In order to develop effective policies and avoid certain pitfalls, it is helpful to review criteria for the establishment of policies:

1. Policies are often used as an excuse for not taking action or not making decisions. ("That's not company policy.")
2. Policies are sometimes so weakly worded or generally written that they have no real meaning and call for no real commitment to anything.
3. Policies and rules are sometimes confused. Policies express a general thought that guides management and leaves room for discretion. Rules are meant to support policy by giving specific directions. Neither policies nor rules are meant to be broken. If they are incorrect or inconsistent with actions, either the policies/rules or the actions need to be changed.

4. Safety and health policy may be designed to defer to public opinion but is neither realistic nor convincing. For example, "Safety first" is a noble slogan, but overlooks the obvious fact that the company must actually say, "Profits and mission first!" Employees and the public can see straight through this transparent contradiction.
5. A safety and health policy is only one of numerous policies the company will have. Too often these clash with each other or are not coordinated. A policy of "Safety first," for instance, is obviously out of step with another policy calling for "efficient production at the lowest cost." Safety and health policies should be designed to complement and support the company mission as well as other company policies and should never do otherwise.

What Is a Good Policy?

A timely, well-prepared policy will avoid most of the problems listed above. A good policy will establish the character of a company internally, instead of letting it be fashioned by external and fortuitous events. Effective safety and health policies allow managers to delegate safety and health activities without having to consider repetitious problems which have been previously solved. Strong policies also promote unified thinking among managers, so they can see safety and health problems in the context of the total corporate structure. It makes them members of the team, instead of merely isolated individuals with problems. Policies also allow improved decision-making. Policy crosses functional lines and provides a common base for solving problems not in a particular manager's area of expertise. They also encourage managerial teamwork in solving common problems, since all are equally affected by the policy and equally responsible for following it. A safety and health policy is a communication tool, as well as a corporate statement. The more clearly it communicates, the more effective it will be. The following are points to remember in creating an effective policy that does communicate:

1. *Be brief.* Explain as succinctly as possible what we will do or try to do in safety and health. Think in terms of a text message rather than a dissertation.
2. *Be inclusive.* State the policy in the broadest terms possible so that it includes all that needs to be done.
3. *Think ahead.* Take a long-range view, so that it won't be out of date too soon. Policy shouldn't have to change every time a new objective or direction is determined.
4. *Involve others.* Involve other managers in framing policy to represent consolidated opinions and to cut down resistance to the policy.

5. *Write it out.* To be effective and stable, a policy must be committed to writing. Verbal policies quickly get vague, distorted, or engender disagreement. A written policy can readily be referred to for clarification.
6. *Communicate it.* Copies of the written policy should be distributed to every manager in the organization. This will assure a greater degree of consistency and application.
7. *Go to the top.* Get the policy approved by the highest corporate authority to give it the power and acceptance it needs.
8. *Manualize it.* Include it in the corporate manual to become an integrated part of the organization's direction.

Policy states the "will" of management, but does not give details. Fuller explanation of the policy is usually needed, but *it should not be part of the policy itself.* Other documents, such as implementing directions, reports, or manuals may be used. They can be used to explain:

1. *Why* there is a policy?
2. *How* the policy is going to be carried out?
3. *Who* is covered by the policy, *who* has responsibility and authority in connection with it, and *who* will give it timely support and direction?
4. *When* the policy will be implemented?
5. *Where* the policy will be in effect in the organization?

A policy statement is a broad plan of action. Good policy formulation is not easy, and it must be understood by those who are responsible for formulating it, as well as by those who are expected to act according to it. The statement of policy and the policy objectives must be related to practical, achievable results. Actual achievement requires clearly defining the responsibilities of every person and function involved in the safety and health program. A clear allocation of responsibilities will assure that

1. Policy and objectives are integrated into the overall organizational functions.
2. The legal basis for programs is clearly defined, and someone will take care of them.
3. Information is collected and analyzed for management decision and control purposes.
4. Roles and responsibilities are translated into training needs.
5. Accountability for safety and health, as well as awareness of the program, is promoted.
6. Recruitment and selection of new personnel will conform to appropriate criteria for safety and health purposes.

Planning and Organizing the Safety System 41

7. A clear channel or line of communication for all safety and health problems is defined and known to all.

Review the policy statements and ensure that the policy has enough scope to address all the major problems likely to occur in the course of operations. It should be comprehensive enough to address the major concerns of human outlook, cost efficiency, legal compliance, and so on. For these reasons, it is essential that the policy be written in a lucid and simple style, so it can be implemented without conflict or confusion. Without a sound policy, the remaining functions of managerial control—responsibility, authority, and accountability—cannot operate effectively, either separately or as an interacting triangle.

RESPONSIBILITY, AUTHORITY, AND ACCOUNTABILITY: LINE AND STAFF

The policy should clearly define who is responsible for what to help assure responsibility is accepted at the proper level, particularly so line management can accept risks. Guidelines and rules should assure the needs, capabilities, and responsibilities of supervisors are addressed also. Top management must make it absolutely clear where and with whom responsibilities are located in the organization. Performance should be regularly reviewed against these responsibilities. The authority and responsibility at any level are directly related to the authority and responsibility at adjoining levels. When an accident or injury occurs, it can always be traced to a higher level of responsibility. Conversely, if appropriate procedures and plans have not been put into effect, the result may be an accident or injury at a lower level. All of this is done within the context of budget and budgetary constraints. In order to carry out an effective safety and health program, a strong business case will have to be made and the safety and health professional should be able to demonstrate the cost benefits of establishing and maintaining an effective program. Safety and health operations complete with every other operation within the organization for resources. Allocation of resources is usually based on expected return on investment.

Specific safety and health responsibilities are largely a result of position and where the position lies within the organization. Most organizations can be divided into line and staff. Line is charged with carrying out the main functions of the organization, whereas staff is responsible for advising line. Line positions usually include production or service as it relates to what the organization does. For example, in a manufacturing facility the line is responsible for making the product. In an architectural firm, the line is responsible for producing the plans. Staff, on the other hand, supports the line with functions such as human relations, accounting, and safety. The role of safety is to monitor and advise line on corrective actions. The role of the line is to implement and enforce the actions they deem critical or important to overall operations. Safety, as a staff function, does not have the authority to implement the actions. The safety professional makes the recommendations to the line and the line implements and enforces any changes. The chief executive officer (CEO) or the designee of the CEO serves as the accountable executive. Implementations generally occur at that level and are then enforced throughout the organization. If recommendations only affect a single department or operation, they may be made at a lower level, but if they are rejected, they can be carried to the accountable executive. Neither the safety professional nor his or her staff should generally have the authority to implement change within the line functions. The only exception to this would be in emergency situations, but the changes would ultimately be the responsibility of the accountable executive in charge of the line.

If the overall program is to fulfill the intention of the policy statement, authority must be properly assigned. Safety and health programs and functions cannot be implemented without authority. Authority to correct system failures, for example, must be clearly delegated, and published in a form and style easily grasped. Special care should be taken not to leave interface gaps when authority is delegated.

Through proper delegation, top management can assure the effective control of lower-level programs. While this delegation does not, of course, relieve top management of its overall responsibility, it assures top-level policy is actually carried out at the operating level. Moreover, vesting authority at the proper level increases program credibility, as well as effectiveness. Authority should rest with the line and all decisions affecting safety and health is ultimately delegated to line management. Changes are recommended by the safety professional who resides on the staff side. The safety professional must make the case for needed changes, but line management must ultimately implement those changes.

Responsibility and authority are only two of the necessary elements that allow top-level programs to be carried out at the procedural and performance levels. There is a third, indispensable element: accountability. Simply stated, top and middle management can never really lose their accountability for the operations under their purview. They can delegate some authority and hold those who have been delegated accountable for their particular portion of the program. Ultimately though, top management remains accountable for the overall safety and health program.

It becomes immediately apparent that the safety professional and staff must not only be able to recognize and make recommendations regarding safety and health issues, but must also be capable of presenting a strong, convincing case to line management to make necessary changes. As mentioned above, this is typically in terms of return on investment or ROI.

Full implementation of the safety and health program occurs largely at the front-line, supervisor level. The supervisor is actually in charge at locations where accidents and injuries most often occur. This important role is often reduced and his diminished by several factors. Nowhere is the difference wider between what the responsibilities and roles of a position *should be,* and what they *actually are.* This is a result of the following. The supervisor:

1. May lack the status of being considered part of the management team.
2. Has priorities are generally set by management pressures.
3. Has responsibility and accountability seldom backed by sufficient authority to do the expected tasks by not being able to hire, fire, or discipline his workers, nor having the authority to promote or demote them. The supervisor does not have a hand in making the rules he and his workers must follow and is not involved in developing the work standards for the jobs he supervises.
4. Must give full support to the company policy or position, whether right or wrong.
5. Is constrained by cost/benefits considerations on safety equipment.
6. Has only restricted control over working conditions.
7. Often lacks adequate communications channels parallel or upward through management.

These are all roadblocks to effective safety and health programming, yet most of the burden of workplace safety is placed directly on the supervisor.

ROLE OF A SAFETY PROFESSIONAL

The professional must engage continual measurement and appraisal of the safety system and provide guidance to management regarding how to improve it while minimizing risk. A safety professional's role will often include aiding in the development of the following:

1. *Safety administration.* Management must develop and implement a strategy to involve others in the program.

2. *Safety standards.* Appropriate rules, regulations, standards, and procedures must be developed and disseminated.
3. *Facility inspection.* Audits, inspections, and feedback from every level should be regularly scheduled.
4. *Accident investigation.* Proper investigation includes analysis, reporting to management, and recordkeeping.
5. *Safety education and training.* Deciding who needs training and why, are vital parts of the safety professional's job. The safety professional may also provide the training.
6. *Safety and health oromotion.* Management may need advice about public relations, consumer liability, writing speeches, and promotional materials.
7. *Safety research and engineering.* This function not only recognizes hazards, but proposes solutions, including where and how to solve them.
8. *Development of a written safety program.* This function involves recognizing and incorporating all relevant safety regulations into an overall, comprehensive safety program.

THE SAFETY PROGRAM

The safety and health program should then be incorporated into existing operating instructions. It should provide clear guidance as to how to accomplish safety goals. There must be processes in place to identify hazards and, depending on the level of risk presented by each, mitigate those hazards outside of an acceptable level of risk. This process of hazard identification and mitigation must become integral to aspects of the organization. It must be done systematically. This is done with the support of managers and their staffs, since it involves their functions. The following are some effective guides to developing a written safety program:

1. *Establish the authority* for the safety and health program, whether it is company policy or governmental regulation.
2. Publish the safety and health policy over the *chief executive's signature.*
3. Establish *safety and health responsibilities.* Normally the chief executive officer (CEO) assumes overall responsibility. The CEO should direct implementation and participation concerning the program, and should personally participate to reduce operating errors and managerial oversights that could lead to accidents and injuries. At some level, identifying hazards and mitigating them become a part of this process.
4. Specify *line responsibilities,* being careful to point out where the buck stops, so that the responsibilities don't continually get passed on until

the supervisor is left holding the bag. While operating responsibilities are passed down the line, higher management levels remain accountable and responsible for safety and health matters.

5. Clearly spell out *supervisory responsibilities*, such as seeking out and correcting work errors, hazards, mechanical defects, malfunctioning equipment and property, substandard work conditions, and environmental problems. The manual should *state what channels are open* to the supervisor if the problems are beyond his resources or control.
6. Discuss and define the *employee's role* in observing and following safety and health rules for his or her own and fellow workers' protection.
7. Discuss the *safety manager's function and role.* It should be clear that the appointment of a safety manager does not relieve the line manager, staff person, or supervisor of any safety responsibility, but does furnish a resource to assist in solving safety and health problems.

The safety and health program can cover much more, but it should definitely contain statements about the following areas:

1. *Organization.* The program should be presented as an integral part of the entire management system, on an organization-wide basis.
2. *Scope.* The program's full scope should be spelled out. It should continuously identify the causes and costs accrued due to operating errors.
3. *Safety program areas.* It should identify all safety program areas, including employee safety, vehicular safety, off-duty safety, public safety, employee health, emergency response, hazardous materials operations, fire protection, etc.
4. *Committees.* Safety committees should be established at each major organizational unit. This item should address the following:

    a. Composition of each committee
    b. Appropriate activities to be undertaken
    c. Procedures for reporting and recordkeeping.

5. *Nonsupervisory Employees.* The manual should address the participation of nonsupervisory employees to assure their engagement in the safety and health program. In addition, the safety program should include some definition of the safety roles and functions at their level as well.

Proper planning and organization are essential to the ultimate effectiveness of any safety and health program. The senior executive should be aware of the recommended steps outlined here, and should review thoroughly *before* attempting to develop a new safety and health program or make radical revisions to an existing one. It can potentially save the organization millions of dollars in lost time, wasted materials, and worker injuries and provide the ROI needed to survive.

REFERENCES

Federal Aviation Administration (FAA). (2014). *Advisory Circular: 120-92b, Safety Management Systems for Aviation Service Providers.* Retrieved from: http://www.faa.gov/documentLibrary/media/Advisory_Circular/AC_120-92B.pdf.

Ferry, Ted S. (November, 1976). "3 P's in Safety: Policies, Procedures and Performance." *Professional Safety*, Park Ridge, IL: American Society of Safety Engineers, pp. 26–29.

Findlay, James V. and Kuhlman, Raymond L. (1980). *Leadership in Safety.* Loganville, GA: Institute Press.

Friedman, Milton and Friedman, Rose. (1980). *Free to Choose.* New York: Harcourt, Brace & Jovanovich, pp. 229–34.

Manualizing Your Safety Function. (1975). Alexandria, VA: Safety Management Information.

Pope, William C. (1977). *Staffing the Safety Function.* Alexandria, VA: Safety Management Information Systems.

Pope, William C. (1973). *Practice of Safety Responsibility.* Alexandria, VA: Safety Management Information Systems.

Pope, William C. (1973). *Safety Management Information Systems "Establishing an Effective Safety Committee."* Alexandria, VA: Safety Management Information Systems.

Pope, William C. (1973). *Systems, "How to Develop a Safety Policy."* Alexandria, VA: Safety Management Information Systems.

Pope, William C. (1976). *The OSHA Act of 1970: A Guide to Safety Program Elements.* Alexandria, VA: Safety Management Information Systems.

Pope, William C. (1977). *The Problem of Locating the Safety Function.* Alexandria, VA: Safety Management Information Systems, *The Gut Issue: Managerial Missions*, (1980). Alexandria, VA.

Chapter 4

Safety Management Framework

Kathy S. Friend and Mark A. Friend

CASE

Scott Fleetwood is unhappy with the safety program he built and administers. When he reviews accident records for the last five years, nothing seems to have improved. The same problems are occurring this year that occurred five years ago. Scott decides it is time for a change, but he doesn't know where to turn. After a long discussion with one of his mentors, former director of a large safety consulting company, Scott determines he is going to look at some different models offered by various government agencies and other organizations. After carefully reviewing them, he decides his best approach is to follow a voluntary standard and adjust it slightly to suit the needs of the team where he works.

OSHA REGULATIONS

Managing the safety system is addressed in numerous guidelines and regulations. Some of the more important approaches are set forth in safety standards proposed or required by government agencies, but there are also some very strong voluntary programs offered by some organizations. Of course, the Occupational Safety and Health Administration (OSHA) sets forth a number of requirements in 29 CFR 1910, Occupational Safety and Health Standards. These are supplemented by 29 CFR 1926, Construction and Industry Regulations. Part 1910 is broken into subparts and each subpart into sections. Basic requirements under 1910 include the following (Friend and Kohn, 2014):

Subpart A addresses the purpose and scope of the act, definitions, petitions, amendments, and applicability of the standards. It also explains individual roles under OSHA and where the standards apply.

Subpart B applies the standards to the employers, employees, and places of employment covered by the Act.

Subpart C provides employees and designated representatives' rights of access to relevant exposure and medical records. It also provides legal access to these records by OSHA compliance officers.

Subpart D addresses the requirements for maintaining walking and working surfaces and further addresses ladders, scaffolding, railing, and other working surfaces.

Subpart E addresses means of egress and provides information on employee emergency and fire prevention plans.

Subpart F covers powered platforms, manlifts, and vehicle-mounted work platforms.

Subpart G deals with air quality, noise exposure over 85 decibels, and radiation exposure.

Subpart H contains information on various hazardous materials and also includes process, safety management requirements of highly hazardous chemicals, hazardous waste operations, and emergency response.

Subpart I has general requirements for minimum personal protective equipment (PPE) used by employees in their tasks.

Subpart J addresses general environmental controls in permanent places of employment to include such items as toilet facilities, washing facilities, sanitary food storage, and food handling. It also covers issues dealing with temporary labor camps.

Subpart K covers medical and first aid to provide the employee with readily available medical consultation as it applies to the on-plant health.

Subpart M applies to compressed gas and compressed air receivers and related equipment in cleaning, drilling, hoisting, and chipping operations.

Subpart N covers materials handling and storage.

Subpart O addresses machine guarding for equipment exposing employees to hazards exposed to moving or rotating parts during operations.

Subpart P describes required guarding, inspection, and maintenance requirements for hand and portable tools including lawn mowers and other internal combustion engine tools.

Subpart R applies to special industries as defined by OSHA, including certain mills, bakeries and laundries

Subpart S establishes electrical safety requirements in the workplace.

Subpart T covers commercial diving operations.

Subparts U-Y are reserved.

Subpart Z addresses toxic and hazardous materials in the workplace. It covers Permissible Exposure Limits (PELs) for contaminants including gases, vapors, and dusts.

The specific standards do not address a management approach to addressing safety; however, in 1989 OSHA published voluntary safety and health program management guidelines. The OSHA (2005) Fact Sheet suggests an effective safety management system goes beyond legal requirements in order to be effective. OSHA calls for major elements to be included in any effective safety management system. In November, 2015 OSHA set forth a draft of new safety and health management guidelines that provides additional guidance. The guidance below incorporated elements from both approaches. Citations from 2015 indicate updates provided by the newly proposed guidelines.

OSHA APPROACH TO SAFETY MANAGEMENT

Management Commitment and Employee Involvement (OSHA, 2005). This requires

- A clearly stated worksite policy on safe and healthful work and work conditions so that all personnel understand the priority of safety and health protection.
- Top management demonstrates its commitment to continuous improvement in safety and health, communicates that commitment to workers, and sets program expectations and responsibilities (OSHA, 2015).
- A clear goal for the safety and health management system and objectives for meeting that goal. All members of the organization must understand the desired results and the measures needed to achieve them.
- Top management involvement in implementing the system.
- Employee involvement in the structure and operation of the system and in decisions affecting their safety and health.
- Assignment of responsibilities for all aspects of the management system, so managers, supervisors, and employees in all parts of the organization know their responsibilities.
- Managers at all levels make safety and health a core organizational value, establish safety and health goals and objectives, provide adequate resources and support for the program, and set a good example (OSHA, 2015).
- Provision of adequate authority and resources so everyone can meet assigned responsibilities.

- Holding everyone at the site—managers, supervisors, and employees—accountable.
- Annual reviews of the system's operations to evaluate success in meeting the goals and objectives, so deficiencies can be identified and the program and/or the objectives can be revised as needed.

Worker Participation (OSHA, 2015):

Workers and their representatives are involved in all aspects of the program—including setting goals, identifying and reporting hazards, investigating incidents, and tracking progress.

All workers, including contractors and temporary workers, understand their roles and responsibilities under the program and what they need to do to electively carry them out.

Workers are encouraged and have means to communicate openly with management and to report safety and health concerns without fear of retaliation.

Any potential barriers or obstacles to worker participation in the program (for example, language, lack of information, or disincentives) are removed or addressed.

Worksite Analysis (OSHA, 2005), including:

- Procedures in place to continually identify workplace hazards and evaluate risks (OSHA, 2015).
- An initial assessment of existing hazards and control measures is followed by periodic inspections and reassessments to identify new hazards (OSHA, 2015).
- Analyses of planned and new facilities, processes, materials, and equipment.
- Analysis of hazards associated with jobs, processes, and/or phases of work.
- Regular site safety and health inspections to identify new or previously missed hazards and failures in hazard controls.
- A reliable system to encourage employees, without fear of reprisal, to notify management personnel about conditions that appear hazardous and to receive timely and appropriate responses.
- Investigation of accidents and "near-miss" incidents to determine their causes and create prevention strategies.
- Analysis of injury and illness trends over extended periods to identify patterns and prevent problems.

Hazard Prevention and Control, calling for:

- Employers and workers to cooperate to identify and select options for eliminating, preventing, or controlling workplace hazards (OSHA, 2015).

OSHA encourages adoption of the following measures: engineering controls, the most reliable and effective; administrative controls that limit exposure to hazards by adjusting the work schedule; work practice controls; and personal protective equipment (PPE)—in that order. In other words, engineering controls are the most desired control and use of PPE is the least.
- A plan developed that ensures controls are implemented, interim protection is provided, progress is tracked, and the effectiveness of controls is verified (OSHA, 2015).
- Systems adapted to meet each workplace's particular characteristics.
- Planning and preparing for emergencies and conducting emergency training and drills.
- A medical program including first aid and emergency medical care.

Safety and Health Education and Training to:

- Ensure all workers are trained to understand how the program works and how to carry out the responsibilities assigned to them under the program (OSHA, 2015).
- Ensure all workers are trained to recognize workplace hazards and to understand the control measures that have been implemented (OSHA, 2015).
- Ensure supervisors and managers understand their responsibilities and the reasons for them.
- Ensure periodic refresher training for all employees.
- Establish a medical program for first aid and emergency medical care.

Program Evaluation and Improvement (OSHA, 2015):

- Control measures are periodically evaluated for effectiveness.
- Processes are established to monitor program performance, verify program implementation, identify program deficiencies and opportunities for improvement, and take actions necessary to improve the program and overall safety and health performance.

Coordination and Communication on Multiemployer Worksites (OSHA, 2015):

- The host employer and all contract employers coordinate on work planning and scheduling to identify and resolve any conflicts that could impact safety or health.
- Workers from both the host and contract employer are informed about the hazards present at the worksite and the hazards that work of the contract employer may create on-site.

The OSHA website at www.OSHA.gov provides additional information on implementation of the updated Safety and Health Program Management guidelines.

THE SAFETY MANAGEMENT SYSTEM AND STANDARDS SUPPORTING IT

In recent years a number of organizations have considered and written standards addressing safety management systems (SMSs). The International Organization for Standardization (ISO) (2015, July) passed ISO 14001 in 2004, addressing environmental management from a system perspective. Safety was a consideration for inclusion in the ISO standards, but was not addressed. ISO attempted to develop a separate safety standard, but the effort did not succeed. Several European groups examined the British Standards Institution standard (BS 8800) (1996), *Guide to occupational health and safety management systems*, and with minor modifications it became OSHAS 18001 (2002). The ILO shortly thereafter published *Guidelines on Occupational Safety and Health*. In 1999, the American Industrial Hygiene Association proposed a standard for health and safety programs, and in conjunction with the American National Standards Institute and the American Industrial Hygiene Association (ANSI/AIHA, 2012) published the ANSI Z10 standard. Their approach was one of continuous improvement or the plan-do-check-act approach.

ANSI Z10

Based on earlier research, there were certain key elements deemed necessary for the achievement of successful safety outcomes (Friend, 2012). These include the following:

- Strong management support. Management must assume total responsibility for safety with accountability at the top.
- Employee engagement in safety processes. Employees must have the time and resources needed to engage in the process from the very beginning. They must also participate in all stages of the process.
- Proactive hazards identification processes. The system must have an ongoing method of identifying hazards before they are actually manifest or cause problems. Ideally, the system will consider any changes occurring and the ramifications those changes will have on safety.
- Risk assessment and mitigation. Risk can never be completely eliminated, but it can be assessed in terms of the potential costs and probability of given

events. Once assessed, those areas with the greatest risk can be systematically identified and the risk can be mitigated.
- Safety training. Managing safety and operating safely at every level requires initial and ongoing training for all employees exposed to hazards.
- Contractor safety. Outsiders interacting with or working inside of a company's controlled environment must abide by the same standards maintain the same levels of safety as employees.
- Home safety. Off-the-job losses, in terms of safety and health, can be very costly to an organization. Every attempt must be made to minimize employee losses while they are away from work.
- Management review and feedback. Continuous engagement by top management is necessary.

The Z10 approach follows a systematic approach to managing safety by engaging in the following steps:

1. Review all relevant management systems and look for gaps.
2. Establish a process to assess and prioritize the issues on an ongoing basis.
3. Establish plans and objectives based on risk-reduction opportunities.
4. Implement the plans.
5. Monitor, measure and assess the process.
6. Implement and incident investigation plan.
7. Periodically audit the system.
8. Take corrective and preventive actions (Friend, 2012).

The standard contains:

- Examples of policy statements
- Outline of roles and responsibilities
- System for encouraging employee participation
- Lists of information to gather
- Guidelines for and examples of objectives
- Risk assessment techniques
- Process for managing change
- Guidance for:
- Procurement
- Contractor safety and health
- Incident investigations
- Audits
- Management reviews
- A useful bibliography (Friend, 2012)

Z10 is an extremely useful document outlining a systematic and relatively thorough approach to implementing a safety management system. It can be ordered online at http://www.techstreet.com/products/1842775?product_id=1842775&sid=goog&gclid=CKrasP3BzMkCFcUXHwodC_0E2w.

Another Approach to SMS

The FAA (2015) suggests a key aspect is *safety culture*, a product of the values and actions of the leadership and the results of organizational learning. This culture isn't created, but it evolves over time as a result of related experiences. It continues to improve as management of safety throughout the organization evolves. The safety culture and accident prevention are strongly correlated. The FAA suggests that the management framework has certain characteristics that lead toward a safer culture. A few of these are as follows:

1. Open reporting. Employees can report errors without fear of reprisal. This occurs within a context of accountability by all employees, including members of management.
2. Just culture. There is an engagement in finding and identifying errors in the system and implementing preventive and corrective actions. Recklessness and willful violations are not tolerated.
3. Personnel involvement. All line personnel at all levels are included in the safety system (FAA, 2015).

A critical piece of the FAA SMS is visible management commitment and involvement to the overall safety processes. Active engagement of leadership is essential to the success of SMS and the development of an effective safety culture. The FAA proposes building the SMS around four essential components: safety policy, safety risk management, safety assurance, and safety promotion.

1. Safety policy. Objectives are set forth, responsibilities are assigned, and standards of performance are clearly stated. Management will clearly communicate its dedication and commitment to safe performance by all employees. This commitment includes provision of necessary resources and definition of requirements for employee reporting of hazards. It also defines behaviors qualifying for disciplinary action. A critical part of the safety policy is the identification of an accountable executive who is the final authority over all matters related to the safety system. The accountable executive will hold managers responsible for identifying hazards, conducting risk assessments, and developing risk controls for their respective areas of responsibility. Managers with the authority to implement changes must be clearly identified

(FAA, 2015). The accountable executive is generally considered the representative of senior management who controls and allocates resources for the safety and health program, supervised by the safety professional.

Walls' (2013) research suggests the following are all essential components of safety policy: accountability for safety, established safety standards, zero tolerance for unsafe acts, senior management review of safety performance, and contractual language binding all contractors to compliance with all safety rules. According to ICAO (2013), all safety standards and safety performance objectives must be linked to safety performance indicators, safety targets, and SMS mitigation actions.

Within this first component the FAA requires an emergency response plan (ERP). ICAO (2013) also promotes the development and testing of an ERP within the organization to ensure appropriate actions are taken by individuals to make certain there is an orderly transition from normal operations to those required in an emergency. It is critical that thorough documentation is maintained to include descriptions of SMS components and elements as well as current, related SMS records and documents (ICAO, 2013).

2. Safety Risk Management (SRM). The SRM is a formal system for identifying and mitigating risk by providing a decision-making process to do so risk. The SRM component is the organization's way of considering risk in operations and reducing it to an acceptable level. SRM is a design process, a way to incorporate risk controls into processes, products, and services or to redesign controls where existing ones are not meeting the organization's needs. This occurs through the consideration of the overall system in terms of its function and purpose; operating environment; processes and procedures; and the personnel, equipment, and facilities necessary for operations. One goal is to identify and address problems before new or revised systems are operational. The system analysis considers activities and resources necessary for the system to function and frequently includes representatives from management, employee groups, subject-matter experts, and other work groups such as safety committees or safety action teams. The hazard identification process involves asking, "What could go wrong that could cause an accident under both typical and abnormal operations?" (FAA, 2015). Possible approaches to making this determination include the following (Transportation Research Board [TRB], 2015):

Checklists: Checklists are prepared for self-inspections and they usually involve reviewing equipment or procedures to assure everything is in place and/or being done properly. For example, before a machine is activated at the beginning of a shift, the operator may pick up a checklist designed to address the major safety components of the machine and review them to be certain everything is intact, in place, and working as it should. Items to check could include guarding, placement of materials, in-place switches, PPE, or other pertinent features. Checklists may be prepared by competent operators and periodically checked to be sure they are thorough and current.

Observation and experience: Managers, operators, and staff members must continuously observe and be aware of circumstances or problems that may pose a safety risk to operations, even when not listed in the checklists. Examples of hazards in this category include industrial trucks speeding in or around the facility, appropriate PPE not begin worn, and access to fire extinguishers blocked by items being stored.

Brainstorming: A group of stakeholders meet to identify hazards and analyze risks. During the brainstorming session, the group develops a list of hazards associated with the issue undergoing assessment. Brainstorming is a common basis for a preliminary hazard analysis (PHA). The what-if game can be very useful in brainstorming. For example, what if a leak of this hazardous gas occurred during operating hours? Sometimes even related events can be very useful in anticipating hazards.

Job safety analyses: this is a technique that uses job tasks to identify hazards. A job hazard analysis explores how the worker, the specific task, the required tools, and the work environment relate. The job is typically broken into eight to twelve tasks. For each task, the potential hazards associated with it are identified and then protective measures for each of those hazards are outlined. Measures may include mitigating the hazard through engineering, implementing specific procedures or utilizing PPE. The JSA is useful in identifying hazards and is often reviewed before a job is undertaken. It is especially useful for jobs infrequently performed.

Preliminary hazard analysis (PHA): based on the safety issue or activity, a preliminary list of potential hazards can be prepared. The goal is to identify the hazard or combination of hazards that may cause an unsafe condition and ultimately result in an accident. This task is normally performed by a group of people familiar with the job or equipment with the goal of identifying and mitigating unacceptable levels of risk.

Once hazards and their associated risks are identified, it becomes necessary to eliminate or control them systematically. Before that occurs, they are evaluated for both severity and probability of occurrence, so the level of risk can be estimated. Calculations may be compared to a risk matrix to help determine problems to be addressed first. Risk controls are then put into place for those problem areas deemed unacceptable, based on the level discerned (FAA, 2015) As mentioned previously, OSHA encourages the use of appropriate measures as follows (OSHA, 2005):

- Engineering controls, the most reliable and effective.
- Administrative controls that limit exposure to hazards by adjusting the work schedule.

- Work practice controls.
- Personal protective equipment (PPE)—in that order.

In other words, engineering controls are the most desired control and use of PPE is the least desired control method for mitigating risk. Not all risk can be eliminated, so the goal is often to lower it to as low as reasonably achievable (ALARA). This is accomplished by either lowering the probability of the risk or the potential severity of the outcome to an acceptable level (TRB, 2015). This often requires a cost-benefit analysis to determine whether or not intervention methods are, in fact, justifiable. This analysis is followed by a decision as to whether or not to implement the appropriate measures.

Santos-Reyes and Beard (2002) noted that prompt and complete communication is necessary in this process and it requires each part of the organizational system to maintain its own, somewhat autonomous level of safety. The implication is that authority to make safety-critical decisions must be maintained at every level in the organizational hierarchy. At the same time, a high level of cohesiveness within the safety management system is maintained throughout the enterprise. A strong organizational structure will permit decision-making to be performed throughout by recognizing the independence of decision-makers at lower levels; yet, will not forego their interdependence with higher levels and throughout the enterprise. Decisions made at every level and within numerous channels of the organization all play a role in creating or minimizing risk. This is particularly true among those designing, managing, or operating the system (Santos-Reyes, 2002).

3. Safety Assurance (SA). SA provides the processes to ensure the system is meeting the organization's safety objectives and that mitigations or risk controls, developed under SRM, are working. In SA, the goal is to observe and review to ensure objectives are being met. SA requires monitoring and measuring safety performance of operational processes and continuously improving the level of safety performance. Strong SA processes will yield information used to maintain the integrity of risk controls. SA processes are a means of assuring safety performance of the organization, correcting it where necessary, and identifying needs for rethinking existing processes. SA essentially involves monitoring the system and making corrections, when necessary. Processes must be in place to ensure this occurs on a continuous basis and that data is regularly reviewed so adjustments can be made. Tools used in this process include inspections, audits, and other means of systematically obtaining feedback on system operations. Employees must have a mechanism for confidential reporting of issues and concerns, as well as possible solutions to safety problems. Regular reviews are required so that when performance is unacceptable or objectives are not being met, the root of the problem is sought and appropriate corrections are made. This must occur on a continuous basis (FAA, 2015).

Mathis (2014) suggests early metrics for measuring the effectiveness of programs were failure metrics; that is, they were reactive and measured accidents or incidents after they occurred. As the enterprise fails less, statistical significance deteriorates until there is no longer direction provided as to where to focus efforts. There is a serious lack of evidence as to how to prevent future problems from occurring. This has caused organizations to seek what are referred to as *leading indicators*, to better predict and prevent problems from occurring. Mathis (2014) categorizes these leading indicators into five topics including leadership, measured in terms of criteria such as the percentage of official communications featuring safety topics, reinforcement of safety topics in interactions and performance appraisals, contributions to safety strategy development, and attendance at safety meetings and training sessions; supervision, measured in terms of safety-coaching training session numbers, numbers of safety feedbacks to employees, numbers of specific worker behaviors addressed, and employee perceptions regarding supervisor practices; conditional control of safety issues measured by safe versus unsafe conditions in the workplace, percentage of unsafe conditions actually addressed and resolved, the discovery of new or previously undetected risks, or projected probabilities the risk could actually cause harm; onboarding practices to include the selection, screening, training, and mentoring of new employees; and knowledge/skill-building activities to include coaching and training activities. Performances are typically evaluated on a numerical scale with higher numbers representing better performances (Mathis, 2014).

Cambon et al. (2006) suggest there are three ways to measure the effectiveness of SMS: the results-based approach (analyzes numbers of accidents, injuries, incidents, etc.), the compliance-based approach (audits the degree of compliance of the SMS using a standard), and the process-based approach (measures the performance of each management processes that constitute the SMS independently). The results-based approach is commonly used because it is easy to implement, inexpensive, and not very time-consuming. The problem is that it is limited in scope and does nothing to assess important elements of SMS (Cambon, 2006). At the same time the Occupational Safety and Health Administration (OSHA) prohibits the use of programs that intentionally or unintentionally provide incentives discouraging workers to report hazards or other safety-related problems (OSHA, 2012). Cambon, et al. (2006) suggest the compliance-based approach, dependent upon audits to determine level of compliance, provides the appearance of performance, but does not adequately address the way the SMS influences the working environment or conditions created by the organization influencing safety at work. The process-based approach actually measures the component processes within

SMS to determine the effectiveness of SMS within the organization, whether the SMS is resilient, and if it is truly embraced by the enterprise. The priority in this approach is performance, and according to Cambon et al. (2006), actual compliance with prescribed practices is not a significant consideration. These authors suggest the Tripod method is the most effective evaluation system. It is based on the idea that the most effective way to control risk is to control the environment. The Tripod method recognizes that risks occur as a result of latent errors, as suggested by Reason's Swiss-Cheese model. Understanding the environment; anticipating latent errors, in place prior to any accident; and forecasting ways latent errors can lead to an accident may be performed through various techniques, such as brainstorming, accident scenario inquiries, and field studies (Cambon, 2006). In their work Cambon et al. (2006) identify the ten basic risk factors and discuss how the Tripod can be used to identify areas of weakness that may lead to potential accidents. The Tripod will also indicate whether management is truly committed to the overall SMS operational performance. The main element of the Tripod is a survey that is administered anonymously to workers in an organization; there are 1,500 validated questions in the central database. The tools provide an assessment of the level of compliance of the company's SMS with a specified standard as well as an "assessment of its influence on the working environment and people's working practices." The results measuring the structural performance and the operational performance are displayed on a radar graph using established SMS criteria.

Following a series of studies of audits and their apparent usefulness based on safety incidents, Dan Petersen (2001), stated there is little correlation between accident rates and audit results. There was virtually no correlation between the two and he found a couple of cases where there was actually a negative correlation. The only categories where strong, positive relationships were found were monetary resources and hazard control. Petersen (2001) concluded that a better measure of safety program effectiveness is a survey of the organization about the quality of the management systems, which effect human behavior as it relates to safety. Among other instruments, he proposed surveys to measure employee perceptions to determine strengths and weaknesses of the safety program. In addition to perception surveys, Petersen (2001) suggests behavior sampling results, percentage to goal on system improvements, and dollars (claim costs, total costs of safety, etc.) may also be useful indicators of the effectiveness of the safety program.

In terms of evaluation of the overall program, Janicak (2009) suggests using the audit tool. The audit process consists of gathering data on the physical condition of the workplace and on the safety performance. A deficiency tracking system should be devised involving accountable personnel, appropriate standards for correction, hazard rating by potential consequence and probability, corrective actions required, and timetable showing estimated and actual dates of correction. Audit findings should be disseminated throughout the enterprise. Findings should also be compared against performance standards (Janicak, 2009).

Ford (2004) points out that a key component in any performance measurement is establishing a baseline before the intervention begins. He states this provides the most valid and reliable form of measurement. It is the only way to demonstrate the intervention had any impact on the organization. Data gathered through the evaluation process should be used to improve the system by eliminating errors and waste. Ford (2004) states the measurement of error and waste reduction will point to future system improvements. Key questions the evaluation should address include the following:

1. Are we addressing the true root causes of the problems?
2. Are we staying on budget and schedule?
3. How can we improve acceptance of the interventions?
4. How can we improve the outcomes?
5. Are we following a defined process?
6. How can we improve the process (Ford, 2004)?

Ford (2004) also points out the effects of training should be considered, using pre- and post-intervention measures. Other factors to review, as pointed out by Brinkerhoff (1988):

1. Goal setting to determine if intended organizational and individual benefits are identified in advance.
2. Program design to ensure the design for complete, feasible, and organizational compatibility.
3. Program implementation to determine if the proposed interventions are working as planned and if needed revisions are being made.
4. Immediate outcomes to determine if participants achieved the intended level of skill, knowledge, and attitude (SKA).
5. Usage outcomes to determine if participants are correctly applying new SKA on the job.
6. Impacts and worth to determine if the organization is benefiting and if interventions produced a positive ROI.

In any case, stakeholders, including customers, should be queried to determine if they are also satisfied with the implementation process, interventions, and outcomes. When there are others with a stake in the process, they should also be embracing the changes implemented (Ford, 2004).

In terms of risk management, Yu and Hunt (2004) suggest that a strong safety record and experience may be poor predictors of future performance. A better measure may be the safety culture. "Safety is culture-driven, and management establishes the culture" (Manuele, 2008, p. 82). The culture comprises numerous factors, including values, beliefs, rituals, mission, and performance measures (Manuele, 2008). Janicak (2009) suggests the culture can be measured through the use of various assessment tools, including perception surveys to measure attitudes and acceptance of the program by employees. They are particularly useful for establishing baselines and monitoring improvements over time. Quizzes are useful for employee demonstration of safety knowledge. Discrete observation of performance may be an indicator of knowledge, ability, and/or attitude. Economic analysis is also useful in determining gains or losses due to investments in safety (Janicak, 2009). Cooper (2000) defined safety culture as consisting of psychological, behavioral, and situational or corporate aspects. Psychological aspects include employee perceptions of safety and the safety management systems in place (Janicak, 2009). Petersen (2004) stated there are six elements necessary to achieve a safety culture:

1. A system must be in place to ensure daily proactive supervisory (or team) activities.
2. The system must actively ensure that middle management tasks and activities are conducted in three areas: ensuring subordinate (supervisory or team) regular performance; ensuring the quality of that performance; engaging in certain well-defined activities to show that safety is so important that even upper managers are addressing it.
3. Top management must visibly demonstrate that safety has a high value in the organization.
4. Any worker who chooses to do so should be able to be actively engaged in meaningful safety-related activities.
5. The safety system must be flexible, allowing choices of activities at all levels to obtain ownership.
6. The workforce must see the safety effort as positive.

4. Safety Promotion. The last component, safety promotion, is designed to ensure employees have a solid foundation regarding their safety responsibilities, the organization's safety policies and expectations, reporting procedures, and a familiarity with risk controls. Training and communication are the two key areas of safety promotion. The overall goal is to assure competence among all employees so they thoroughly understand and can effectively perform their safety duties. During training sessions, competency should be assessed through testing processes to assure competencies are in place (FAA, 2015).

The FAA (2015) provides an implementation plan by first mapping and analyzing the organization. This begins with detailed descriptions of all departments and the management personnel responsible for specific operations within the organization. This will also provide a record of personnel authorized to make modifications and to accept risks within each department. A gap analysis is then performed to compare the existing operations to those proposed and identify the gaps that need to be filled. An implementation plan is then constructed to address milestones in the development of SMS. It is periodically reviewed and updated as progress toward that development takes place. The overall objective is to develop and implement a comprehensive, integrated SMS for the entire organization. The FAA (2015) proposes this is done through implementation levels that include planning and organization, developing and implementing basic safety management, applying the SRM process to various parts of the system, continuously monitoring and improving the SMS, and then a high-level implementation.

Krause (2001) reports the use of visual and verbal feedback provides reinforcement for safety-related behaviors and improvement efforts. Historically, causes of accidents have been divided into two major categories—worker-related where the employee made an error or caused the problem or facilities-related where maintenance, equipment, or design problems resulted in an accident. Despite popular thinking, Krause (2001) suggests that rarely is a problem caused by the employee. Following an extensive causal analysis of injuries in various organizations over multiple years, Krause (2001) reports in most cases actual causes of injuries are typically due to interactions between the worker and the facility. In order to reduce these injuries and improve safety, Krause (2001) maintains the enterprise must systematically define and improve these critically important interactions between the worker and facility. Engaging the employee in the improvement process will also serve as a catalyst in this process. Krause (2001) suggests that employees value safe performance and it is reinforced as it improves and feedback regarding behavior is obtained. Feedback is based on behavioral observations used to plan improvements, and the focus is on improving the system and its components rather than on the individual. System components include facilities, equipment, design, maintenance, and other, less obvious mechanisms such as purchasing and decision-making (Krause, 2001).

According to Krause (2001), certain behaviors, identified as *enabled*, are within the employee's control. For example the employee may or may not use an appropriate safety device. The behavior is non-enabled if the employee has no access to the device. This behavior may be difficult if obtaining the use of the device is time-consuming or challenging. This would be the case if the employee must travel a long distance to obtain the device or if it is otherwise not readily available. Krause (2001) determined that non-enabled and difficult behaviors often occur more frequently than enabled behaviors. By identifying critical behaviors, Krause (2001) suggests employees who work unsafely are not necessarily doing so because they are at fault or to blame, but because barriers that interfere with their efforts to work safely are in place. Exposures to technology and other variables having an adverse effect on safe performance may actually be signal indicators to overall problems in the system. Potential problems can be pinpointed by identifying and studying critical safety behaviors within the employee population (Krause, 2001).

Janicak (2009) points out the five safety indicators of leadership: two-way communications; employee involvement, that is, participation in risk identification, assessment, and hazard control; learning culture or applying lessons learned from previous accidents and analysis of unsafe acts; and attitude toward blame or focusing on the cause of accidents rather than placing blame.

Many view safety as the absence of accidents or incidents or operation at an acceptable level of risk where few things go wrong (DNM, 2013). According to traditional beliefs in safety, termed by this article as Safety I, things go wrong due to failures—technical, human, or organizational. When things go right it is because the system functions and people work as planned. Things go wrong when something has malfunctioned or failed. The two modes are assumed to be different, and the purpose of safety management is to ensure the system remains in the first mode and never wanders into the second. Humans are often viewed as a liability or hazard. These beliefs were formulated when organizational systems were simpler and could easily be broken into component parts and diagnosed as working or not working (DMN, 2013). In a more complex world where systems cannot be meaningfully separated and diagnosed individually, this model no longer works. Safety I does not explain why things go right, only why things go wrong. Specifically, it does not account for human performance behind things going right. This performance is not necessarily a result of individuals being told what and how to do things, but may be a result of people adjusting their work to the conditions within which they work. As systems develop and change, these adjustments become increasingly important. The challenge is understanding the adjustments and why safety performance goes right (DMN, 2013).

Safety management should move away from attempting to eliminate things that go wrong and toward ensuring that as many things as possible go right (DMN, 2013). This concept is termed Safety II, and it relates to the organizational ability to operate successfully under various, evolving, and changing conditions. The challenge in the management of safety is to attempt to anticipate various developments and events and their resulting conditions. Thus, an investigation considers how things usually go right, as a basis for understanding how things went wrong. Risk assessment attempts to understand conditions where performance variability can be difficult or even impossible to monitor and control (DMN, 2013). Understanding and embracing Safety II does not eliminate the need for practices that have historically been based on Safety I, but it does require the service enterprise to consider the incorporation of additional beliefs and techniques not previously incorporated into the safety management system (DMN, 2013).

The whole concept moves toward ensuring that safety management investigates why things went right and then works toward ensuring it continues to happen (DMN, 2013). Failures should not be considered unique, but rather an expression of everyday performance variability. It is likely that something that goes wrong will have likely gone right many times before and will likely go right multiple times in the future. When something does go wrong, the emphasis should be on finding how it usually goes right instead of focusing on the anomalies that only explain the cause of one, specific incident. Ensuring that things go right cannot be accomplished by simply responding to problems, but requires interventions to occur prior to the problems manifesting themselves (DMN, 2013).

Events are usually explained by tracing back from event to cause(s) (DMN, 2013). The causes are then associated with some component, function, or event that typically failed. Although the adverse outcome is, in fact, real, it may be due to transient conditions particular to that time and space. The observed, final outcome is real, but the conditions leading to that outcome may be gone. It may therefore be impossible to identify, eliminate, or control those conditions (DMN, 2013).

CONCLUSION

The SMS should be scaled to the size and complexity of the organization and incorporate a mechanism for maintaining and evaluating its effectiveness based on the four safety management pillars or components: (1) safety policy and objectives; (2) safety risk management; (3) safety assurance; and (4) safety promotion (ICAO, 2013). Service providers should seek inputs from key stakeholders as they develop their SMS. Those providing input may include safety or industry professionals, regulatory and administrative authorities, industry trade associations, professional associations and federations, international organizations, subcontractors or principals of a service provider, and/or representatives of the public (ICAO, 2013). Once programs are implemented, evaluation becomes critical, as it is an integral part of the overall continuous improvement process.

The authors suggest beginning the implementation process of SMS through the use of the ANSI standard. It can then be enhanced with additional guidance from OSHA and the FAA.

REFERENCES

Brinkerhoff, R. (February, 1988). An Integrated Evaluation Model for HRD. *Training and Development Journal*, pp. 66–68.

British Standards Institution (BSI). *BS 8800: 1996-Guide to Occupational Safety and Safety Management Systems*. London, UK: British Standards Institute.

Civil Aeronautics Administration. (2001). *Airport Risk Self-superintending System*. Taipai: Civil Aeronautics Administration.

Cooper, M.D. (2000). Towards a Model of Safety Culture. *Safety Science*, 36(22): pp. 111–32.

DNM Safety. (2013). *From Safety-I to Safety-II: A White Paper*. Brussels, Belgium: European Organisation for the Safety of Air Navigation (Eurocontrol).

Edwards, E. (1988). Introductory Overview. In E.L. Winner and D.C. Nagel (Eds.), *Academic Press Human Factors in Aviation*. New York: Academic Press.

Federal Aviation Administration (FAA). (2015). *Advisory Circular 120-92B: Safety Management Systems for Aviation Service Providers*. Retrieved from: http://www.faa.gov/documentLibrary/media/Advisory_Circular/AC_120-92B.pdf.

Friend, M.A. (2012). *Safety Management Systems-a History and Simple Guide*. Presentation for National Safety Congress, Orlando, FL.

Friend, M.A. and Kohn, J.P. (2014). *Fundamentals of Occupational Safety and Health*. Lanham, MD: Bernan Press.

International Civil Aviation Organization ICAO. (2013a). *Safety Management (SMM)*, 3rd edition (Doc 9859). Retrieved from: http://www.icao.int/safety/SafetyManagement/Documents/Doc.9859.3rd%20Edition.alltext.en.pdf.

International Organization for Standardization (ISO). (2015, July). *ISO 14000-Environmental Management*. Retrieved from: http://www.iso.org/iso/iso14000.

Janicak, C.A. (2009). *Safety Metrics: Tools and Techniques for Measuring Safety Performance*. Blue Ridge Summit, PA: Roman & Littlefield Publishing Group.

Krause, T.R. (2001). Moving to the 2nd Generation in Behavior-based Safety. *Professional Safety* (46.5): pp. 27–32.

Manuele, F.A. (2008). *Advanced Safety Management: Focusing on Z10 and Serious Injury Prevention*. Hoboken, NJ: John Wiley & Sons, Inc.

Mathis, T.L. (2014, July 3). *Safety Drivers: The First Level of Leading Indicators. EHS Today*. Retrieved from: http://ehstoday.com/training/safety-drivers-first-level-leading-indicators?NL=OH-04&Issue=OH-04_20140716_OH-04_452&YM_RID=mark.friend@erau.edu&YM_MID=1476922&sfvc4enews=42&cl=article_1.

Occupational Safety and Health Administration (OSHA). (2005, July). *OSHA Fact Sheet. Voluntary Safety and Health Program Management Guidelines*. Retrieved from: https://www.osha.gov/OshDoc/data_General_Facts/vol_safety-health_mngt_.pdf.

Occupational Safety and Health Administration (OSHA). (2015, November). *OSHA Safety and Health Program Management Guidelines*. Retrieved from: https://www.osha.gov/shpmguidelines/SHPM_guidelines.pdf.

Occupational Health and Safety Advisory Service (OHSAS). (2002). *OHSAS 18001*. Retrieved from: http://www.osha-bs8800-ohsas-18001-health-and-safety.com/ohsas-18001.htm.

Petersen, D. (May 1, 2001). The Safety Scorecard: Using Multiple Measures to Judge Safety System Effectiveness. *EHS Today*. Retrieved from: http://ehstoday.com/safety/best-practices/ehs_imp_34484.

Santos-Reyes, J. and Beard, A.N. (2002). Assessing Safety Management Systems. *Journal of Loss Prevention in the Process Industries* (15): pp. 77–95. Retrieved from: http://158.132.155.107/posh97/private/SafetyManagement/assessment.pdf.

Transportation Research Board (TRB). (2015). *Airport Cooperative Research Prgram Report 131: A Guidebook for Safety Risk Management for Airports*. Retrieved from: http://onlinepubs.trb.org/Onlinepubs/acrp.

Walls, D.B. (2013). *World-class Safety Program. (Doctoral Dissertation)*. Retrieved from: http://media.proquest.com.

Yu, S. and Hunt, B. (2004). A Fresh Approach to Safety Management Systems in Hong Kong. *The TQM Magazine*, 16(3): pp. 210–15.

Chapter 5

Directing the Safety System

Mark A. Friend

CASE

Michael Allen is the safety director for a company that manufactures and distributes pest-control chemicals. He regularly checks on application procedures to be certain employees are correctly following the rules regarding personal protection and protection of others. Larry Maze, an employee recently reported that, as part of a prank on a fellow employee, he sprayed his colleague with a pest-control chemical. Larry was apologetic, but reported his actions as part of the voluntary-reporting system now in place. Michael believes Larry reported simply because he feared the victim was going to tell on him, and he could stay out of trouble by conveying the information first. Michael is now in a quandary as to how to proceed.

INTRODUCTION

Management of the safety system is ultimately the responsibility of the accountable executive within the organization. In the United States, there are both privately and publically owned organizations. Authority within public organizations originates with the public agency that oversees it. The agency selects and appoints members of top management and authorizes those managers to run the organization. The managers then delegate responsibilities throughout. Authority flows from the taxpayers to the elected representatives and then to the agency itself

Private organizations are proprietorships, partnerships, or corporations. A proprietorship is owned by a single person who derives authority from the fact that he or she has all rights to the assets and organization and can therefore direct all operations. Authority flows from the ownership of the company and its assets. Partnerships are similar, except they have more than one owner. A corporation is owned by stockholders, who appoint a board of directors to run the company. The board typically hires a president or chief executive officer (CEO) to run the company and authorizes this person to exercise authority in running the company. Regardless of the type of organization, authority flows from ownership of the property. Property owners appoint and authorize leaders to manage the organization within specified guidelines.

The line-staff organizational structure is typical of many organizations. Authority flows through the line portion of the organization. The line is charged with carrying out the major function or objectives of the enterprise. The line begins with the president or CEO and flows downward through production-type personnel. Staff supports the line by providing assistance to it. The staff role of safety is usually one of support in terms of monitoring the program and advising management. From a pure safety management perspective, the remoteness of safety operations can make motivation of employees to do the job safely difficult. All employees have direct supervisors, and unless employees are members of the safety staff, the direct supervisor is someone other than the safety professional. It is the direct supervisor's responsibility to provide the leadership needed to get the job done and done according to established policies and procedures—including those related to safety and health. The supervisor may have a wide span of control with many direct reports under him or her, and those direct reports may be located out of sight or geographically remote.

PERSONNEL NEEDS

Regardless of the management position, line or staff, managers want to know, "What is it that motivates employees?" What motivates people in general? People are motivated by unmet needs. The needs may be physical, psychological, or social. There are different ways to consider needs. Abraham Maslow, an American psychologist, categorized needs as follows (Chruden and Sherman, 1976):

Physiological needs to include food, water, air, rest, etc. These needs are required to maintain the body.
Safety needs to include safety and security from both a physical and a psychological perspective. People need to be protected from external danger, including harm to both the body and the psyche.
Love and affection to encompass the desire to have affectionate relationships and have a place for a comfortable fit in the group.

Self-esteem and esteem by others to include a desire for respect, achievement, and competence. This encompasses the desire for a strong reputation and prestige.

Self-actualization or the desire for fulfillment in one's life—to realize one's full potential.

Maslow placed these needs in a hierarchy that started with the physiological needs and finished with self-actualization. Typically, the lower-level needs are satisfied before the higher-level needs become prominent. As the lower-level needs are met, higher levels begin to emerge (Chruden and Sherman, 1976).

David McClelland suggested people are motivated by the needs for affiliation, achievement, and power. Different people have different needs and McClelland sought to identify those needs among individuals by testing them (Internet Center for Management and Business Administration, Inc [NetMBA], 2010). His conclusions were somewhat different from Maslow's. Regardless, needs represent the physical or psychological requirements that must be met and people are motivated by attempting to fill unfilled needs. *Attempting to fill* implies that actions and the resulting rewards do not always lead to fulfillment of needs, but the effort put forth is often based on the *perception* of likely fulfillment.

Frederick Herzberg theorized a motivation-hygiene theory where he suggested influencing factors fell into one of two categories. First, there are factors that may lead to dissatisfaction if removed or adversely impacted. He referred to these as *hygiene factors* (Chruden and Sherman, 1976). Hygiene factors include (Management Study Guide [MSG], n.d.):

- Pay and fringe benefits—reasonable and appropriate
- Policies—fair, clear, and somewhat flexible, including work hours, dress code, vacation, and breaks
- Working conditions—safe and clean
- Job security
- Relations—no conflicts or problems with supervisors, peers, or subordinates

With no problems in the above areas, there tends to be less dissatisfaction. There is a clean or hygienic workplace—one with little dissatisfaction (Chruden and Sherman, 1976). Herzberg suggested these are not motivators. Motivators include (MSG, n.d.):

- Recognition or praise for a job well done
- Sense of achievement

- Growth and promotional opportunities
- Responsibility—employees have ownership of their work, but are held accountable for it
- Meaningful work—job is interesting and challenging

The implication is that although improving the hygiene factors may tend to reduce the level of dissatisfaction, they may not positively affect the level of motivation. Raising motivation is more likely a result of addressing Herzberg's motivators.

Much work has been done on the subject of what it takes to improve motivation and this discussion has only addressed a few of the points. The thinking of employers is often that employees work to fulfill needs, and they perceive those needs are being fulfilled through various aspects of the job, including salaries, benefits, relationships, recognition, the work itself, the above factors, and more. As long as employees perceive their actions result in appropriate levels of fulfillment, they should continue to do their job appropriately. With increased perception of greater fulfillment, they may increase productivity, work longer hours, and perform their duties in a safer manner. Effectively managing the workplace is essentially reinforcing employees' perceptions that the work requested by management will provide the returns desired by employees. It should be immediately evident that different people are motivated by different rewards, and it may be difficult or impossible to determine what rewards people perceive are appropriate for them and the relationship between the reward and the behavior expected. This can be especially challenging in a relatively large organization where one manager has a wide span of control or many employees. From the perspective of the safety professional who has limited or no direct authority over personnel directly involved in operations, even if understanding and helping to meet needs is feasible, it may be outside of the realm of his or her authority. In either case it may be difficult or impossible to gain a complete understanding of individual needs and even more difficult to meet those needs.

Although the needs of specific individuals may be indeterminable, there are certain common elements among nearly all employees. According to Horowitz (2014), open communication is the number one priority of employees. Employees want to have an awareness of future events that may affect them. As the safety program begins and evolves, employees should be given a heads up on anticipated changes and have the opportunity to provide input on those changes. Participation should be encouraged in all phases of changes with employees throughout the organization. Horowitz (2014) reports that 65 percent of employees don't think their employers show genuine concern toward their employees, and 62 percent do not trust employers to keep their promises. He suggests asking employees what they want to know, both privately and in meetings. Employees

Directing the Safety System   71

should have the opportunity to provide inputs and see a demonstration of consideration toward each. A strong, closed-loop system can aid in this effort.

The safety program is usually built around accomplishing objectives and meeting specified goals. Behaviors will occur more frequently when followed by perceived positive consequences or rewards. Those not followed by positive consequences or that have negative consequences will occur less frequently. *Reinforcement* is the process of changing behaviors by affecting the consequences that follow it (Skinner, 1984). Positive reinforcement occurs when behaviors are followed by perceived meeting of needs. Negative reinforcement occurs when behaviors are followed by perceived lower levels of meeting of needs or perceived punishment. Perceived positive reinforcement tends to strengthen or increase behaviors and perceived punishment tends to weaken or decrease those behaviors (Williams, 2015). There are numerous variations of reinforcement to be considered, but management behaviors tend to improve or worsen depending upon perceived needs and how well they are or are not met.

ASSURING SAFE PERFORMANCE

What steps should managers and supervisors take to assure conformance with safety policies and procedures? What is the role of the safety professional in this process? Since the safety professional is not in a position of direct authority over employees, his or her responsibilities include monitoring the safety process and providing advice, based on results of the monitoring process, to the supervisor or manager. Although the safety manager cannot be present to supervise all activities, various approaches can still be effective in helping to motivate employees to work safely:

1. Periodically sample the work and provide feedback to management on the results of that sampling. The safety professional can show up unannounced and sometimes even observe the work unnoticed. Videos and snapshots of both safe and unsafe behaviors can be provided to management to reinforce observations and encourage appropriate feedback to employees. Of course, anytime unsafe behavior that may threat the safety or health of the employee is observed, the safety professional and/or supervisor should immediately intervene.
2. Provide training to all employees regarding safe and unsafe behaviors. Encourage employees to intervene on behalf of fellow employees anytime unsafe behaviors are observed. This can be accomplished through group interactions in training sessions where each member of the group is asked to discuss the advantages and disadvantages of speaking out when unsafe

behaviors are observed. Discussions with the informal group leader ahead of time to encourage his or her cooperation and support in this endeavor can be very useful. Since members of the management team will not be available to observe all behaviors, group reinforcement of safety can be very effective, if all members of the group are onboard. The informal group leader can be very useful in helping to make this happen.
3. Utilize the job hazard analysis for critical jobs or those not often performed. Have all affected employees participate in the development of the JHA for their respective jobs and require them to review it periodically or prior to the performance of any infrequently performed job.
4. Ensure that encouragement of safe job performance is not simply lip service. In order to effectively influence members of any working group, management must provide a role model to not only promote safe performance, but to personally engage in it and make every effort, under all circumstances, to encourage safe performance on the part of employees. When time becomes an issue or the job constraints encourage shortcuts, the safety and health of individual workers must come first.
5. The safety and health program must be a closed-loop system. Information is the lifeblood of any safety system, so feedback must be provided to any individuals providing input regarding safety. Employees should feel free to report problems they see or even create. There is a balance between self-reporting that results in no discipline and a self-reporting of clear violations. When an employee is aware of rules and clearly violates them, there must be repercussions; otherwise, the reporting system simply becomes a get-out-of-jail-free card.

Discipline

Some aspects of the job may require negative feedback. There are two parameters that should be considered:

1. Discipline should be progressive. When discipline is necessary, it may begin with a call into the office and a warning not to let this happen again. If that doesn't work, a brief suspension without pay may be necessary. As a last resort, the employee may have to be terminated. The point is that discipline should go from less harsh to more harsh. In any case, all disciplinary action should be done in private.
2. Discipline should follow the *Red Hot Stove Rule.* Douglas McGregor suggested discipline is like a red hot stove in the following ways (Sison, 1991):
   - The stove immediately burns when touched.
   - The stove is glowing and everyone can see it is red hot. It provides warning that, if touched, it will burn.

- The stove is impersonal. It doesn't matter who touches the stove, they get burned.
- The stove is consistent. Every time it is touched, the burn occurs.

Consciously breaking the safety rules, or any rules, should lead to the same consequences. Employees know the rules and when they are broken, there are quick, impersonal, negative consequences. The application of the rules is consistent.

Trust

Effective management of the SMS in the organization begins with a climate of trust. Employees at every level must perceive management "has their backs." This begins with a total commitment at the top of the organization and continues through the encouragement and enforcement of a strong safety climate throughout. The safety professional monitors the safety program through traditional feedback in the form of reports, but also relies on personal observations of individuals to provide additional insights into the overall effectiveness of the safety program. When problems are detected, input is provided to the line-management portion of the organization and appropriate changes are made. The line-management part of the team is fully committed to maintaining a safe and healthful working environment because it provides a positive return and is in the best interests of all personnel.

Effective management of a safety system requires a commitment on the parts of management and employees throughout the organization. It begins at the top of the organization and is delegated and authorized throughout. Every individual in the organization has responsibilities for safety, knows those responsibilities, and is held accountable for properly executing them. Generally, there is one accountable executive (AE) responsible for safety within the organization. That person may be the CEO, but is more likely an executive who reports to the CEO. The AE oversees and authorizes resources utilized within the safety function.

EFFECTIVE SAFETY MANAGEMENT

The safety function supports the organization by helping achieve the overall mission. A major share of this support is accomplished by managing risk and minimizing losses that ultimately result in higher costs and lower profits. The safety professional who oversees the safety function has two major responsibilities: monitoring risk throughout the organization and advising the accountable executive on actions to be implemented within the organization to minimize that risk and resulting losses. In advising management it is critical that a convincing case be made; otherwise, less than optimal safety performance may occur. The main purpose of most commercial organizations is to make a profit; therefore, the most effective case to be made is usually in terms of return on investment. Safety is *not* first in all things. Safety is performed in conjunction with production or service and the safety department competes with every other organizational function for resources. The safety professional must demonstrate a return on investment comparable or superior to that of other functions. Safety competes with marketing, design, HR, and others for resources. The safety department must contribute to accomplishment of the organization mission and to the overall well-being of the organization. Within that requirement is the moral imperative to protect the well-being of all employees. It is the responsibility of the organization to ensure employees finish the workday as safe and healthy as when the day began. This responsibility begins with top management and is carried out by the line portion of the enterprise. The safety department and professionals within it advise line management on how that should occur. At any level, a line manager or line worker may reject that advice. Owing to the design of most organizations, the worker and any manager, all the way to the top, have one and only one superior to whom they report. It is up to the line portion of the organization to carry out the safety directives initiated at the top. Safety begins and ends with line management. All safety success and failures are a result of management actions. All safety incidents and safety problems are indicative of problems within the overall management system. Management is in control of everything related to safety.

In the event of an accident, the company is held responsible. OSHA cites employers, not workers. The public opinion is nearly always on the side of the worker and when an accident kills employees, the media and public opinion nearly always rest on the side of the worker and against the employer. Following an accident or an incident, an investigation takes place to determine the root cause of a problem. The root cause is always a fault in the management system. Anytime the worker is blamed, root causes cannot be determined. Management must own the safety problem and take full responsibility for it.

Safety programs must be initiated at the top but must always include input from all levels. This requires a closed-loop system whereby all employees can provide information and expect feedback on that information. If employees offer suggestions or point out problems, they should always expect a response to that input. It may be a yes, no, maybe, or later, but the closed loop is always required. The safety professional must listen for problems among employees at all levels. Those closest to the problems often have the best understanding, as they are typically the employees actually performing the job.

The safety function ultimately involves reducing risk to its lowest, cost-efficient, effective level. Risk is always present and cannot be completely eliminated, and although zero accidents are not theoretically possible, this level may be achieved through a practical reduction of probabilities and severities. Loss control and risk reduction are key roles of the safety professional. These are always within the scope of cost-benefit, an important rationale for safety. It is up to the safety professional to look for problems in the management system. The safety professional must devise methods to zero in on problems in the management system and recommend adjustments accordingly. The goal is to help develop an overall, proactive safety culture capable of anticipating and mitigating hazards before they become problematic, rather than react to them after-the-fact. To accomplish this, the safety professional must depend on a management system that will actively support a fully integrated safety and health system into the overall organization.

It is obviously impossible for top management to maintain expertise in all of the workplace safety and health issues, but it is vital for management to have a strong grasp of the problems, face them squarely, and utilize available resources to develop solutions. The executive can function as an expert by realizing the sources of costs and resource depletion and developing an appreciation of new developments in workplace health and safety. This knowledge alone makes him or her something of an expert, since so few recognize the magnitude of the situation. Knowing that the cost of workers compensation increases as quickly as or more quickly than inflationary costs (due to increased benefits) will help in strategy sessions. Being aware of the increasing role of employees in safety and health issues and supporting safety within the organization from a line-management perspective will help ensure safety is effectively working throughout the organization. The *line* part of an organization is responsible for carrying out the major function of it. This may be production, sales, or another function the organization carries out. The *staff* provides support and advisory functions to include human resources and accounting. The CEO, the production worker, and all of the links between them are considered a part of the line. Appreciating the degrading effects of accidents—not just to the injured worker, but to the company's own goals—will make for more efficient management, particularly in terms of allocating resources. This is an integral part of the line function.

The senior manager—within the line—does not have to become a safety and health professional, yet can act with expertise in the area by being fully aware of all of the implications of safety and health matters, being aware of current techniques and resources available for dealing efficiently with them, and fully supporting the safety and health professionals and programs within the organization.

Worker Involvement

There is a growing trend for government, industry, and all workers to come together on safety and health problems. Employees' roles and activities can be strengthened through emphasis on the rights of the worker to refuse to work in an unsafe or unhealthy environment and through the establishment of joint safety and health committees and other working groups. These rights strengthen the employee's role in safety and health matters and can benefit the company as a whole.

There are now routinely, nonmanagement members of safety and health committees who take part in company accident investigations or conduct their own and who receive safety and health training paid for by their company. For example, companies often pay for worker participation in OSHA ten-hour and thirty-hour training programs that address 29 CFR 1910 and 1926, as well as other safety standards. Management must embrace worker involvement in safety and health activities. Safety and health participation and advocacy on the part of all employees will increase the pace of workplace safety and health improvements.

CONCLUSION

Safety and health participation at all company levels is well established and has an economic and social impact that involves management from the chief executive down. Management is feeling safety and health pressures from government, society at large, the workforce, and the marketplace, as well as from their own co-managers and staff. New technologies are creating new safety challenges and new health concerns; a better educated and trained workforce has a greater concern for safe working conditions. Management must realize that safety and health integration can be a viable and profitable means of helping accomplish the overall organizational mission.

REFERENCES

Chruden, H.J. and Sherman, A.W. Jr. (1976). *Personnel Management*, 5th ed. Cincinnati, OH: Southwestern Publishing.

Horowitz, Shel. (2014). *Hiam: What Motivates Employees? UMass Amherst Family Business Center*. Retrieved from: http://www.umass.edu/fambiz/articles/nonfamily_managers/motivate_employees.html.

Internet Center for Management and Business Administration, Inc. (NetMBA). (2010). *McClelland's Theory of Needs*. Retrieved from: http://www.netmba.com/mgmt/ob/motivation/mcclelland/.

Management Study Guide (MSG). (2012). *Herzberg's Two-Factor Theory of Motivation*. Retrieved from: http://www.managementstudyguide.com/herzbergs-theory-motivation.htm.

Sison, P.S. (1991). *Personnel and Human Resource Management*. Quezon City, Philippines: Rex Printing Company.

Skinner, B.F. (1984). *A Matter of Consequences*. New York: New York University Press.

Williams, C. (2015). *MGMT: Principles of Management*. Mason, OH: South-Western.

Chapter 6

Controlling the Safety Program

Mark A. Friend

CASE

Melanie Sanders, the SMS director for Oxypro, wants to evaluate her SMS, but isn't exactly certain how to do so. She wants input from the employees, but also wants to be sure her system is on par with others in the industry. After much contemplation, she considers soliciting feedback from internal safety and health committees and also input from a safety consulting firm. She is still a little unsure as to where to begin, so she turns to the American Society of Safety Engineers by attending a local meeting. For lunch that day, she not only hears a speaker from OSHA, but also meets other safety professionals who give her insight based on their experience.

INTRODUCTION

Once the planning and organizing phases of management's tasks are complete, there remains the ongoing business of controlling the system that has been set in operation. The most meticulously planned and organized safety program will rapidly become ineffective unless adequate attention is given to the control function. Control essentially involves monitoring the safety system and adjusting according to the findings. Controlling the safety program is no different from controlling any other part of a company's operations; however, there are some areas that require particular attention where safety and health is concerned. Most effective safety and health programs rest heavily on the following control devices:

1. Communications and information systems

2. Safety and health audits
3. Safety and health committees

COMMUNICATIONS AND INFORMATION SYSTEMS

Most of management's ongoing safety and health responsibilities become apparent only after the planning and organizing stages are complete, and the program is actually put into effect. Controlling and directing the program then becomes management's task. Achieving coordinated action, monitoring performance, evaluating feedback and results, and revising the program to meet developing needs are only a few of the tasks. The control process rests solidly on a base of effective communications: relaying information down the line, gathering needed feedback, and receiving information from lower levels. There must always be a two-way system stretching all the way from the president down to the employee, and from the employee back up to the top. Each level of information flow from the top to the bottom and bottom to top is critical, because information is often filtered and management or sometimes workers may not always receive a true picture of the safety system. Two particular actions will help minimize the problems related to filtering. First, a closed-loop system whereby those providing input receive relatively prompt feedback on that input helps assure a continuous flow of information. When an employee or supervisor reports a problem, someone reports back to that person to provide a response as to whether the problem was addressed and how. If for some reason management determined no action will be taken at this time, a rationale should be provided to the person who originally reported the problem. This typically occurs through channels. Also, every employee should also have a protected privilege to report problems directly to the safety department—anonymously if they prefer. Although it may be difficult to close the loop on this type of reporting, every effort should be made to respond to all reports and make employees aware of the response.

Communicating Roles and Responsibilities

Every employee must know and understand his or her role and responsibilities for safety and health. In addition, they must be aware of the roles and responsibilities of other functions and persons, since these may overlap or depend on each other. Rarely is any manager effective in isolation. Moreover, the manager can only fulfill responsibilities insofar as he or she has been apprised of and understands them. How can these roles and responsibilities be communicated? Among other ways, they can be:

1. Included in job and position descriptions
2. Made a part of entry training and briefings
3. Integrated into appraisal forms and periodically reviewed during appraisals
4. Manualized in company handbooks
5. Referred to in company statements
6. Specifically mentioned in job assignments
7. Incorporated in reviews of job and position definitions

As new problems develop or new safety measures are instituted, this communication process must be repeated to make sure all levels of management are aware of changes in their roles and responsibilities as they relate to safety and health. In briefings, meetings, and reviews, higher management should seek to find out if lower management levels grasp and understand communications and clarify them where necessary.

Securing Proper Input and Information

Communicating down the line is only half the process; perhaps even more vital is the need to secure proper input and information from lower levels. An effective system requires a good deal of information and input, including

- Death and injury rates
- All risks, including environmental, personnel, and property
- Public or political pressures in safety and health matters
- The potential loss from the worst event that could happen
- Average annual losses (by division or department, as well as company-wide)
- Both frequency and severity rates of accidents and injuries
- Identification of departments or functions involving special risks
- Impending legislation or regulations
- Comparison of company safety performance with industry-wide performance
- Items having significant or unusual cost impact
- Notable changes in safety performance (either better or worse)
- Proactive means currently employed to prevent problems in the future.

These are often recognized as a function of the management commitment to safety and health and its willingness to put in the necessary time and resources.

Leadership and visible commitment to safety is indicated by periodically reviewing and tracking trends of the following (Mathis, 2014):

- Percentage of official communications featuring safety topics
- Reinforcement of safety in appraisals and interactions

- Attendance at safety meetings and training sessions
- Contributions to safety strategy development
- Numbers of safety training/coaching sessions
- Numbers of employee feedback
- Numbers of specific behaviors addressed
- Employee perceptions of supervisor practices
- Safe versus unsafe conditions/practices in the workplace
- Percentage of unsafe conditions actually resolved

Trends of this nature tend to indicate future directions of the overall safety system. Performances are typically evaluated on a numerical scale with higher numbers representing better performances (Mathis, 2014). Intelligent and effective management decisions will require this type of input, allowing the manager to compare actual performance with goals and objectives, review the resource allocations and priorities for dealing with the situation, and consider alternative actions. Management can then either accept the situation as satisfactory, or take steps to remedy an unsatisfactory condition.

Obtaining Feedback

Dan Petersen (2001) concluded a strong measure of safety program effectiveness is the response from the entire organization to questions about the quality of the management systems. These affect human behavior as it relates to safety. Petersen proposed the use of surveys to measure employee perceptions regarding the strengths and weaknesses of the safety program. This can be done through the use of perception surveys whereby employees are simply asked about how they view the safety system. Employee queries may address items, such as perceptions as to how well management is doing its safety job, how safety aligns with the overall mission of the organization, adequate resources to do the job safely, and appropriate safety training. Employees often perceive their own, direct experiences relative to safety as characteristic of how the organization overall is performing. By asking about perceptions related to critical areas, employers can get useful feedback on how employees view the safety system and whether or not they believe it is effective.

Petersen (2001) also suggests behavior sampling results may also be useful indicators of the effectiveness of the safety program. This form of indirect feedback comes about through both formal and informal observations of employee job performance. The job performance of the employee may be reviewed periodically to assure conformance to safety standards on either a scheduled or random basis. The observation may take place during a spontaneous walk through the facility or the task may be observed during a scheduled period. Tasks not done frequently or repeated often may be of higher risk, because employees sometimes fail to remember and implement all safety procedures. Although it may be difficult to observe these randomly, determining a time when the task will be performed and observing it may be particularly worthwhile.

The Australian Institute of Management (AIM, n.d.) applies a systematic approach toward assessment as a part of continuous improvement. Evaluators are particularly concerned with satisfaction rates, competency completion rates, outcomes of complaints and appeals processes, outcomes of management processes, and opportunities for improvement by staff and stakeholders. Data are collected using a variety of methods to help ensure validity. Their primary method is through written surveys of participants and employers. Performance indicators from a learner or employee perspective include training quality, work readiness, training conditions, and learner engagement. Qualitative data is collected from evaluations, complaints and appeals, and audit results. Findings are used to identify opportunities for improvement. Once improvements have been identified, they are clearly defined with responsibilities and guidance. Data continue to be collected to ensure improvements as delineated are made. A continuous improvement reporting procedure remains in place throughout in terms of regular (at least monthly) continuous improvement meetings and continuous improvement forms collected and reviewed at the periodic meetings (AIM, n.d.).

With the amount of information and feedback involved, it would not be possible or effective for the manager to attempt to assimilate and deal with every single item personally. Intelligent screening and allocating are necessary. The safety staff should separate the items and send each to the appropriate manager for action. It can eliminate those with little significance and pass the rest through the organization for routine action. In some cases, a safety systems specialist will be needed to furnish detail before information is provided to management. Perhaps a risk-loss analysis or some other information will be required. Constant attention is needed to alert the manager to any noticeable trends in hazards or risks or to any new situations since the last analysis. Watching for trends is particularly critical, because these indicate areas of change. The change may be for better (as when a safety system is being effectively managed), or it may be for worse (indicating a problem area that may need review). When specific problems do occur or trends begin to appear, the additional or more detailed information can then be secured, as needed.

The information and feedback system is, of course, a means of furnishing the data needed; however, raw data are very rarely useful by itself. A variety of safety and health analysis techniques are available for this purpose; however, these techniques are often quite sophisticated, requiring the services of an expert. Moreover, they are sometimes limited because

1. Some risks cannot be detected by the worker or first-line supervisor on the site.
2. Some hazards and risks can *only* be identified by the worker or first-line supervisor.
3. Hazards not directly associated with the worker may be detectable only by an expert or specialist (as when the hazard cannot be detected by the physical senses).

Realizing the limitations of the feedback system is essential to providing an adequate information base. Given the above limitations, adequate input requires regular feedback from inspections or audits. These are conducted by someone other than the first-line supervisor—often the worker or specialists who provide specific information from items such as air quality sampling, and noise monitoring.

ESTABLISHING STANDARDS FOR MEASUREMENT

One of the most vital roles in the safety and health mission is to establish standards against which performance can be measured. Only then can safety performance be meaningfully assessed, and corrective action taken. The standards themselves are influenced by many factors, some of them outside the control of the manager. Among these are

- Regulations from OSHA or other federal agencies depending on the organization type.
- Recommendations set forth by agencies such as the American Conference of Governmental Industrial Hygienists (ACGIH), the National Fire Protection Association (NFPA), and the American National Standards Institute (ANSI). Management must be aware that standards set forth by these and other agencies may be interpreted as minimums in the event of OSHA inspections. If not adhered to they can also be used against the company in the event of litigation following an accident.
- Technological changes.
- Product liability trends.
- Sociopolitical pressures.
- Increased insurance and compensation costs.

Some standards are mandated by law; others are set by industrial associations or labor unions, and still others are dictated by the state-of-the-art itself. The manager will want to be fully aware of these and integrate them into measurement and control systems. Certain other standards of measurement can be directly controlled and determined by management. In addition to the measures listed above, as related to leadership, these include

- Acceptable frequency of lost time and other injuries. More important is a measure of what is being done to lower these as well of rates of other losses.
- Acceptable costs of insurance and compensation claims.
- Damage to property and equipment.
- Results of inspections, environmental monitoring, and mishap investigations
- Lost production data.
- Departmental budget averages which reflect production and maintenance costs.

Other pertinent inputs, capable of measurement and integration into an overall control standard, are

- Labor turnover
- Tardiness and absenteeism
- Productivity level
- Grievances and morale
- Budget overruns
- Waste and rework levels

Extensive research and examination may be needed before management can set generally acceptable standards. There is no single rule that can be applied, as every organization will differ in its definition of what is "standard" or "acceptable." The above listings are intended simply to serve as a guideline to managers in reviewing and establishing standards. An adequate standard will carefully consider *all* the items mentioned above as a bare minimum, plus whatever other items the individual company considers pertinent, such as customer complaints, returns or rejections of finished goods, and quality control reports.

SAFETY AND HEALTH AUDITS

One of the keys to informational feedback, as well as one of the major control devices, is the safety and health audit. Properly conducted, it can provide much of the necessary input on the state of the safety and health program. Audits are different than inspections. During inspections, no quantitative evaluations are made. Inspections typically determine whether operating standards are being met or not. For example, an inspection will determine whether a machine guard is in place before the machine is operated. The statement on an inspection form for a given machine, will require a yes/no response, such as, "The guard is in place and operational." Audits are used to determine whether the tools are in place to assure inspections and other necessary actions are occurring, as well as to determine the effectiveness of those tools (Friend and Kohn, 2014). Audits are much more thorough and in-depth than inspections. An audit will determine whether processes are operational to assure the machine guard is in place before the machine is operated. It looks at the management system and evaluates it on a quantitative basis. For example, it may state, "Management has designated an employee who has responsibility and accountability for assuring all machines and related safety equipment are inspected, monitored, and maintained according to standards at all times." The auditor will answer on a 1–5 Likert-type scale. Anything less than a top score will require additional attention, but critical items with lower scores are usually addressed first.

The Audit Process

The audit is ultimately a managerial responsibility, requiring managerial follow-up to assure correction of deficiencies. Management at every level may be involved, both on the giving and receiving ends of audits. An audit involving safety and health is no different from any other type of audit. Managers should not wait for safety expertise before conducting an audit, nor should they wait until an outside consultant (such as an insurance inspector or OSHA team) requests one. Outside auditors of this sort are generally looking for different things, and will advise on those particular things, such as fire safety or regulatory compliance. This may be quite different from what the manager wishes to achieve. What the top manager is really after is the most efficient operation. Granted, management objectives and those of the outside auditor may be related, but in most cases they are not identical. There should be an established schedule for auditing. The line manager can, with instruction, make audits a part of the regular control system for monitoring performance and self-assessing. A problem with this is that the line manager cannot objectively grade his or her own performance. This can be resolved by having collateral managers with no conflicts of interest make the audits.

Audit Management

Basically, audit management involves the following five key functions:

1. *Description.* The audit process should be thoroughly described, preferably through the safety and health manual. The description should include a

listing of responsible parties, their various duties, the audit practices, procedures and performance expected.
2. *Definition.* Audit management should define *what* needs to be audited (i.e., what items will be especially looked at and examined).
3. *Designation.* Audit management should clearly designate *who is going to conduct the audit.* It should consider and define the respective roles of line, staff, and middle management in the audit process.
4. *Communication.* Audit management must work out reporting procedures to ensure two-way communication and keep employees at every level in the loop. All inspection forms should go to the appropriate manager, with copies to safety and health personnel.
5. *Recordkeeping.* It is vitally necessary to make audits a matter of permanent record. These can be used as benchmarks to evaluate the overall, continuous improvement of the safety process.

Authority to Shut Down?

Should the audit team or individual inspector have the authority to shut down a process or to stop work when an unsafe condition is discovered? If the group or individual represents appropriate line management, certainly. Shutdown actions by staff or safety personnel should be taken only in a case of imminent danger, where death or serious injury could result if prompt shutdown action are not taken. The action might then be justified, but *such authority must be used only with great care.* Including the line manager on the audit team eliminates this problem.

Types of Audits

Audits can be divided roughly into three different types: (1) periodic, (2) intermittent, and (3) continuous. Each type has its particular uses and applications.

1. *Periodic audits.* These are conducted on a recurring basis *at regular intervals.* They may review the same or different items each time. These are particularly valuable for processes or functions where changes occur frequently. Some companies develop an audit cycle for all safety components and review each at least annually. Those areas where difficulties have been encountered may be reviewed more frequently. Audits may be done based on subject areas, such as fire, industrial hygiene, and environmental issues. They may also be based on operational areas or specific facilities.
2. *Intermittent audits.* These are conducted at irregular intervals—as needed. When work areas are having ongoing difficulties and need prompt attention, audits may provide useful insight into the problems.

3. *Continuous audits.* Some operations must undergo continuous audit, particularly where legal standards exist. Routine quality control sampling is one type of continuous audit. In some operations, continuous monitoring is required of air quality, radiation or toxic exposures, temperature or humidity levels.

Basic Steps in Making an Audit

The audit process consists of gathering data on the physical condition of the workplace and on the safety performance. A deficiency tracking system should be devised involving accountable personnel, appropriate standards for correction, hazard rating by potential consequence and probability, corrective actions required, and timetable showing estimated and actual dates of correction. Audit findings should be disseminated throughout the enterprise. Findings should also be compared against performance standards (Janicak, 2009). Proper planning and proper procedure are essential to conducting a successful audit. It is essential that the critical components are identified and the tool properly measures the effectiveness of those components. Most audits will benefit from the following guidance:

1. *Plan ahead.* Determine the area(s) to be audited and prepare all details well in advance. Be prepared for known or specific problems. Have all the materials needed for the audit. This will include records, writing materials, checklists, safety equipment, meters and other measuring devices, reference copies of regulations and company procedures, and records from the last audit (particularly recommendations of corrective actions suggested at that time). Develop or acquire an audit instrument, based on the needs, but be aware that any instrument acquired from another source will need adjusting to fit the needs of the company (Friend, 2014).
2. *Communicate.* Set up a pre-audit conference with all collaborating personnel to discuss the purpose and scope of the audit (Friend, 2014). Contact the proper line manager and ask for someone to accompany the audit team. Go over the findings with the supervisor and/or manager before leaving. Answer any questions or clarify any confusion that may arise at the time.
3. *Follow up.* Following the audit, written reports of the findings should be forwarded to management and safety as soon as possible. Other reports and forms that may be required for files should be completed promptly and filed or forwarded as required. Follow-up should include the paperwork such as recommendations on any subsequent activity needed to see that corrective actions are taken. In some cases, this may involve a subsequent visit or inspection. All changes must be tied to the organizational mission and objectives (Friend, 2014).

Assistance in making audits can often be obtained outside the organization, through the insurance carrier, federal or state consultants, the nearest OSHA office, or professional consultants. An online search will reveal numerous auditing instruments. When the necessary expertise for conducting a thorough audit does not exist in-house, it may be vital to seek outside help. A perfunctory or haphazard audit is hardly better than none and does not give top management the control it seeks, or the information it really needs.

SAFETY AND HEALTH COMMITTEES

If committees are skillfully constituted and properly administered, they can become one of the manager's most effective operational and control devices. Depending on management's approach, the safety committee can be a valuable two-way communication device with broad powers for action, or simply a forum for small and insignificant talk.

Advantages and Disadvantages

Most senior managers are more than familiar with the various advantages and disadvantages of the committee as an operating tool. There are peculiar features of safety and health committees that make a discussion of their specific advantages and disadvantages appropriate.

Advantages

A properly constituted and administered safety and health committee can offer the manager several advantages:

1. It can serve as a communication channel for safety and health policy and practices through every level of the organization.
2. It can give management the benefit of multiple judgment and group opinion on proposed or actual safety practices. It can foster understanding and cooperation, where safety and health matters cross departmental lines.
3. It can accomplish more than one safety and health expert working alone.
4. It increases active participation and involvement in implementing safety programs and solving problems.
5. It can lead to consensus which may be very necessary in implementing new safety practices.
6. It multiplies safety and health efforts, allowing members from various departments to share their knowledge or expertise with each other.

7. It can bring the skills of several different people to bear on a single department's problem while calling on those actually involved in the jobs and processes to provide their input.
8. In small operations, it is sometimes the only device needed for handling all safety and health needs.

Disadvantages

There are also disadvantages in relying on a safety and health committee. The manager should be fully aware of these, since most of them can be easily remedied.

1. Management may try to use the committee to shift safety and health responsibility, avoid making decisions, and evade taking action on safety matters.
2. It may be difficult to provide adequate leadership with the requisite safety and health expertise, and to provide adequate management support.
3. Committee actions may not be recognized, appreciated, or properly valued, in terms of their effect on overall safety and health.
4. Committee recommendations may not be (or sometimes cannot be) acted upon or responded to by the appropriate level of management.
5. Members may not be representative of either the organization or its problems. (For instance, workers or union members may be excluded, or a department with an unusual number of safety and health problems may not be represented).
6. Scheduling difficulties may keep the safety committees from meeting on a timely basis. (As some safety and health matters are urgent, postponing a meeting even by a few days could be crucial.)
7. The safety committee may not have adequate guidance on what to do or how to proceed without someone from management taking the time to prepare and relay it.
8. Proper records of meetings may not be kept, nor passed to the proper management level for action or recognition.
9. The safety committee, if not carefully organized and chaired, may simply waste the participant's valuable time on trivial issues that could more appropriately and efficiently be solved by one person.
10. Safety committees take time and time is money.

If the committee is properly mandated, organized, recognized, and directed, and if two-way communications are kept open between the committee and top management, none of these problems need develop. In short, these are all factors that management can control.

Organizing the Safety Committee

There are numerous ways of defining and organizing a committee. They may be authoritative, advisory, or informative committees; they may be *ad hoc* committees, informally convened to solve one particular problem, or they may be permanent in nature, formally constituted as part of the organizational structure.

Membership

On an advisory committee, properly selected members can be quite effective even without any special knowledge or qualifications. Expertise in management decision-making may, of course, be helpful, but is rarely available in operating-level committees. It is probably more essential that the committee member simply be thoroughly familiar with the sector of the company represented, knowing its operations, problems and particular needs.

How many people should be on the committee? Most companies tend to select representatives of all major departments and staffs. This often entails too many people for effective action or good communication. The larger the group, the more diffused and ineffective it may tend to become. *The ideal size of a committee for effective exchange of information and decision-making ability is from five to seven persons.* This allows maximum two-way participation and input of valuable views. The question may become, for some managers, how to adequately represent, say, fifteen departments on a committee of seven. Here are some suggestions for dealing with this problem:

1. Select someone from the department with the most potential for accidents and injuries (they are going to have a problem).
2. Select someone from the department with the poorest accident/injury experience (they already have a problem).
3. Select someone with a demonstrated interest and enthusiasm for safety and health matters (they will work on the problem).
4. Select someone from the department with an outstanding safety and health record (they will have experience that the others can learn from).
5. Select someone who represents the interests of the workers, such as a labor organization representative (they have a vested interest in preventing problems).
6. Supplement these members as needed with experts or temporary members who are needed to solve a particular problem. Such participants do not have to be regular members to be used effectively, and can add much to the committee's resources.
7. If resources are available, utilize multiple committees to involve employees from all phases of the organization.

Chairperson

The designated or elected chairperson should be a respected leader who will keep the meeting on track, properly direct and follow up on problem-solving actions, and handle people skillfully. The chair should be able to draw out quieter people, keep more vocal elements from dominating the meeting, and encourage all to contribute. A major problem can be the inclusion of a higher-level manager, in that lower-level participants may be unwilling to challenge or divulge problems related to their boss. The system should always include a method of permitting individuals to input information outside the normal committee meetings.

The regular safety and health person should not be the chairperson. He or she may, at some levels, be the secretary/recorder, or merely an advisor or observer. From that position, the safety professional can ensure the proper items are considered, acted upon, and that recommendations are followed up. If the safety person is the chair, much as is the case with a higher-level manager, there may be the tendency to dominate the meeting, when the whole idea is to get others involved as participating members of the safety team. In the ideal situation, there will be a committee at every level of operation. These will interrelate on a regular basis, help solve each other's problems, bring lower-level problems to the attention of the next higher level, and finally bring top management's attention to those things that cannot be solved at a lower level.

Where this ideal situation is not possible, some provision must be made for disseminating the committee's findings or recommendations to all levels and some means provided so that all levels can give the committee input, bring problems to its attention, or pass on information from higher up.

In a larger organization, the ideal would be a committee composed of top management, which might be called the "policy" safety and health committee. At the next lower level, there should be a management "advisory" safety and health committee. Finally, at the operating levels, there will be committees representing the different major functions, departments or divisions. The minutes of the lower-level committees would be fed upward through the other two, until finally the accountable executive receives them. Thus, safety and health problems not solvable at a lower level will be passed upward. Should the problems exceed the ability of this next higher level to solve, they will again be passed upward, until they reached a level at which they can be solved. The higher levels can then pass recommendations down the line, through these levels. Moreover, the accountable executive can see if his or her mandate on safety and health has been clear, if committees understand their responsibilities and tasks, and if the safety and health program is working as expected.

Some Practical Considerations

All meetings should be scheduled in advance for a convenient time. Adequate time should be allowed between meetings to follow up on actions and decisions. Once per week is probably too often in most organizations; every other month not often enough. Monthly meetings are probably the most frequently used. This assures the committee members do not have to give an unreasonable amount of time to attending meetings, and that recommended actions can be well undertaken since the last meeting.

Roll call should be taken at all meetings, and minutes kept, with the names of those who make motions or who were assigned to take action. Making this a matter of record is essential for follow-up, and will also aid management in assessing individual performance of committee members.

If there are not enough agenda items to make a meeting worthwhile, it should be cancelled. Nothing destroys an effective committee faster than the thought that they are wasting their time. If a majority of committee members are absent for any reason, the meeting should also be cancelled. In either case, the meeting should be rescheduled if possible.

When to Use a Committee

As a rule of thumb in determining when a safety committee should be established, here are some general guides. Use a safety committee when:

1. Information or input is needed from several different sources.
2. The full support and understanding of different departments is needed to implement policies.
3. The action to be taken involves three or more management functions for effective coordination.

On the other hand, there are times when handing a matter to a safety committee may *not* be practical. This is true when

1. Speed and time are vital to a decision.
2. The safety problem can be handled by one or two people who are directly concerned.
3. The problem is clearly the responsibility of a single person or function to decide.
4. The decisions required are of a very routine or trivial nature.

A carelessly organized or badly handled committee may not only be a waste of time, but a positive liability in effective safety and health action.

REFERENCES

Australian Institute of Management (AIM). (n.d.). *Continuous Improvement Policy and Procedure.* Retrieved from: https://www.aim-nsw-act.com.au/sites/content.aim-prod.sitbacksolutions.com.au/files/policies/AIM_Continuous_Improvement_Policy_and_Procedure.pdf.

Briscoe, G.J., Lofthouse, J.H. and Nertney, R.J. (1979). *Applications of MORT to Review of Safety Analysis.* Idaho Falls, ID: System Safety Development Center.

Findlay, James V. and Kuhlman, Raymond L. (1980). *Leadership in Safety.* Loganville, GA: Institute Press.

Friend, M.A. and Kohn, J.P. (2014). *Fundamentals of Occupational Safety and Health.* London, U.K.: Bernan Press.

King, R.W. (1979). *Industrial Hazards and Safety Handbook.* London: Newens-Butterworth.

Mathis, T.L. (July 3, 2014). Safety Drivers: The First Level of Leading Indicators. *EHS Today.* Retrieved from: http://ehstoday.com/training/safety-drivers-first-level-leadingindicators?NL=OH-04&Issue=OH-04_20140716_OH-04_452&YM_RID=mark.friend@erau.edu&YM_MID=1476922&sfvc4enews=42&cl=article_1.

Nertney, R.J. (1976). *The Safety Performance Measurement System.* Idaho Falls, ID: System Safety Development Center.

Petersen, D. (May 1, 2001). The Safety Scorecard: Using Multiple Measures to Judge Safety System Effectiveness. *EHS Today.* Retrieved from: http://ehstoday.com/safety/best practices/ehs_imp_34484.

Pope, W.C. (1973). *Authority and How to Use It.* Alexandria, VA: Safety Management Information Systems.

Pope, W.C. (1973). *Practice of Safety Responsibility.* Alexandria, VA: Safety Management Information Systems.

Pope, W.C. (1978). *The Process of Safety Inspection.* Alexandria, VA: Safety Management Information Systems.

Chapter 7

Staffing and Job Processes— A Safety Perspective

Mark A. Friend

CASE

James Gary, supervisor of maintenance for a large manufacturing company, was recently approached by the director of safety and health. "James, I'm unhappy with the number of accidents and incidents reported by many of your workers lately. It seems like there is something new that comes up every week. What is the problem?" After some consideration, James responds that he perceives the problem is that his crew addresses different situations on almost every repair call. Also, the typically go out on the job without the aid of a supervisor. With so many unfamiliar circumstances, the crew isn't as prepared as they should be. James is stumped. They have a broad range of safety training, but these unfamiliar jobs often lead to injuries.

INTRODUCTION

An effective and profitable safety management system is far more than just a matter of theory or paperwork, properly-worded policy statements, or formal objectives. Its real-life arena is the actual day-to-day operation of the job process, what people do, and how they go about doing it.

Effective job process and procedure control is the responsibility of top management and involves three main elements within the system:

1. People
2. Hardware
3. Procedures

These elements are in not independent; they interface and radically affect each other and the whole job process. In establishing an effective staffing and job process control system within SMS, the senior manager must consider three primary interfaces:

1. Do the people match the hardware?
2. Does the hardware match the procedures?
3. Do the procedures match the people?

All too often, when things go wrong, organizations treat the symptoms, rather than the causes. This is especially true when a workplace problem arises, or a safety and health event has adverse effects. When production falls off in a department, pressure is put on the supervisor. If the accident level increases, the supervisor may simply lecture the workers. Management may overlook the fact that there may be good reasons for the drop in production, an obvious cause for the accidents or the worker's error. It is the symptom which catches the eye.

A sound SMS seeks to discover the *cause* of the trouble, trace it to its source, and correct it. This is the only way to keep the event from being repeated. It will tackle the cause, not the symptom. As suggested earlier, the system is examined for failures allowing the event to happen. The drop in production may be due to material shortages or critical equipment breakdowns. Too many accidents can point to sloppy housekeeping, poor supervision, or an inadequate training program. Perhaps the worker's error was due to poor machine design, inadequate instruction, or poor maintenance. All can indicate a management deficiency at some level of operation, such as an inefficient purchasing system, inadequate training program, or a failure to establish proper job procedures.

It is up to the principals of an organization to review and approve its long-range SMS objectives. In a small corporation, this may rest on one or two persons who determine the broad policies. More commonly, a committee representing top management and heads of functional departments, approves the plans and programs, and that includes safety and health planning and programming. These top management people, more than anyone else in the organization, should be deeply concerned with discovering the root causes of workplace problems, and evolving sound systems and processes to recognize, diagnose and correct them.

THE MAIN ELEMENTS OF JOB PROCESS CONTROL

Safety and health considerations are not extraneous to day-to-day operations. They are an integral part of every process and procedure in the organization.

Establishing job processes and procedures and effectively incorporating safety and health considerations are management's responsibility. The astute manager knows this requires more than just a properly worded policy statement in the corporate manual or a formal list of objectives. In the area of job processes and procedures, management is not dealing with abstractions, but with actual workers, the things they do, and how they go about doing them. This is the real-life arena of safety and health, where every facet of the on-paper safety program will be proved out.

Basically, job process and procedure control involves three main elements:

1. People
2. Hardware
3. Procedures

These three factors cannot really be separated; they are very interdependent, each affecting the others and being affected by them. In order to judge the adequacy or effectiveness of job processes and procedures, the manager must ask three basic questions:

1. Do the people match the hardware?
2. Does the hardware match the procedures?
3. Do the procedures match the people?

These represent the main element of job process control.

People Requirements

A large hydraulic press is a powerful, dangerous piece of equipment involving several types of energy in its operation. It cannot be made 100 percent safe. Its safe operation does not depend solely on the hardware. It depends on the hardware, the use of safe procedures, and the proper actions of people. In addition, it requires a sound evaluation of the entire process to do the job safely and properly. The first considerations are the staffing requirements and what management should do to assure the best and safest use of the organizational staff.

Providing adequate control of personnel involved in the job process system rests on several basic requirements:

1. Establishing an appropriate employee selection process.
2. Providing adequate training.
3. Establishing an adequate testing and qualification program.
4. Evaluating the current status of each employee.

Staffing

The steps in establishing a well-rounded personnel selection and placement program consist of:

1. Adopting a clear selection policy and making it known to all supervisory personnel.
2. Determining the actual job requirements, with special regard for the psychological and physiological demands.
3. Knowing the applicant's interests, experience, abilities, and attitudes.
4. Determining the applicant's physical capabilities through medical examination.
5. Placing the applicant by matching job requirements with his or her abilities, interests, motivation, background, and physical capabilities.
6. Acquainting the worker with the job, using a planned procedure.
7. Regularly following up.

It is necessary to establish *criteria for the selection* of personnel before determining the *methods of selection*. The safety-related job requirements must be defined in order to determine physical and mental capabilities needed for the job. Outside assistance, such as someone who knows the job and/or safety requirements extremely well, may be needed and should be consulted before establishing criteria. Employees should be selected based on their ability to match the work environment and interface with the procedural and managerial controls. For example, a hard-of-hearing person should not be placed in a situation where keen hearing is absolutely necessary to carry out the job safely. If the actual task calls for good eyesight and no particular use of hearing, a hard-of-hearing person with 20–20 vision matches the requirements very well. These examples only apply to the safety and health aspects of personnel selection. Consideration must also be given to the Americans with Disabilities Act (ADA). According to the Department of Labor (n.d.) website:

The ADA prohibits discrimination against people with disabilities in employment, transportation, public accommodation, communications, and governmental activities. Employers with fifteen or more employees are prohibited from discriminating against people with disabilities by Title I of the (ADA). In general, the employment provisions of the ADA require equal opportunity in selecting, testing, and hiring qualified applicants with disabilities; job accommodation for applicants and workers with disabilities when such accommodations would not impose "undue hardship;" and equal opportunity in promotion and benefits.

Any evaluations done must comply with ADA requirements and should be vetted to assure conformance with the standard. Tests for occupational skills, coordination, strength, interests, and aptitudes may build a useful profile of the person's real innate abilities, but care must be exercised not to violate the law. The manager is best advised to consult a human relations professional, who can give expert advice on available testing instruments, practices, and legal constraints where needed.

Personnel Training

At the time of selection, personnel should be screened to see if they are already suitably trained and skilled or if they will need a training program. The training program should be adequate in scope, depth, and detail. Applying the personnel training needs to the question of job process control raises particular questions. These are as follows:

1. Is the training kept current with changes in hardware, procedures, and management control? How is this done?
2. Do personnel know what risks it can accept and the ones it must refer to a higher management level?
3. Is the training directly related to procedures to be followed on the job?

Making training work, even when it is well designed and presented, calls for the active support of supervisors and managers. Supervisory and management personnel should be well trained in safety and health, should understand the principles of risk management, should be able to identify hazards, and should know what risks are acceptable and what risks are not. Mere training of the new employees is not adequate; thorough training of supervisors and management must be accomplished first.

Testing and Qualification

A testing and qualification process, regularly monitored and verified, will assure proper training is being conducted. This might call for simulators, tests, or examinations. A system of verifying the person's current qualification status is needed, particularly where workers are hired as already trained or skilled. Testing and qualifying should not be performed just at the time of hire, but on a continuous and periodic basis. If task criteria are changed, new testing and qualification are needed for all employees. The new examination should address the new criteria, and procedures should be established for retraining and requalifying the existing workforce. The manager should assure this testing is conducted on the worksite wherever possible, so that individuals can demonstrate their actual skills under realistic conditions.

Evaluation

Organizations need to establish training processes to evaluate not only the current status of personnel training, but also the support functions of supervisors, staff, and management in the training process. For example, supervisors must understand their responsibilities in order to assess their employees. They should know what performance indicators might reveal problems. In some cases, the program may involve supervisory backup for medical programs and employee assistance programs for drug and alcohol problems. How effectively does the supervisor give on-the-job training? How alert is the supervisor to the existence of training deficiencies? How closely does the supervisor follow established and approved training materials or procedures?

Hardware Requirements

The next basic component involved in job process control is hardware, or more properly, the hardware and plant (sometimes referred to as "equipment"). The elements involved in the total hardware and plant are complex. They may consist of hundreds or even thousands of separate machines, tools, support systems, and buildings. For the sake of simplicity, it is easiest if the manager views hardware in terms of its life cycle; thus, a thorough review of the hardware picture can be accomplished by considering each stage of the life cycle.

From the simplest standpoint, effective plant and hardware control in the job process system involves four main tasks:

1. Determining the need and design requirements. Making a life cycle study and safety analysis of the product.
2. Actually constructing and acquiring the product, inspecting it as a quality control function and testing it for suitability.
3. Operating, maintaining, and inspecting the equipment.
4. Decommissioning and disposing of it at the end of its useful life.

It is during the design phase human factors are considered and built into the hardware. This calls for an expert, even when relatively expensive. It is the best opportunity to design out problems of hardware-human interface, providing for safer operation throughout the entire life cycle of the equipment or hardware.

Life Cycle Study

The thorough safety analysis in an SMS must necessarily involve a life cycle study. Plans must assure this study will be initiated *during the planning phase*. The study will review each stage of the life of the *hardware*, from planning and design through fabrication, operation, maintenance, and disposal. It may also include an analysis of the environmental impact of the hardware. These reviews are carried out at predetermined milestones during the life of the hardware, preferably by independent reviewers, rather than by staff directly associated with the use of the hardware. The following life cycle stages represent the stages often referred to in the development of hardware. Life cycles of facilities are addressed in chapter 11.

Fabrication

Once the design is complete, hardware must then be fabricated. Here, too, proper controls must be established and rigorously followed. One aspect of fabrication control involves following well-established guidelines, such as accepted engineering practices, quality assurance practices, engineering and design codes, standards, regulations, and the use of standardized parts or components. The other aspect deals with the introduction and use of the fabricated parts or components in the system itself. Fabrication controls should be placed on both in-house and outside work, with special attention to in-house operations. These may require separate identification, quality assurance, and documentation controls. Specifications are to be kept up-to-date for in-house fabrications and checked at every stage.

Installation

The actual installation of hardware calls for considerable advance preparation. Care must first be taken to assure there is no damage or degradation of the hardware between fabrication and installation. Next, an installation plan and description is prepared and reviewed to assure clear and precise information for installation personnel. Post-installation inspection is performed before the equipment is tested or put into use to assure compliance with all directions and specifications. This includes checking all clearances between the new equipment and other parts of the system, using proper grounding and bracing, and employing adequate electrical wiring and shielding where specified—particularly to see if any damage was done to the hardware during installation.

Occupancy-Use

The preparation of the hardware for facility use involves applying stringent criteria to assure operability and maintainability for meeting safety and health needs. This may involve inspecting for operational readiness, as well as startup of functional operation. A suitable plan to determine occupancy-use readiness should be developed well ahead of time.

Operation

Before hardware is actually placed in operation, all safety requirements must be specified, available, and ready for use. Requirements should include operating instructions, manuals, maintenance documents, emergency instructions, and records to be kept with the hardware. Documentation of the requirements should be up-to-date and available to the user at all times. If the new hardware replaces old hardware, the latter should be removed from the system, along with its documentation.

The workforce should be given an adequate pre-task briefing to consider the effect of changes, maintenance, and any new or potential hazards. Pre-operational training should cover all these factors, as well as actual operating instructions for the new equipment. Adequate technical support should be provided at the worksite, along with careful scrutiny of the interface between operations and maintenance personnel.

An analysis should be made for each task, involving a potential for error, injury, damage, or unwanted energy flow. Task safety should be exercised through a job safety analysis (JSA). Operational control requires adequate maintenance support and inspection of equipment, as well as a plan broad enough to cover most aspects of operation. Gaps in the coverage should be called to management's attention. The plan should also provide for failure analysis, particularly to determine its real causes. As covered in the previous chapter, the plan should also provide for minimizing maintenance-related problems. Finally, there should be dry runs or tests to for all hardware and procedures, adjustments of oversights, and any final modifications prior to the first actual hands-on operation.

Application of the life cycle stages and the controls necessary for each in facility construction are discussed in chapter 11. Either approach may be useful, depending on the specific application.

Job Safety Analysis

The JSA has been addressed in previous chapters with a brief description, but a more detailed explanation is provided here. It is a well-thought-out safety and health program with emphasis the system, rather than the individual. Both job- and system-safety analyses are based on three facts:

1. A specific task or operation can be broken down into a series of relatively simple steps or operations.

2. The hazards associated with each step of the operation can be identified.
3. Solutions can be developed to control or eliminate these hazards.

Job- or system-safety analysis can be used to prevent accidents at any level of operation. For purposes of simplifying this, consideration will be given on three levels: (1) policy level, (2) procedure level, and (3) performance level. The first, or policy level, involves the work of top management; the second, or procedural level, is that of middle management and staff; and the third, or performance level, is where the work is actually carried out.

The JSA can point out the need for changes in safety and health policies, thereby involving the top management or policy level. Through JSA, one can also detect problems in safety and health procedures and involve middle management and staff in procedural corrections. Finally, at the working or performance level, the JSA can call attention to the need for worker protective equipment for handling hazardous materials. This level of JSA is often used as a basis for developing emergency plans and procedures.

JSA is a simple process involving four steps. In their proper sequence, these are as follows:

1. *Job Selection.* The first step is to select the task to be analyzed. The analysis is done for a single task at a time; thus, several JSAs may be required to cover a complex task or operation.
2. *Job Breakdown.* Next the task is broken down into its basic steps. For example, if the task is to load a cart with waste material, the steps might be: determine the amount of waste, select an appropriate cart, bring loading tools, position cart, etc. While this may seem very detailed and time-consuming, most tasks can actually be broken down into eight to twelve steps.
3. *Hazard Identification.* Next the actual or potential hazards associated with each step are identified. There is a definite procedure to follow in doing this. Typically, one might first ask if the worker can slip; trip; fall; or overexert by lifting, pushing, or pulling. One can then ask if the worker may come in contact with energy sources, chemicals, excessive heat, etc.
4. *Hazard Control.* In this step procedures or equipment are stipulated to reduce or eliminate each hazard listed in Step 3. If Step 3, for instance, indicates that the worker might be exposed to toxic fumes, one could specify that he or she must wear breathing apparatus. If one of the hazards is having to lift heavy objects, one might specify that two workers perform the task, that a forklift be used, or some other useful alternative.

As the JSA is performed, a chart should be developed, as shown in the accompanying diagram. This chart will list the basic job steps, the corresponding hazards, and the recommended safe procedures and equipment for each step. The completed JSA chart can then be used as a training guide for new personnel or as a safety review guide for experienced hands. It also makes an excellent guide for reviewing new or infrequently performed tasks. The JSA can also be included in the supervisor's safety manual, and can be extremely useful in day-to-day operations or during the safety inspection or audit. The JSA, at least initially, is done in the workplace with employee participation, perhaps by the supervisor or a committee, with subsequent management review.

Modification

There must be a formal program to ensure modified procedures and hardware do not present new hazards that may catch workers off guard. An adequate review process is undertaken whenever modifications are made in the system. Each change is carefully reviewed in terms of its immediate and long-term effects on personnel and the environment. Additions or changes in chemicals require special consideration and the effects must be carefully evaluated. Even a minor process change can have effects that ripple through the whole process and impact the outcomes in ways not anticipated, if no review occurs.

Decommission and Disposal

When hardware reaches the final stage of its life cycle and the decision is made to discard it, safety and health are prime considerations. Definite plans should be made to minimize disposal problems and hazards by considering hazards involved in dismantling and moving equipment, safe disposal areas or processes, potential environmental impacts where machinery is simply junked or discarded, scrapping operations, and so on. If the hardware is involved in complex operations or is very large, plans should be made for sectional removal, without completely redesigning or shutting down during the replacement process. This should take into account the impact of changed or modified operations, temporary arrangements, and so forth. These detailed disposal plans should also be subjected to a thorough JSA to detect and eliminate hazards.

Procedural Requirements

The next major element in the job-process-control system is the development of procedures. By now it should be clear that as carefully as personnel and hardware are selected and as effectively as policy is framed and authority is delegated, the operation will only be as safe and efficient as the procedures actually used. There are three essential factors that must be included in developing safe procedures: participation, testing, and feedback.

Participation

Establishing safe procedures does not happen in a vacuum or even on a drawing board. Engineers, designers, and managers may write the safety and health procedures, but they are not the ones who will carry them out. Their involvement is important, but participation of the on-site worker is absolutely essential. The worker is the person actively involved in performing a task: he or she knows what the job requires in the way of personal actions. In addition, a seasoned worker has established personal routines, perhaps discovered the most efficient motions or arrangement of tools, is likely to have developed short cuts—often at the expense of safety, and knows exactly what difficulties the procedures present. An actual worksite study may be extremely helpful, as in developing the JSA or reviewing existing JSAs against actual workplace practices. When possible, the worker should be asked for input, verbally, in terms of demonstrating work practices, and in making suggestions based on actual experience. Not only will this participation enable the manager to develop a more realistic picture of actual workplace needs and practices, but it can also have the effect of making the resulting official procedures more acceptable to the workforce. Where safety and health practices are concerned, this morale factor can be a tremendous asset.

Testing

Even when procedures have been developed and committed to paper, they should only be considered tentative. The only way to determine the validity of procedures, check the flexibility of plans, and verify the clarity of instructions is through actual testing. Both regular and emergency procedures should be tested in actual workplace conditions by the workers who will actually be using them. This may require a period of monitored activity with direct observation by management or by an appropriate committee of engineering safety and management personnel. Emergency procedures and drills should be tested up front and periodically. Findings should be thoroughly reviewed and, where needed, corrections made. The resulting revised procedures should then be tested in exactly the same way, with further review and revision. Only after thorough testing and review should the procedures be regarded as firm and incorporated as a part of the permanent job process system.

Feedback

Change is a constant factor in the workplace, and procedures may require periodic adjustment. A vital part of this adjustment process is the existence of an adequate feedback system, running all the way from top management to the worker and back again. There must be a *closed-loop system* that entitles everyone who provides input to receive feedback. Also, input which originates with the worker, such as with an employee suggestion, must find its way to management. Management must indicate the message was received and will be fully considered. Whether the suggestion is to be acted upon or not, it should be acknowledged. The feedback chain is not complete until the worker confirms reception of management's acknowledgment. Likewise, feedback originating with management, such as with a revised procedure, must find its way to the worker. A system should be established to verify affected workers have received the message. This can be done through initialing a copy of the new instructions, signing an acknowledgment form, or having the supervisor verify.

Failure to give feedback can completely destroy an effective safety and health program, as when employees become convinced management has not even seen their suggestions, or when management assumes everyone has received the new directive when, in fact, they have not. Feedback is an indispensable link making the whole system work.

Interfacing the Elements

There are three major elements involved in the job process control system: people, hardware and procedures. Each of these three interfaces with the others. The interfaces are

1. People to hardware
2. Hardware to procedures
3. Procedures to people

Each of these interfaces, the points at which one element of the system meets and affects another, represents a critical facet of the system. For instance, we cannot know how safe hardware really is by considering the hardware alone; what happens when it interfaces with people and procedures must also be reviewed. System control involves considering each of these interfaces and asking some pertinent questions.

People to Hardware

- Are the people basically compatible with the hardware? (This includes aptitudes, skills, as well as physical and mental traits.)
- Is personnel training designed for existing hardware? Is it adequate?
- Has hardware been thoroughly tested to determine its safety for personnel?
- Are personnel requirements reviewed when hardware is changed, and is retraining provided?

Hardware to Procedures

- Were procedures developed with existing hardware in mind?
- Was hardware designed or selected to conform with existing procedures?
- Have procedures been tested on the actual hardware?
- Has new hardware been introduced since procedures were written?

Procedures to People

- Are the contents and language of the procedures appropriate to the people? Highly technical instructions may be beyond the grasp of some workers or bilingual manuals may be needed.
- Do existing procedures match what workers have actually been trained to do?
- When new procedures are developed, is training modified to match them?
- Do procedures match the actual capabilities of current personnel?

These questions should be used as a checklist, particularly before starting a new or modified job process or the introduction of new procedures into an existing job process. They can greatly aid the manager, both in pinpointing potential trouble spots and in maintaining an adequate job process control system.

CONCLUSION

The heart of the job process control system within the overall SMS is the relationships among man, machine, and environment—or, as termed here, people, hardware, and procedures. While each involves distinct and complex questions that must be asked and requirements that must be met, no part of the system can be considered as entirely separate from the others. Effective job process control must include continued attention to the interfaces among these elements. Senior management, more than anyone, must be heavily involved in this task, using its overview to develop job processes that are not only adequate and efficient, but include sound safety and health practices from the planning stages on.

REFERENCES

Bird, F.E. Jr. and Loftus, R.G. (1976). *Loss Control Management.* Loganville, GA: Institute Press.

Briscoe, G., Lofthouse, J.H., and Nertney, R.J. (1979). *Job Safety Analysis,* Idaho Falls, ID: System Safety Development Center.

Bullock, M.G. (1979). *Work Process Control Guide*. Idaho Falls, ID: System Safety Development Center.

Department of Labor. (n.d.). *Disability Resources*. Retrieved from: http://www.dol.gov/dol/topic/disability/employeerights.htm.

Grimaldi, J.V. (1975). *Safety Management*. Homewood, IL: Irwin.

Hendrick, K. (1979, June). USCG, "Job and Systems Analysis." *USAF Study Kit*.

King, R.W. (1979). *Industrial Hazards and Safety Handbook*. London: Newens-Butterworth.

Nertney, R.J. (1976). *Occupancy Use Readiness Manual*. Springfield, VA: National Technical Information Service (NTIS).

Nertney, R.J. (1978). *Management Factors in Accident and Incident Investigation*. Idaho Falls, ID: System Safety Development Center.

Pavlov, P.V. (1979). *Job Safety Analysis*. Idaho Falls, ID: System Safety Development Center.

Pope, W.C. (1978). *Managing a Housekeeping Activity*. Alexandria, VA: Safety Management Information Systems.

Chapter 8

Safety and Health Training and Education

Mark A. Friend

CASE

Roberta Williams requested noise monitoring at her work station, since noise levels seemed so high. She was afraid that over a period of time, she might suffer hearing loss. After a brief discussion with her supervisor, he issued ear plugs and she returned to work. No conversation regarding this matter occurred between the supervisor and the safety professional, and no review of the noise levels or the intervention strategy occurred. Following a subsequent OSHA inspection, the organization was cited for not providing her with any training on the PPE use and for not determining if a more effective method of addressing problem with engineering or administrative solutions would have been more appropriate than PPE.

INTRODUCTION

An effective SMS must have an ongoing and efficient training program. While senior management rarely needs to involve itself directly in the training, it is up to higher management levels to help determine the ultimate scope and extent of training programs. The particular tasks are as follows:

1. Determining existing and projected training needs.
2. Developing overall training program strategies.
3. Assigning training responsibilities.
4. Determining in-house training capabilities.
5. Determining the need for outside training services.
6. Encouraging professional growth of the in-house safety and health staff.
7. Selecting suitable and reliable consultants when needed.

THE NEED FOR TRAINING

In any SMS there is a need for every person in the organization to have some level of safety and health training in order to maintain a safe and healthful working environment. There should also be emphasis on discovering inter- and multi-disciplinary solutions to the problems of identifying and controlling workplace hazards. This complexity makes training and education indispensable. Regarding training and education, there are two further considerations in work situations:

1. Today's workforce generally has a higher level of education than before and can usually handle a higher level of training.
2. In order to develop specific solutions for specific problems, training is useful for *all* levels of management.

GOVERNMENTAL REQUIREMENTS

In many cases the manager will have no choice as to whether to provide safety and health training. OSHA does not require a written safety and health program, but they strongly encourage it (Friend and Kohn, 2013). The OSHA standards do, however, require worker training in a number of areas, depending on the circumstances and the jobs in which employees engage. Standards affecting most employers include the following:

Hazard Communication Program. When employees handle hazardous materials as part of their jobs, the company will need a HAZCOM program. The standard provides some exceptions, for example, if the material used is only maintained in household quantities. For most hazardous materials, a few basic requirements exist, including required training for employees.

Emergency Action Plan. Employees should be trained on standard operating procedures in the event of fire, weather emergency, or other likely adverse situations. Standard operating procedures to account for all employees following evacuation or emergency notification.

Emergency Response Plan. Companies involved in emergency response operations to the release of or substantial threat of release of hazardous substances must develop an emergency response plan and provide training for those likely to encounter the emergency, as well as responders to it.

Personal protection equipment (PPE). A company is required to have a written PPE program if employees use personal protective equipment in conjunction with their duties. *PPE* refers to equipment to protect eyes, face, head, and extremities; protective clothing; respiratory devices; and protective shields and barriers. The employer must select and have each affected employee use the types of PPE necessary for protection. Training for each affected employee must be done on when PPE is necessary; what PPE is necessary; how to properly don, doff, adjust, and wear PPE; limitations of the PPE; and proper care, maintenance, useful life, and disposal of the PPE. Workers needing PPE must be trained on its application and use. Any PPE assigned requires training on its application and use. Prior to utilizing PPE, engineering controls should be considered. If these are not feasible, administrative controls are evaluated. PPE is generally considered the last and least desirable option for worker protection. OSHA can and will cite employers who don't assure engineering controls and then administrative controls are carefully reviewed before assigning PPE.

Permit-required Confined Space Plan. Companies are required to identify confined spaces throughout their facilities. A confined space is large enough and so configured that an employee can bodily enter and perform assigned work; has limited or restricted means for entry or exit (e.g., tanks, vessels, silos, storage bins, hoppers, vaults, and pits are spaces that may have limited means of entry); and is not designed for continuous employee occupancy. For the confined space to be considered permit-required, it contains or has a potential to contain a hazardous atmosphere, contains a material that has the potential for engulfing an entrant, has an internal configuration such that an entrant could be trapped or asphyxiated by inwardly converging walls or by a floor which slopes downward and tapers to a smaller cross section, and/or contains any other recognized serious safety or health hazard. Companies permitting employees to enter confined spaces must establish a training program.

Lockout/Tagout. Companies are required to comply with the Lockout/Tagout standard if servicing and/or maintenance takes place during normal production operations in which:

—An employee is required to remove or bypass a guard or other safety device, or
—An employee is required to place any part of his or her body into an area on a machine or piece of equipment where work is actually performed upon the material being processed (point of operation) or where an associated danger zone exists during a machine operating cycle.

Exceptions to the standard include minor tool changes and adjustments, and other minor servicing activities, which take place during normal production operations. These are not included if they are routine, repetitive, and integral to the use of the equipment for production, provided that the work is performed, using alternative measures which provide effective protection.

Depending on the overall operations of the company, there are numerous other types of OSHA-required training. A careful review of training requirements in specific circumstances surrounding operations is necessary.

TRAINING AND HUMAN BEHAVIOR MODIFICATION

Most of the training available in safety and health is oriented toward modifying human behavior. This approach is based on conclusions about the causes of unsafe actions. It assumes that people do unsafe things because of the following:

- They don't know they're doing a task incorrectly.
- They misunderstand instructions.
- They don't consider the instructions important.
- They have not received specific instructions for the task.
- They may have forgotten the correct and safe way to perform a task.
- They perceive safety is second to another organizational or personal priority.
- They may find it awkward to follow instructions.
- They deliberately disregard instructions.
- They may never have learned the proper way to perform the task.
- They may be unable to overcome bad work habits, despite instruction.

While training is a partial solution to the problem of unsafe acts, management error and oversight plays a part too. Top management has a vested interest in safety and health training.

Getting Started

The many levels and types of activities involved in a complete SMS depends largely on how clearly the program objectives are stated. The objectives must include the entire organization and all functions. While approval and sometimes actual development of the training program objectives comes from the top, it is the staff and management levels that bear most of the training burden. To be effective, there are certain basic requirements that apply to any type of training:

1. It must relate to identified needs of individuals and groups within the organization.
2. It must be related to organizational policies and objectives as a whole.
3. It must be both practical and acceptable to those involved.
4. Its various levels and aspects must be interrelated. (For example, new employee training meets an identified need, but would take place only after training needs for supervisors in that area had been met.)
5. It must be assessed periodically to determine how effectively it is meeting its objectives.

Program Development

The National Safety Council has identified six steps in the development of any safety and health training program:

1. *Identifying training needs.* The biggest problem in training is determining when there is a need. Signs of need include high accident and injury rates, many "near-misses," worker grievances, unusual labor problems, frequent equipment breakdowns, wastage and excessive scrap, customer complaints, changes in equipment and processes, and so on. The exact nature of the training problem must be identified, or training may be provided in the wrong area or on the wrong operational level.
2. *Formulating training objectives.* Organizations must develop specific training objectives to solve specific problems. Beyond that, the development of training objectives, course objectives, testing, and evaluation methods required of the training staff.
3. *Gathering materials and developing course outlines.* This step involves a commitment of resources from top management and staff. Is in-house training the most economical or are there more suitable training materials and resources available from outside sources?
4. *Selecting training methods and techniques.* Depending on the learners and material covered, certain training techniques are more effective for different types of learners dealing with certain kinds of tasks or behaviors. A training specialist may be in the best position to assess specific needs.
5. *Conducting the training program.* This involves the actual application of curriculum, timetables, teaching environment (the proper place, lighting and seating, if the training is not actually done on the job), supplies, instructors, certificates, and incentives. If there is no training staff for these activities, management may want to hire a competent, outside, training consultant.
6. *Evaluating the program.* To assure the training program serves company's objectives, even when mandated, it is essential to provide a method for evaluating its effectiveness. Tests, on-the-job evaluations, evidence of favorable changes in the accident or wastage rates, and many other methods of evaluation may be applied. Management must ensure evaluations are made periodically, and higher management receives the results.

Training Program Content

Exactly what should a training program for safety and health contain? This will vary according to the person, group, or level being addressed, but the organization's overall safety and health training program should contain the following elements, depending on specific job functions and assignments of individuals being trained:

- The safety management system (SMS) and its components
- Concepts and philosophies of environmental health
- Concepts and philosophies of accident prevention
- Company safety and health policies
- Plant processes
- Job safety analysis (JSA) and job instruction training (JIT)
- Accident investigation procedures
- Relationship between safety and health and company objectives
- Accident, incident, and injury reporting
- Legislative and reporting requirements
- Personnel selection and appraisal techniques
- Job placement and training activities
- Hazard detection techniques
- Operating procedures and rules
- Gathering and interpreting safety and health information
- Personal protective equipment
- Human aspects of safety and health
- Emergency planning and procedures

These subjects have wide application, as safety and health functions are involved in any company function; however, efficient application requires specialized training and approaches. Different operating levels may need different knowledge about a given area. For example, the assembly line worker does not need detailed information about personnel selection or information-gathering techniques but does need to know how standards of performance apply or whom to inform of an unsafe condition. The senior executive needs to know what governmental regulations require of the company; the worker needs to know what rights he or she has under the same legislation. The manager needs to know how to develop an effective plan for dealing with emergencies; and the individual worker needs to know what specific actions to take in an emergency.

Training Responsibilities by Operating Function

The exact training needs of an individual or group in an organization are related to its function. The following list attempts to summarize responsibilities of each operating function.

Executive and Senior Management

1. Understand and accept responsibility for the overall safety management system.

2. Develop, agree on, promulgate, review, and evaluate safety and health policy.
3. Determine safety and health program objectives.
4. Allocate resources to achieve those objectives.
5. Assign responsibilities and establish accountabilities for safety and health objectives.
6. Develop, disseminate, and periodically review and evaluate the SMS.

Staff (Non-Safety and Health)

1. Develop:
   a. Operating rules and standards
   b. Accident investigation procedures
   c. Personnel selection and training procedures
   d. Useful information and reporting formats
   e. Emergency planning procedures
   f. Operating objectives
   g. Training requirements and resource allocations
2. Ensure safety and health considerations are included in phases of the product life cycle
3. Interpret policy for line management
4. Provide information and advice for management decisions
5. Review and evaluate safety performance
6. Provide input on safety and health problems observed

Line Mid-Management

1. Understand the SMS
2. Communicate objectives and goals
3. Work with staff in developing objectives, rules, emergency plans, processes, and equipment design
4. Develop strategies to comply with rules and regulations
5. Develop job procedures with staff and supervisors
6. Interpret information for preventive application
7. Implement procedures for use of personal protective equipment and other safety devices
8. Provide feedback to senior management and staff on objective performance as a basis for review and change in policies and procedures
9. Investigate, evaluate, and report on accidents and injuries
10. Take corrective action on conditions uncovered through accident investigation
11. Coordinate training needs and make personnel available for training
12. Provide input on safety and health problems observed

Supervisor (Front-Line)

1. Understand the SMS and reasons for accident prevention and health measures as part of the SMS
2. Understand, adopt, and ensure compliance with safe practices
3. Implement government requirements, company policies, and procedures
4. Develop safety awareness in workers
5. Supervise workers to assure use of safe practices
6. Brief workers on and discuss safe work practices
7. Inspect to identify unsafe practices and conditions
8. Inform management of unsafe practices and conditions beyond own control
9. Ensure use and maintenance of personal protective equipment and all safety devices

Employee (Non-Management)

1. Understand:
    a. Overall safety management system
    b. Concepts of accident causation and prevention
    c. Program objectives and personal application
    d. Need for implementation of safe practices
    e. Communication system for safety and health items
    f. Emergency plans and personal roles
    g. Employees' OSHA rights
    h. Hazard reporting system
    i. Environmental hazards of the workplace

2. Comply with regulations and company rules
3. Use personal protective equipment and safety devices as required
4. Report accidents, injuries, and hazards

Safety and Health Personnel

1. Have a strong understanding of the safety management system
2. Know:
    a. Basics of accident investigation, analysis reporting
    b. Policies and programs
    c. Organizational levels and functions
    d. Safety and health concepts and practices
    e. Accident reporting and investigation systems
    f. Line and staff roles in accident and injury prevention
    g. Applicable government regulations
    h. Rationale for safety and health throughout the organization

3. Understand and make recommendations on programs
4. Monitor adherence to government regulations and company rules
5. Monitor and assist line and other staff in carrying out their functions
6. Serve as a resource on safety and health matters
7. Interpret safety information for senior management
8. Monitor accident investigation and reporting activity
9. Promote interest and participation in programs
10. Inspect to identify hazards and unsafe practices
11. Establish supplemental programs to attain objectives
12. Make recommendations to management, line, and staff
13. Utilize information and feedback for accident prevention purposes.

Given its functions, what sort of basic or extra training will any operational level need? Proceeding in this way has the advantage of pinpointing the safety and health functions of any particular position. Additional training can then be added to meet new facility occupation, changing procedures, installation of new or modified equipment, and the activities of safety and health committees or accident investigation teams.

Internal Safety and Health Training

Almost every organization already has some provisions for in-house training. However, when higher management contemplates a major revision of its safety and health training program, the question naturally arises: How much training should the company attempt to do in-house, and what kinds of training can be handled more effectively by contacting an outside firm or consultant? To assist the manager in making this decision, following are some points to keep in mind about in-house training.

Safety and Health Staff as Training Resources

The organization may already have one very good resource for in-house safety and health training: the safety and health manager or staff. They should be involved in any in-house training scheme, but there is always a question as to the extent of that involvement. Some organizations with regular safety and health staffs may include staff members hired specifically as trainers. Training can become such a time-consuming task that it prevents the safety and health personnel from performing much more urgently needed functions. The safety and health person rarely has first-hand knowledge of actual work processes, and safety and health activities should not be separated from the regular daily activities in the workplace.

The safety and health professional is needed to provide an extensive advisory service affecting all activities. To also be a training director as well dilutes overall effectiveness. This is particularly true when speaking of routine areas like first aid, firefighting, emergency rescue, and the proper way to perform specific tasks. These training segments are better performed by others. The maximum benefit is achieved when the safety and health manager merely acts as an advisor to whoever gives the actual training by helping select the right training methods and get the right material together. The safety professional should usually serve in an advisory capacity than as an actual instructor.

Training Strategy

A sound training strategy begins with company policy. The policy should make a clear reference to safety and health training. Next, the objectives of the training should be clearly stated, and the responsibilities for different training should be clearly located and spelled out. This does not relieve top management of its responsibility but properly delegates part of it to the management levels who can most effectively direct it. The overall responsibility should be assumed by a senior manager.

In terms of training objectives, strategy should be based on required changes in present performance levels, and changes expected in the future. Some objectives will be based on statutory requirements. In all cases, strategy should be based on

1. The main safety and health problems of each functional area.
2. Problem areas where improved performance is desired.
3. Realistic and attainable goals.

In order to make intelligent decisions about training strategies, as well as other aspects of the safety and health training program, the manager should pose several questions.

1. What are the expected benefits of training?
2. Who does the company need to train?
3. What levels of performance are required?
4. How long will training take? How much time can we allocate to training?
5. How urgent is the training need? Does someone's life depend on it?
6. How many people need training?
7. Who will be expected to conduct the actual training?

Since line management has the responsibility for implementing the program, steps should also be taken to assure the program is acceptable and practical. The senior manager must present the training objectives, training content, training methods and responsibilities to line management, making sure that they are clearly understood. Line management's feedback may point out needed revisions in approach. When these have been addressed, the senior manager can secure a firm commitment from everyone involved.

Arrangements are also made for monitoring the training and assessing its value when completed. The training is periodically reviewed for pertinence, quality, and timeliness, and changes are made as required.

Special Safety and Health Training: The New Employee

More seasoned employees may require "spot" training. They may already be familiar with safe work practices, the use of personal protective equipment, and how to report a workplace hazard. The new employee, however, represents a special case for training; he or she still has everything to learn about the organization. In training the new employee, it is not enough simply to demonstrate how to perform the specific task he or she has been hired to do. Training must include a much broader range of safety and health information, including

- Company safety and health philosophy
- Safety and health requirements as a condition of employment
- Employee safety and health responsibilities
- Safety rules and regulations
- Requirements for reporting injuries, accidents and hazardous conditions
- Location of emergency exits, alarms, first aid kits, etc.

These subjects require the understanding and backing of top management and should be presented to all new employees. This early training period is an ideal time for top management to demonstrate its sincerity about safety and health, discussing these subjects with new employees as part of their induction briefing, either before or during the first week of employment. A few words on the subject from top management add viability and credibility to the program. Being given a message of welcome encouragement by the top executive (whom they might not see again for weeks, months, or years) is a great way to start off a new job. It gets things off on the right foot and gives management a chance to emphasize programs they feel strongly about. Certainly safety and health emphasis means a lot more coming from the top executive than it does coming from safety personnel or a lower-level personnel clerk.

This is not enough, however, and there is a time when safety and health become specific and applicable to the job being performed. This emphasis becomes the responsibility of the immediate supervisor, preferably right in the workplace. The supervisor's task is made much easier when the pace has been set by top management.

SAFETY AND HEALTH CONSULTANTS

It can be wasteful of management time and resources to develop a new internal safety and health training program, only to find an equally suitable program already exists and is readily available at a reasonable cost. Similarly, investing in developing an in-house safety and health staff, only to learn later that an outside consultant is available to accomplish the same goals at a more reasonable price, can also be a problem. Many safety and health consultants can offer or design specialized programs and present them on-site; others can work on particular safety and health problems, but do not provide direct training. There are also consultants who work full-time for a business (such as an insurance company), providing services for the business's clients. Any of these may prove suitable to a company's particular needs.

If the decision is made to seek help from an outside consultant, the manager is still faced with the problem of finding a reliable professional who can provide exactly the services desired. Following are points for consideration:

1. What is the alternative besides using the consultant?
2. Why pick this particular consultant? Cost should not be the prime consideration.
3. What is the reputation? Seek references for any outside consultant.
4. How was the name of the consultant obtained? Is the source reliable?
5. Who are the consultant's references? Verify those references and contact them before contracting with the consultant.
6. Who are the consultant's associates or instructors?
7. Does the consultant have previous experience in your type of business? Is this kind of experience necessary?
8. How does the consultant self-evaluate effectiveness?
9. What can the consultant provide in the way of handouts, manuals, slides, etc.? Will those cost extra?
10. Will the consultant furnish necessary facilities and equipment?
11. Is the consultant willing to present training courses on-site?
12. Is the fee competitive?
13. What are the contractual arrangements? This is particularly important if the training is complex or long-term.
14. What is the consultant's attitude about working with and through the existing safety and health staff?

Even the finest consultant cannot work effectively without the active backing of top management. Learn what the consultant needs from top management in order to create and implement effective programs. It is management's task to decide whether an outside consultant will be needed or whether in-house training is the best answer. This question may have to be reviewed from time to time, as training needs change, new staff is added, or as existing programs appear to be ineffective.

CONCLUSION

Training is an ongoing need; it simply cannot be accomplished on a one-time basis. For that reason, the senior management must be continually informed of the ongoing success of training programs, keep them under continual review, and ensure they are being conducted according to plan. As changes are made in the system, affected personnel must be informed and trained on any necessary functions in which they may be engaged.

REFERENCES

Anton, Thomas J. (1979). *Occupational Safety and Health Management.* New York: McGraw-Hill.
College and University Safety Courses. Chicago: National Safety Council, Annual.
Findlay, James V. (1979). *Safety and the Executive.* Loganville, GA: Institute Press.
Friend, M.A. and Kohn, J.P. (2013). *Fundamentals of Occupational Safety and Health.*
General Materials Catalog. Chicago: National Safety Council, Annual.
Good, Carter V. (1973). *Dictionary of Education.* New York: McGraw-Hill.
Grimaldi, John V. (1975). *Safety Management,* 3rd ed. Homewood, IL: Irwin.
Hammer, Willie. (1980). *Product Safety Management and Engineering.* Englewood Cliffs, NJ: Prentice-Hall.
King, Ralph W. (1979). *Industrial Hazards and Safety Handbook.* London: Newens-Butterworth.
ReVelle, Jack B. (1980). *Safety Training Methods.* New York: Wiley.
Supervisor's Safety Manual. Chicago: National Safety Council, Periodic Revisions.

Chapter 9

Ergonomics

Mark D. Hansen

CASE

Mal Kaufman worked in a brickyard, supervising teams of five people who moved bricks from a flat car onto a device where they were then stacked, banded, and palletized. Each employee had seven seconds to move approximately twenty-two bricks from the flat car to the machine right behind them, and then the machine moved forward. Before beginning the operation, employees donned heavy-duty, leather pads on their hands. Each pad had been covered by rubber patchwork; otherwise, the leather would be disintegrated in a few hours. At the first position, the worker picked up the bricks, turned 180 degrees and placed the bricks on the bottom of the rack. After each row of eleven bricks, the next row began. The fifth position filled in the last two rows, so there would be a stack of bricks ten rows high. The conveyer holding the bricks moved every seven seconds. The person working position number five could stop the conveyer, if anyone fell behind. Employees were given a five-minute break every hour and a half-hour break for lunch. Production typically ran for 40 hours, plus a half day on Saturday. Owing to the relatively quick work cycle and the types of movement, Mal worried about worker injury, but he wasn't sure how to determine whether this job was safe or not.

INTRODUCTION

As a way of introduction, the word "ergonomics" comes from the Greek, "ergo" which means work, and "nomos" which means to study. Simply put, *ergonomics* is the study of work. In practice, ergonomics means fitting the job and the work place to the individual capabilities of people. For safety and health professionals it means to reduce employee exposures to risk factors associated with work-related ergonomic injuries and illnesses and to improve the overall efficiency of operations.

This chapter will address the typical work-related injuries, developing an ergonomics program within the SMS and SMS-assessment tools used to eliminate and/or mitigate ergonomics exposures in the workplace.

TYPICAL ERGONOMIC INJURIES AND ILLNESSES

The most common types of ergonomic injuries and illnesses are musculoskeletal disorders (MSDs). MSDs are injuries/illnesses that affect muscles, nerves, tendons, ligaments, joints, or spinal discs. Employees may suffer ergonomic injuries/illnesses when work tasks include reaching, bending over, lifting heavy objects, using continuous force, working with vibrating equipment, and/or performing repetitive motions. MSD injuries/illnesses are determined by the part of the body affected. If employees suffer an ergonomic injury/illness, occupational physicians may diagnose the injury as an MSD. Some common MSDs are discussed below.

De Quervain's Tenosynovitis

De Quervain's tenosynovitis, the most common disorder with tendon sheath swelling, occurs in the abductor (moving away from the mid-line) and extensor tendons of the thumb. These tendons share a common sheath, and swelling can affect both. Impingement on the tendon by swollen sheath and the production of excess synovial fluid can lead to loss of tendon function. De Quervain's tenosynovitis, which is commonly recognized as an occupational disorder, may be precipitated by forceful grasping and turning, particularly of hard objects such as vials. Symptoms include swelling, pain, and tenderness at the base of the thumb. Pain is aggravated by attempts to extend the thumb. Flexion and adduction may produce a "trigger" effect or popping sensation.

Carpal Tunnel Syndrome

Carpal tunnel syndrome is the most common nerve-entrapment disorder. Symptoms result from compression of the median nerve as it passes through the wrist within the carpal tunnel, a narrow, confined space formed by the eight carpal bones and the transverse carpal ligament. Within this limited space, swelling of any of the components can increase pressure in the tunnel. Because the median nerve provides both sensory (feeling) and motor (muscular movement) innervations to both the thumb and middle three fingers, damage can result in pain and disability.

Although it has been extensively studied, carpal tunnel syndrome still remains a subject of controversy in terms of etiology, relation to work, diagnosis, and treatment. However, its association with cumulative, occupational trauma is becoming more generally accepted. Carpal tunnel syndrome can result from any process that leads to increased pressure on the contents of the tunnel. The nerve is the most vulnerable component. Nonspecific tenosynovitis (swelling of the tendon sheaths) of the flexor tendons within the tunnel may be the most common cause of increased pressure. Decreased capacity of the tunnel may result from prior fracture or discoloration of the hand or forearm, rheumatoid arthritis, or congenital anomalies. Early symptoms include numbness, painful tingling, and burning pain and weakness in the thumb and first three fingers. These symptoms are often precipitated by repetitive hand or finger actions, but nocturnal symptoms that wake the patient are common. Shaking the hand may bring relief. Sensory loss in the ring finger frequently occurs only on the lateral (away from the body) side.

Numbness rarely radiates proximal to the wrist, but sometimes the forearm and shoulder aches. Typically the dominant hand is most affected, with involvement of the other hand being evident only upon electro-diagnostic studies. Weakness eventually develops in the muscles that abduct and oppose the thumb. In advanced cases, the thumb cannot move properly in opposition to the other fingers, and the worker may drop objects.

Tendonitis and Tenosynovitis

Two of the best documented work-related disorders are tendonitis and tenosynovitis. Both are painful and disabling and play a role in nerve compression syndromes. Tendonitis-inflammation of the tendon (fibrous band connecting skeletal muscle to bone) can affect any extremity, but in the workplace, it is most common in the wrist and fingers. Tenosynovitis is inflammation of the synovial sheath that encloses tendons running within fibro-osseous (having both connective tissue and bone) tunnels. The inner membrane of the sheath secretes a viscous fluid that lubricates the joint, and this secretion increases during an inflammatory process. Tendonitis, which is often accompanied or preceded by tenosynovitis, is common among people who perform repetitive work, especially when a tendon rubs against other structures as it passes through a fibro-osseous tunnel. Occupational risk factors include repetitive tension and motion, bending, and vibration. Risk increases with age, due to tendon stiffening. Tendonitis is diagnosed by tendon swelling found on physical examination, with localized pain on palpation or resisted movement. Common disorders are rotator cuff tendonitis, biceps tendonitis, lateral epicondylitis ("tennis elbow"), and medial epicondylitis ("golfer's elbow").

Rotator Cuff Tendonitis

Rotator cuff tendonitis is a shoulder disorder characterized by inflammation of the supraspinatus muscle to the humerus (upper arm bone). Rotator cuff tendonitis has several causes, which may be classified as extrinsic (due to mechanical impingement from outside of the cuff) or intrinsic (due to changes within the cuff, such as aging or diminished vascular supply). In primary mechanical impingement, elevation of the arm leads to pressing of the supraspinatus tendon against the acromion. When this is repetitive or excessive, the resulting irritation and ischemia (reduction in blood supply) lead to rotator cuff tendonitis. The primary symptoms of rotator cuff tendonitis are shoulder pain, sometimes radiating down the arm. Generally tendonitis pain is exacerbated by movement and relieved by rest, but it may occur at night also, especially if the cause is impingement. Movement may be limited by pain, stiffness, or weakness. Rotator cuff injuries are common among workers who perform repetitive tasks with their elbows above mid-torso height, particularly if their arms are raised overhead.

Thoracic Outlet Syndrome

Thoracic outlet syndrome consists of upper extremity symptoms resulting from pressure on nerves of blood vessels between the base of the neck and the axilla (armpit). The syndrome may involve any of the several structures in the thoracic region, and symptoms make it difficult to distinguish the specific area. Structures that are subject to compression are usually the nerves of the brachial plexus (the braid of nerve fibers formed by the anterior branches of cervical nerves C5-C8 and thoracic nerve T1; it begins in the base of the neck and extends into the axilla, where it divides into the major nerves of the arm). Less commonly (in fewer that 10 percent of cases), the subclavian artery or vein is compressed. Symptoms include neck pain, arm weakness, and numbness extending along the inner forearm into the medial two fingers. Symptoms may be precipitated or aggravated by postural changes, especially arm elevation. Vascular symptoms are aching or throbbing in the arms, coldness, and periodic blanching of the fingers. Diagnosis is difficult because many physical signs are nonspecific.

Wrist Ganglion

The wrist ganglion is considered by many to be a herniation of the joint capsule or of the synovial sheath of the tendon; other authorities believe it to be a cystic structure. It usually appears slowly after a wrist strain and contains a clear, mucous fluid. The ganglion most often appears on the back of the wrist, but can appear at any tendonous point in the wrist or hand. They are often treated surgically.

Trigger Finger

The trigger finger or thumb is an example of stenosing tenosynovitis, a condition where the tendon surface becomes irritated and rough, the sheath becomes inflamed, and the tendon sheath undergoes progressive constriction. It most commonly occurs in a flexor tendon that runs through a common sheath with other tendons. Thickening of the sheath or tendon occur, thus constricting the sliding tendon. A nodule in the synovium of the sheath adds to the difficulty of gliding. The worker complains that when the finger or thumb is flexed, there is resistance to re-extension, producing a snapping that is both palpable and audible. This disorder is usually associated with activities involving the repetitive usage of tools that have handles with sharp edges or hard edges. Examples of laboratory activities involving work with sharp edges can include repetitive work opening vials, closing vials, pipetting, and cover slip applications.

Back Injuries

Back injuries are categorized as ergonomic disorders when they result from chronic or long-term, injury to the back rather than from one specific incident. Once back muscles or ligaments are injured from repetitive pulling and straining, additional injuries are more likely to occur because the back muscles, discs, and ligaments may be scarred and weakened and can lose their ability to support the back.

One of the main causes of low back injuries in the laboratory is the awkward lifting of centrifuge rotors. Overhead lifting of materials off of shelves and frequent rearrangement of laboratory equipment due to lack of space also lead to a significant amount of back injuries.

Ergonomics programs have become just another component of a comprehensive safety program for safety and health professionals to implement and manage. It is often included in employee orientations, annual training, and awareness. Even though there are no Occupational Safety and Health Administration (OSHA) regulations on ergonomics at this time, it still must be addressed as a workplace hazard. Under Section 5(a)(1) of the Occupational Safety and Health Act, OSHA requires that employers furnish to each employee a place of employment free from recognized hazards that are causing or are likely to cause death or serious physical harm to its employees.

DEVELOPING A WORKSITE ERGONOMICS PROGRAM

As part of the SMS, committed management is a significant key to a successful worksite hazard control effort. A typical ergonomics program should include the following:

- Introduction
- Program management
- Training and education
- Ergonomics process
- Surveillance
- Analysis and design of jobs
- Medical management

Program Management

As part of the overall SMS, safety professionals must secure senior management commitment. Management commitment is demonstrated by the provision of organizational resources and the assignment of accountability for the ergonomics program. Employee involvement is necessary not only for identifying existing and potential hazards but also for participating in their own personal protection.

Management commitment provides visible involvement of managers at all levels. It places a high priority on eliminating ergonomic stressors while assigning and communicating the responsibilities for various aspects of the program and requiring accountability for fulfilling those responsibilities in a timely manner. Management provides authority and adequate resources to meet the assigned responsibilities.

Training and Education

An effective worksite hazard control program within the SMS will incorporate an ergonomics training component. The goal of the training program should be to prepare supervisors, managers, and workers to identify job ergonomic hazards, to recognize signs and symptoms of related injuries or illnesses, and to participate in the development of strategies to control or to prevent workplace hazards. The depth and detail of the training will depend on the role you expect each person to play in your ergonomics program.

For example, the safety and health professional should have the most intensive training. This type of training may involve, at a minimum, several weeks of course work. Medical personnel will need training in recognition and treatment of these injuries. Supervisors will need training to help them identify and correct problems on the production lines they supervise. This type of training may involve several days of instruction. Engineers will need training so they can design and redesign equipment and jobs with sound ergonomics principles in mind. Some organizations set up ergonomics teams who try to spot and correct problems that could lead to musculoskeletal disorders. These groups will also need special training.

Employees, in general, should be given ergonomics awareness training. Training should take place in two types of settings—classroom and on-the-job training. Objectives should be as follows:

- Recognition of workplace risk factors and an understanding of control measures.
- Identification of the signs and symptoms of musculoskeletal disorders that may result from exposure to risk factors and familiarity with the company's health care procedures.
- Knowledge of proper reporting procedures, including the names of designated persons who should receive the reports.
- Knowledge of the process the employer is using to address and control risk factors, the employee's role in the process, and ways employees can actively participate.

Types of Training

Training must be developed to meet the needs of the company while addressing the ergonomics exposures. There are six types of training listed below.

General Training. Facilities and Safety and Health personnel are provided an introduction to the general principles of ergonomics and to the ergonomics program. This training may also be accomplished using outside consultants in coordination if deemed appropriate.

All employees are required to take the general training. New employees receive general training during new employee safety orientation. Existing employees are scheduled to receive general training during monthly safety meetings.

Job-specific Training. Every employee (new, old, reassigned) is taught how to use tools and equipment for maximum efficiency and ergonomic comfort, and is responsible for using safe work practices on the job. Training for commonly used tools and equipment takes place in the classroom with interactive teaching methods, including computer-based training (CBT) (student participation and practice). Safety practices for tools and equipment unique to a work area will be demonstrated on the job by supervisors. Trainees are expected to actively participate in their own protection by performing self-assessment of their work habits and implementing basic changes in their work areas.

Management Briefing. Management and some administrative personnel are required to attend tailored ergonomics briefings.

Training for Supervisors. Supervisors are required to attend the jobspecific training for the positions they supervise. In addition, supervisors need briefings similar to those provided for managers in order to gain a complete understanding of their responsibilities.

Support Training. All departments have a responsibility to maintain ergonomic knowledge and skills current based on direction from safety and health professionals to apply ergonomic principles in performing their duties. Appropriate technical training is provided for support employees on an as-needed basis.

ERGONOMICS PROCESS

A process is developed to identify and manage ergonomics injuries and illnesses in the workplace. The process includes identifying signal risk factors, quick fixes, job improvements, design control and changes, and medical management. Examples of the Medical Management process and the ergonomic process are shown below in figures 9.1 and 9.2, respectively.

A *process* is a system whereby necessary components function and interact to produce a desired goal. Similarly, an *ergonomics process* is a plan in which a facility gathers all relevant information on work organization, employee capabilities and limitations, and work-related MSDs, to develop solutions to better accommodate these employees and reduce MSD rates and their associated costs. One important aspect, actually key purpose of the process, is to communicate information among all those involved, so that adequate and feasible solutions to problems having ergonomics issues can be solved. Employee involvement in the ergonomics process, across several layers of an organization, helps to ensure its success. When individuals in a company, from the upper echelons of management to the hourly employee, contribute to making changes in work systems and job sites, they become empowered and more responsible for these changes. This cooperation and employee involvement leads to feasible and successful changes that serve to improve the company.

A successful ergonomics process needs to be implemented and refined across various types of organizations, so that a wide variety of work-related issues can be incorporated. It comprises five fundamental elements:

- The workplace;
- The ergonomics committee;
- Management;

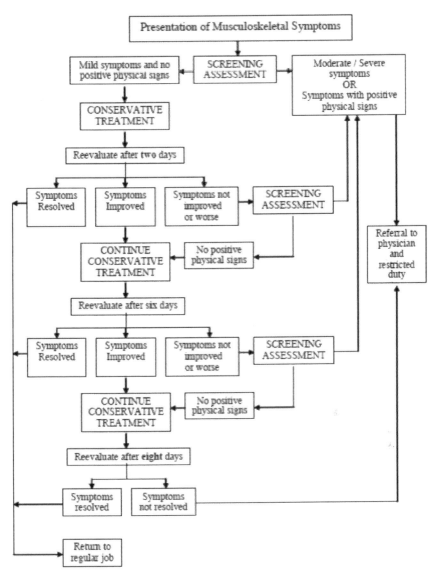

Figure 9.1 Example Flowchart of Medical Management Process, Showing the Steps Necessary to Assess and Treat Symptoms

- Medical management; and
- The ergonomics expert.

The element from which basic ergonomics issues arise is the workplace; that is, the area in which the physical work is being performed and where employees become injured due to musculoskeletal stressors. It is the responsibility of the ergonomics committee to identify these stressors and provide solutions. The remaining three components (management, medical management, ergonomics expert) provide support for the interaction between the Workplace and the Ergonomics Committee. The company's management is responsible for committing to the process itself; that is, emphasizing the value it places on its employees, and for initially providing monetary and personnel resources to sustain the committee and the changes it makes within the facility. The Medical Management component deals with the health of the employees, by providing symptom recognition, providing prompt and appropriate treatment, and returning injured employees back to work as efficiently as possible. Finally, the Ergonomics Expert is tasked with training the Ergonomics Committee, providing ergonomics input beyond the committee's expertise, and assuring that the team progresses in the correct direction.

The description of the ergonomics process contained in this document can be used as a guide for the interested reader or for those developing an ergonomics process within their company. The components outlined and discussed are consistent with the components of ergonomics processes outlined by the Occupational Safety and Health Administration in their document *Ergonomics Program Management Guidelines for Meatpacking Plants.* Many facilities already have an active safety and health program. However, many MSDs differ from acute injuries, in that these injuries develop over time, and there may be many risk factors throughout the entire job that

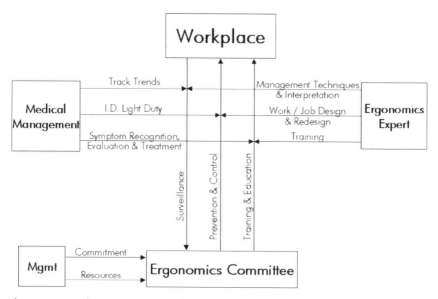

Figure 9.2 **Major Components of a Typical Ergonomics Process and How these Functions Interrelate**

contribute to these types of injuries. Thus, components of an existing safety and health program may need to be enhanced to facilitate the understanding of these types of injuries, as well as their causes and prevention strategies.

The responsibility of these components in dealing with workplace issues within the ergonomics process are outlined below.

MANAGEMENT COMMITMENT

For an ergonomics process to be successful, the company's management must be fully committed to integrating ergonomics into the workings of the organization. This includes understanding ergonomics concepts on a global level—that work systems designed within individuals' capabilities and limitations will improve work efficiency and productivity and will reduce the numbers and costs of MSDs. A progressive management structure with the SMS is one that honestly values its employees and communicates this commitment through corporate policies that tout this dedication to employee health and well-being; this includes the implementation of an ergonomics process. Management must further understand that ergonomics itself is an ongoing process. A successful ergonomics process must be approached like any other SMS process, where there is a commitment to continuous improvement.

This level of commitment by management is necessary because of the types of MSDs often addressed by ergonomics committees. These MSDs usually are cumulative in nature; that is, they do not occur after a one-time exposure, but rather they accrue over a period of weeks, months, or years of activity considered stressful to the body. Because of their cumulative nature, workplace changes may not immediately result in a drop in injury rates. Management must realize that some patience is required when beginning an ergonomics process and that, over time, the benefits of the investment will be returned in terms of reduced numbers of MSDs and their associated costs and an increase in productivity.

ESTABLISHMENT OF THE ERGONOMICS COMMITTEE

The core of this part of the SMS process is the ergonomics committee. It is within the committee that all activities in the facility having ergonomics impact be directed. The formation of the committee and their training requires considerable effort. Before the Ergonomics Committee can be productive, members must be trained on the methods, tools, and expected outcomes. A strong committee foundation results in a more comprehensive understanding of workplace concerns (surveillance), more complete evaluations of workplace stressors and solutions more likely to be technically feasible (hazard prevention and control), and a better understanding and acceptance of the ergonomics process throughout the facility (training and education). The elements of the committee are described below.

1. *Composition of the Committee.* Just as ergonomics itself is comprised of multiple disciplines, an Ergonomics Committee should be composed of members with diverse backgrounds and areas of expertise. It is important to consider what traits each team member can contribute to the ergonomics process. Manufacturing engineers, for example, have an understanding of why existing work processes were designed as such and can give insight as to what potential solutions may or may not be feasible. Experienced production employees from different departments within the facility often have performed many jobs and have excellent working knowledge of the tasks involved in the jobs, as well as insights for potential solutions to problem areas. Other members to consider for the ergonomics committee include the plant manager, process engineer, human resources or safety manager, department supervisors, and maintenance personnel. The committee should be composed of individuals who are known and respected within the facility, since others will be more likely to provide information and accept job changes if they also are supported by these employees.

   It is extremely important to maintain balance within the committee. That is, the team should be composed of half management and half production personnel. This is especially important when the process is being implemented within a union facility. The Ergonomics Committee must be viewed as part of a company cooperative endeavor rather than one strictly created by and under the direction of either labor or management. The number of team members also must be considered. For any type of committee, more than eight to ten members can become unwieldy and unproductive. An ergonomics committee is no exception. However, when implementing an ergonomics process within a large corporation, a team of this size may not be able to address all ergonomics issues that arise. When this is the case, separate subcommittees, having similar compositions, can be formed that periodically report to a main overseeing body.

2. *Training the Committee.* Before beginning to evaluate the facility from an ergonomics perspective, the team first must receive training. Because of the multidisciplinary nature of this field and the diversity of educational backgrounds of committee members, only a general awareness of ergonomics concepts may be possible. That is, the committee must have enough training to understand how work activities can affect the body

and produce MSDs. The training topics covered should be tailored to the needs of the facility. For example, if a majority of MSDs to employees occur in the low back, then the principles of low back injury information and strategies should be thoroughly covered in the training materials. Similarly, if many jobs are performed within noisy work environments, then concepts of perception, communication, and noise abatement should be addressed. These special topic areas need to be addressed in addition to the basic ergonomics principles that can be applied to a wide range of work situations.

3. *Active and Passive Surveillance.* One goal of the ergonomics committee is to adequately survey the workplace for problem areas. This includes identifying both jobs and work practices that have historically caused MSDs and those destined to produce future problems if not corrected.

    To begin, the committee will undoubtedly focus, with the help of medical management personnel, on those jobs and work areas known already to have caused MSDs (passive surveillance). These initial areas of focus usually arise from a review of accident and injury reports (e.g., OSHA 300 logs. This reactive approach is accepted as the method for committees who are beginning to tackle ergonomics issues. However, just as in other SMS approaches, the ultimate goal of the ergonomics process is for the team to become proactive. This occurs over time as the team acquires a better understanding of ergonomics and a familiarity with the process. For instance, more established ergonomics teams may distribute discomfort surveys to all employees periodically. An example discomfort survey is provided later in this chapter. The information gathered from these surveys (qualitative assessments) helps to identify jobs that are producing problems and employee discomfort, but have not yet resulted in injuries or illnesses.

    Once trained, the team will possess the adequate skills to identify and correct ergonomics stressors that currently exist within work processes before actual symptoms and MSDs result. They will also be able to analyze new work processes that are to be designed in the future.

4. *Hazard Prevention and Control.* A second goal of the committee is to prevent MSDs from occurring, or, attempt to control the rates at which MSDs do arise. This goal can be achieved through identifying problem areas, proposing solutions, and implementing those solutions best believed to produce a positive change.

    Proper identification of job stressors likely to produce MSDs can be gained through the ergonomics training previously received by the committee. The training will also provide team members with an understanding of which ergonomics tools and methods are most appropriate for proper risk factor identification. The assessment tools enable the team to asses the extent of the problem.

Potential solutions to these problems are generated by team members (initially in coordination with the ergonomics expert), from discussions within the committee and through testing of ideas, so that the best solution can be determined and then be implemented. Thorough follow-up, consisting of job reevaluation, is necessary to ensure that no additional problems have been created and to determine if the solutions are working as desired.

5. *Ergonomics Awareness.* Another goal of the committee is to generate awareness of ergonomics throughout the facility. This includes the promotion of the team, its duties, and its purpose. The purpose is for the team to interact with the work force and explain how they can change jobs to make them easier and safer for individuals to do. They also need to become part of the environment where employees feel comfortable bringing up issues to team members. This involves, to some level, team members describing ergonomics principles to employees so they see how ergonomics can improve their jobs. One way to accomplish this is through a "train-the-trainer" approach. With this technique, ergonomics principles are taught to personnel at the upper levels of a facility, then those individuals teach others, and this follows through the company's chain of command. This system assures that all employees become familiar with the ergonomics process, and it also increases the likelihood that the process will be accepted by employees on many different levels.

These methods are an integral part of SMS and provide effective communication within the facility. As employees become familiar with the concept of SMS and ergonomics with the SMS and understand that such a process exists within their facility, they intend to see it work. It is the responsibility of the team to continually keep employees updated on the team's activities. For example, if a work site is ergonomically evaluated, affected employees must be kept appraised of the status of the evaluation. Similarly, the team should develop a list that prioritizes jobs based on past injuries or current levels of discomfort. Because not all concerns will be addressed at once, this list can be used to explain why some jobs or processes are receiving more attention than others. Providing feedback on progress is critical to the overall SMS.

D. Medical Management Program. A medical management program addressing injuries and illnesses is essential for the success of a facility's efforts to reduce both the incidence and severity of MSDs. As highlighted in the flowchart of the medical management process, a medical management program includes components that address the following:

- Tracking the trends of MSDs and physical discomfort in the workplace;
- Identifying jobs within the facility that employees with temporary physical limitations can perform; and
- Assisting in educating the work force about what symptoms may develop into an MSD if exposures to job stressors are not corrected as well as effective treatment of MSDs.

Each of these components is elaborated upon below.

1. *Track Trends of Musculoskeletal Disorders.* Companies are required to keep a record of injuries and illnesses that occur within their facility (e.g., medical department visits, OSHA 300 logs). Tracking the type and duration of MSDs found within the company, whether it is performed by the medical staff or the ergonomics committee, provides important information. For example, injury tracking can identify jobs or departments that may account for the majority of MSDs and indicate the types of MSDs most prevalent among employees. The computation of MSD incidence based on these injuries can signal if rates are changing (increasing or decreasing) over time.

   Tracking also can be accomplished with symptoms of MSDs that have not yet developed into reportable or lost-time events. These can serve as leading indicators. Discomfort surveys (an example of which is shown in the Appendix) given to employees by the ergonomics team and medical management personnel allow them to indicate areas of their bodies where they experience pain or discomfort and also to note the extent of this discomfort (e.g., mild or severe). The periodic administration of these surveys also provides the team with trend information, that is, if levels of employee discomfort are changing over time.

2. *Develop a Return-to-Work Plan.* Early return-to-work has been found to reduce the health care expenditures associated with these injured employees and other direct and indirect costs (U.S. Dept. of Labor, 1990), in addition to increasing the likelihood that an injured person will eventually return to gainful employment. The logic behind the use of restricted-duty jobs is that employees recovering from an MSD can return to work earlier if a job or certain task is found that accommodates their temporary physical impairments, identified as restrictions for return-to-work by the treating physician.

   Identification of restricted-duty jobs or tasks requires an assessment of the physical requirements of jobs. These requirements are then matched with the physical restrictions of employees returning to work from an injury, so that the ailment is not aggravated.

   This process requires good communication with the employees' treating physician(s), so that detailed and relevant information about the types of restrictions are passed to the ergonomics committee and the person responsible for the return-to-work. In general, physicians who treat the company's employees need to understand the physical nature of the jobs in the facilities, and the company and Ergonomics Committee need to understand the MSDs that occur to employees.

3. *Recognition and Treatment of Musculoskeletal Disorder Symptoms.* Health care providers knowledgeable about the causes and treatment of MSDs may be called upon to educate the work force about general symptoms that may be related to future MSDs, if work activities go unchecked. This is a part of the committee's long-range goal: to address and remedy workplace stressors before they become more severe problems. The treatment of MSD symptoms also is an important part of the medical management strategy, where the focus initially is on conservative treatment.

E. The Ergonomics Expert. For companies beginning to implement an ergonomics process, the ergonomics expert has the initial responsibility of developing this process within the facility. This individual should have received formal training in the field of ergonomics. This often includes an advanced degree such as a masters degree and a certification from an accredited organization. The most prevalent accredited certification is the Certified Professional Ergonomist (CPE) granted by exam from the Board of Certification in Ergonomics (BCPE), the Board of Certified Safety Professionals (BCSP) which offers the Certified Safety Professional (CSP), once offered a specialty certification in ergonomics; however, that certification is no longer active. Individuals displaying these certifications are usually professionals who have passed or completed a combination of work experience and educational standards (by examination). Safety and health professionals must understand there are many certifications that purport to be accredited, but these are the ones generally recognized.

Given that the process has already been established, the expert is tasked with three general responsibilities to the ergonomics committee:

- Training;
- Merging the ergonomics process with the management style of the company; and
- Providing expertise and assistance when requested by the committee.

These three functions are discussed below.

1. *Training the Ergonomics Committee.* The ergonomics expert must educate the committee regarding those ergonomics principles and concepts relevant within the facility. Because of the multidisciplinary nature of this

field, a comprehensive review of all ergonomics issues can be extremely time-consuming, therefore, topic areas included in the training should be selected carefully.

The goal of the training is not to make team members experts themselves, but to enable them to recognize work-related factors causing or contributing to MSDs and how to resolve the problems.

As mentioned, many topic areas within the field of ergonomics may be applicable. However, it is recommended the following basic concepts be included: understanding body size and strength differences among people (anthropometry); loading of the joints in the body due to both external and internal forces (workplace biomechanics); and evaluating a work site from an ergonomics perspective (task analysis). Other topics may need to be included, depending on the needs of the facility. As discussed earlier, a train-the-trainer approach may be useful, so team members not only understand how the ergonomics process is to function, but have the ability to train others within the facility on ergonomics concepts.

2. *Management Techniques and Interpretation.* Each company manages work processes differently, and an ergonomics process is not exempt from these differences. The ergonomics expert is responsible for understanding the company's work environment and integrating the process within this atmosphere and the overall SMS.

3. *Work/Job Design and Redesign.* A properly trained ergonomics committee will be able to accomplish much within a facility, evaluating and correcting work sites that contain MSD risk factors. There may be circumstances when the issues of a problem work process are beyond the capabilities of the committee. The ergonomics expert can assist the team with these problems, by applying additional knowledge of the field or by evaluating the workplace with more sophisticated or technical equipment. It is the responsibility of the expert to know which advanced tools need to be applied, when they should be used, and how this information will benefit the ergonomics team and the company. In summary, the objective of the ergonomics expert is to direct a fledgling team and its activities. This relationship should continue until the team is able to sustain itself independently, with little or no expert assistance.

ANALYZING WORKPLACE HAZARDS

There are signs that indicate ergonomic problem areas. For instance, the OSHA Form 300 log or workers compensation claims may show cases of musculoskeletal disorders. In addition, safety and health professionals should watch for ergonomic indicators such as

- Jobs that require the same motions every few seconds (three to five) for more than two hours at a time.
- Fixed or awkward work postures for more than a total of two hours, such as bending, bent wrists, kneeling, twisting, and squatting.
- Use of vibration or impact tools or equipment for more than a total of two hours.
- Lifting, lowering, or carrying more than twenty-five pounds more than once during the workshift.
- Piece rate or machine-paced work for more than four hours at a time.
- Workers' complaints of physical aches and pains related to their work assignments.

Ergonomic Risk Factors

There are also a number of ergonomic risk factors that contribute to musculoskeletal disorders. These factors should be addressed by the safety and health professionals as part of the worksite control program. Among the factors of concern are

- Regular repetitive tasks
- Jobs requiring forceful or prolonged exertions of the hands
- Vibration and/or cold temperatures
- Jobs requiring heavy lifting, pushing, pulling, or carrying of heavy objects
- Poor body mechanics
- Restrictive workstations
- Awkward postures
- Hand tools that do not meet the requirements of the job

CONTROLLING ERGONOMIC HAZARDS

After safety and health professionals have identified workplace hazards that contribute to ergonomic injuries and illnesses, the next step is to implement methods of controlling these hazards. As proposed in the mitigation of other types of hazards, NIOSH recommends reducing or eliminating potentially hazardous conditions using engineering controls. One of the goals in controlling worksite hazards is to design problems out of the job. To meet ergonomics challenges, equipment in the industrial environment may need to be adapted to eliminate hazards. It also means that any new equipment or processes must be studied by safety and health professionals in advance to ensure that the same ergonomics guidelines are followed. It is more efficient and economical to make changes during the job design planning stage than after problems arise.

Engineering controls for ergonomics are generally of three types: workstation design, design of work methods, and tool design. An example of workstation design is changing the height of a worktable to make it more efficient, easier, and more comfortable for the employee to use. Design of work methods is a second type of engineering control. An example is putting handles on boxes or providing bins or boxes fabricated with built-in handholds so the employee can more easily lift the load. The third type of engineering control is tool design. An example is providing counterbalancing for a tool that weighs over 200 pounds.

If the hazard cannot be fully controlled by applying engineering principles, the next priority is work practice controls or administrative controls. Changes in work practices and management policies, sometimes called administrative controls, are the second alternative.

Administrative Controls

Safety and health professionals implement administrative controls in an attempt to mitigate exposures by reducing the frequency or length of exposure. For example, employers may provide more rest pauses for a person exposed to excessive tool vibration or excessive heat. Other administrative controls include: rotating workers through physically tiring jobs, training workers to recognize ergonomic risk factors and to learn techniques for reducing stress and strain while performing their work tasks, reducing shift length or limiting overtime, broadening or varying the job content to offset certain risk factors (e.g., repetitive motions, and static and awkward postures), and adjusting the work pace to relieve repetitive motion risks and give the worker more control of the work process. Although engineering controls are preferred, administrative controls are useful temporary stopgaps until engineering controls can be put into place.

Work Practice Controls

If engineering, work practice, or administrative controls don't completely control the hazard, then safety and health professionals may implement the use of personal protective equipment (PPE). Personal protective equipment for musculoskeletal disorders is extremely limited and its effectiveness is arguable. OSHA recognizes these types of PPE to control musculoskeletal problems:

- Gloves to reduce vibration transmission
- Padding to keep hands, wrist, arms, etc., from coming into contact with hard surfaces

- Gloves or clothing to keep hands or body warm or prevent vibration illnesses

Note: Braces, splints, and back belts are not considered PPE to prevent musculoskeletal disorders.

Work practice controls are not as effective as engineering controls, since monitoring is necessary to determine whether the controls are being followed. An example of work practice controls is keeping tools in top condition. For instance, screwdrivers that have become dull, chipped, or warped require more hand pressure to use than ones in good condition. The more hand pressure an operator has to use, the more likely musculoskeletal disorders of the hand, wrist, and shoulder can occur.

Medical Management

The purpose of medical management is to focus on early identification and treatment of musculoskeletal disorders, reduce the severity of functional impairment, and prevent future problems. To reduce the extent of injury or illness, employees should feel encouraged to report early symptoms and not fear retaliation or discrimination.

Employer Responsibilities

To create an environment where employees feel comfortable in reporting early symptoms of ergonomic injuries or illnesses, employers may:

- Train employees to recognize signs of musculoskeletal disorders.
- Provide prompt access to health care providers for medical evaluation.
- Require health care providers to take periodic walk-throughs of the worksite to learn about the workplace and to keep in touch with employees.
- Modify jobs or accommodate employees if health care providers so advise.
- Assure employees that medical findings will remain confidential and private to the extent permitted by law.

Health Care Provider Responsibilities

Health care providers need to be familiar with employee jobs and job tasks in order to play an effective role in matching jobs and the job environment to the needs of the worker. One of the best ways for a health care providers to become familiar with jobs and job tasks is by periodic plant walk-throughs. Other approaches include reviewing job analysis reports, detailed job descriptions, job safety analyses, and photographs or videotapes accompanied by narrative or written descriptions of the jobs.

Treatment

Health care providers are responsible for determining the physical capabilities and work restrictions of affected workers. Employers are responsible for giving an employee a task consistent with prescribed restrictions. Until effective controls are put in place, employee exposure to ergonomic stressors can be reduced through restricted duty and/or temporary job transfer. Complete removal from the work environment should be avoided unless the employer is unable to accommodate the prescribed work restrictions.

Assessment Tools

Assessment tools for the safety professional vary from qualitative to quantitative. Qualitative methods primarily include checklists. Checklists provide basic information and some insight into the hazards presented. Discomfort surveys while largely qualitative can also have a quantitative component. Quantitative methods like RULA, the Liberty Mutual formula, and the NIOSH lifting formula are primarily quantitative. Each is provided/discussed below.

MANUAL HANDLING CHECKLIST

Yes	No	
☐	☐	Are objects able to be grasped by good handholds?
☐	☐	Are objects stable?
☐	☐	Are objects able to held without slipping?
☐	☐	When required, do gloves improve the grasp without bunching up or resisting movement of the hands?
☐	☐	Is there enough room to access and move objects?

Manual Handling Checklist

Yes	No	
☐	☐	Are mechanical aids easily available and used whenever possible?
☐	☐	Are working surfaces adjustable to the best handling heights?

Does material handling avoid:

☐	☐	Movements below knuckle height and above shoulder height?
☐	☐	Static awkward postures?
☐	☐	Sudden movements during handling?

☐ ☐ Twisting of the trunk?
☐ ☐ Excessive reaching while holding/moving the load?
☐ ☐ Is help available for heavy or awkward lifts?

Are high rates of repetition avoided by:
☐ ☐ Job rotation?
☐ ☐ Self-pacing?
☐ ☐ Sufficient rest pauses?

Are pushing/pulling forces reduced/eliminated by:
☐ ☐ Casters that are sized correctly and roll freely?
☐ ☐ Handles for pushing/pulling?
☐ ☐ Availability of mechanical assists?
☐ ☐ Are objects rarely carried more than 10 feet?
☐ ☐ Is there a prevention maintenance program for manual handling equipment?

Does the design of the task reduce or eliminate:
☐ ☐ Bending or twisting of the trunk?

Manual Handling Checklist

Yes   No
☐ ☐ Squatting or kneeling?
☐ ☐ Elbows above mid-torso?
☐ ☐ Extending the arms?
☐ ☐ Bending the wrist?
☐ ☐ Static muscle loading?
☐ ☐ Forceful pinch grips?
☐ ☐ Are mechanical devices used to lift or move objects that are heavy or require repetitive lifting?
☐ ☐ Can the task be performed with either hand?

Manual Handling Checklist

Yes   No
Are the materials:
☐ ☐ Able to be held without a forceful grip (no slipping)?
☐ ☐ Easy to grasp?
☐ ☐ Free from sharp edges?
☐ ☐ Do containers have good handholds?

		Are fixtures and vises used to reduce or eliminate hard grasping forces?
☐	☐	
☐	☐	If gloves are needed, do they fit properly?
☐	☐	Does the task avoid contact with sharp edges or corners?

Is exposure to repetitive motions reduced by:

☐	☐	Job rotation?

Manual Handling Checklist

Yes **No**

☐	☐	Self-pacing?
☐	☐	Sufficient rest pauses?

Is the workstation designed to reduce or eliminate:

☐	☐	Bending or twisting of the trunk?
☐	☐	Squatting or kneeling?
☐	☐	Elbows above mid-torso?
☐	☐	Extending the arms?
☐	☐	Bending the wrist?
☐	☐	Hands behind the body?
☐	☐	Are mechanical aids and equipment available?
☐	☐	Can the work be performed without repetitive (or static) bending or reaching more than 20 inches?

Are awkward postures reduced by:

☐	☐	Providing adjustable work surfaces and supports (such as chairs or fixtures)?
☐	☐	Tilting the surface?
☐	☐	Are all job requirements visible without awkward postures?
☐	☐	Is an arm rest provided for precision work?
☐	☐	Is a foot rest provided for those who need it?
☐	☐	Are cushioned floor mats and foot rests provided for employees who are required to stand for long periods?
☐	☐	Have jobs been reviewed to determined if they are best suited for sitting or standing work?

Manual Handling Checklist

Yes **No**

Are tools selected to:

☐	☐	Reduce exposure to localized vibration?
☐	☐	Reduce hand force?
☐	☐	Reduce/eliminate bending or awkward postures of the wrist?
☐	☐	Avoid forceful pinch grips?
☐	☐	Are tools powered where necessary (e.g., to reduce forces, repetitive motions)?
☐	☐	Are tools evenly balanced?
☐	☐	Are heavy tools suspended or counterbalanced?
☐	☐	Does the tool allow adequate view of the work.
☐	☐	Does the tool grip/handle prevent slipping during use?

Are tools equipped with handles that

☐	☐	Do not end in the palm area?
☐	☐	Are made of textured, nonconductive material?
☐	☐	Have a grip diameter suitable for most workers, or are different size handles available?
☐	☐	Can the tool be used safety with gloves?
☐	☐	Can the tool be used by either hand?
☐	☐	Is there a preventive maintenance program to keep tools operating as designed?
☐	☐	Can triggers be operated by more than one finger to avoid static contractions?
☐	☐	Does the tool design or workstation minimize the twist or shock to the hand (in particular, observe the reaction of power tools after the torque limit is reached)?

QUANTITATIVE METHODS

RULA

The RULA Assessment Tool was developed to evaluate the exposure of individual workers to ergonomic risk factors associated with upper extremity MSD. The RULA ergonomic assessment tool was developed to consider biomechanical and postural load requirements of job tasks/demands on the neck, trunk, and upper extremities.

For more information and for computerized RULA assessments tools are available online from:

- Osmond Ergonomics (http://www.ergonomics.co.uk/rula.html)
- Humanics (http://www.humanicsergosystems.com/rula.htm)
- ErgoIntelligence (NextGen) (http://www.nexgenergo.com/ergonomics/ergointeluea.html)

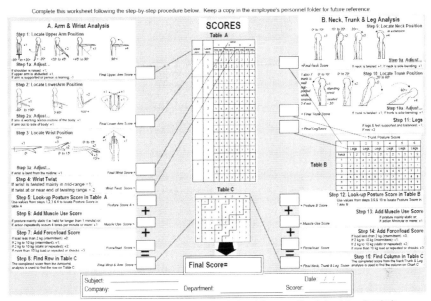

Score	Level of MSD Risk
1–2	Negligible risk, no action required
3–4	Low risk, change may be needed
5–6	Medium risk, further investigation, change soon

Liberty Mutual Lifting Formula

The Liberty Mutual Research Institute for Safety developed and regularly updates the manual material handling tables originally published by Snook in 1978, and by Snook and Ciriello in 1991. The new "Liberty Mutual Manual Materials Handling Tables" provide both the male and female population percentages capable of performing manual material handling tasks without over exertion, rather than maximum acceptable weights and forces. This manual material handling analysis tool is based on the same "Liberty Mutual Manual Materials Handling Tables" and can be used to perform ergonomic assessments of lifting, lowering, pushing, pulling, and carrying tasks with the primary goal of supporting ergonomic design interventions. The detailed tables can be found at: https://safetyresourcesblog.files.wordpress.com/2014/08/liberty-mutual-snook-tables.pdf.

Many jobs requiring a wide variety of manual handling tasks (lifting, lowering, pushing, pulling, and/or carrying) can be assessed as a whole using the

Snook Tables. This can be done by comparing data for each of the specific manual handling tasks against the appropriate table, and then using the 'total frequency for all the tasks' as the frequency value to determine the percentage of the population that would find the task to be acceptable.

For example, if a job requires lifting at a rate of one lift every two minutes, a push every five minutes and a carry every five minutes, the worker would do four and a half "tasks" over five minutes, or approximately one task per minute. The evaluator can then compare the data for the lift, carry, and push against the appropriate table but use the same frequency (one per minute) for each to determine a result.

When a mixture of males and females are doing the task, the task should be designed so it is acceptable to at least 75 percent of the female population; that would make it acceptable to more than 90 percent of the male population. Any task that cannot be performed by at least 75 percent of the total population should be considered for MSD prevention controls and redesign.

Preparation

The evaluator should prepare for the assessment by interviewing and observing workers to gain a complete understanding of the job tasks and demands. Selection of the job tasks to be evaluated are a function of the most difficult manual material handling work duties, based on worker interview and job observation. Equipment needed: A tape measure is required to take distance measures and a weigh scale or force measurement gauge is required to determine the weight of the object(s) being lifted/lowered or carried and the forces required for pushing and pulling.

Data Collection

For each job task analyzed, the evaluator will need to collect relevant data. Measurements and data required for this method include:

- *Weight:* the weight of the object being lifted, lowered, or carried.
- *Lift/lower distance:* the distance of travel of the hands while lift or lower taking place.
- *Hand distance:* the distance from the front of the body to the hands. This will normally be half the width of the object being handled unless the object is purposely held away from the body.
- *Hand height:* the height of the hands on the object being pushed or pulled, or the height of the hands when carrying a load.
- *Push/pull/carry distance:* the distance the item is pushed or pulled, or carried.

- *Frequency:* the number of lifts, lowers, pushes, pulls, or carries expressed in terms of number of activities done in 'x' seconds, minutes, or hours.
- *Force requirement:* For each pushing and pulling task evaluated, the amount of force required is measured to get the item moving (initial force) and then the amount of force it takes to keep the item moving (sustained force) is also measured.
- *Lift/lower zone:* the area of the body in which the lift/lower finishes. The observer notes the position of the hands when the worker has completed the lift/lower (floor to knuckle, knuckle to shoulders, or shoulder to overhead reach).

Using the Snook Tables

The following is an example application for using the Snook Tables for assessing lifting tasks.

Example: Above Shoulder Lift

Variables determined by the assessment:

- **Vertical Location—above shoulder lift (54"+)**
- **Frequency—average of 1 lift every 5 minutes**
- **Horizontal Distance—10" (front of body to mid-line of hands)**
- **Distance of Lift—30" (lifts from cart at 25" to rack height of 55")**

Above Shoulder (above 54 in)		Horizontal Distance (Front of Body to Hands) [in]								
		7			10			15		
		Distance of Lift [in]			Distance of Lift [in]			Distance of Lift [in]		
Frequency		10	20	30	10	20	30	10	20	30
1/8 h	1/8 h	35	31	29	29	26	24	26	24	22
1/30 min	2/1 h	31	26	24	24	22	20	22	20	18
1/5 min	12/1 h	26	24	22	22	20	18	20	20	18
1/2 min	30/1 h	26	24	22	22	20	18	20	20	18
1/1 min	1/1 min	26	24	20	20	20	18	20	18	15
1/14 s	4.3/1 min	20	20	18	18	18	13	18	18	13
1/9 s	6.7/1 min	18	18	15	15	15	13	15	15	13
1/5 s	12/1 min	18	18	13	13	13	11	13	13	11

Design goal = **18 pounds**

These tables are accessible from Liberty Mutual online at: https://libertymmhtables.libertymutual.com/CM_LMTablesWeb/taskSelection.do?action=initTaskSelection.

Niosh Lifting Equation

The NIOSH Lifting Equation is a tool used by safety and health professionals to assess the manual material handling risks associated with lifting and lowering tasks in the workplace. This equation considers job task variables to determine safe lifting practices and guidelines.

The primary product of the NIOSH lifting equation is the Recommended Weight Limit (RWL). The RWL defines the maximum acceptable weight (load) that nearly all healthy employees could lift over the course of an 8-hour shift without increasing the risk of musculoskeletal disorders (MSD) to the lower back. In addition, a Lifting Index (LI) is calculated to provide a relative estimate of the level of physical stress and MSD risk associated with the manual lifting tasks evaluated.

NIOSH Lifting Equation Outputs

The RWL determines if the weight of the lift is too heavy for the task. The LI determines the significance of the risk.

An LI value of less than 1.0 indicates a nominal risk to healthy employees. An LI of 1.0 or more denotes the task is high risk for some fraction of the population. As the LI increases, the level of low back injury risk increases correspondingly. The goal is to design all lifting jobs to accomplish an LI of less than 1.0.

The NIOSH lifting equation always uses a load constant (LC) of 51 pounds, which represents the maximum recommended load weight to be lifted under ideal conditions. From that starting point, the equation uses several task variables expressed as coefficients or multipliers (In the equation, M = multiplier) that serve to decrease the load constant and calculate the RWL for that particular lifting task.

NIOSH Lifting Equation: LC (51) × HM × VM × DM × AM × FM × CM = RWL

Task variables needed to calculate the RWL:

H = Horizontal location of the object relative to the body
V = Vertical location of the object relative to the floor
D = Distance the object is moved vertically
A = Asymmetry angle or twisting requirement
F = Frequency and duration of lifting activity
C = Coupling or quality of the workers grip on the object

Lifting Index (LI): Weight ÷ RWL = LI

Additional task variables needed to calculate the LI:

- Average weight of the objects lifted
- Maximum weight of the objects lifted

The RWL and LI can be used to guide lifting task design in the following ways: (1) The individual multipliers that determine the RWL can be used to identify specific weaknesses in the design. (2) The LI can be used to estimate the relative physical stress and injury risk for a task or job. The higher the LI value, the smaller the percentage of workers capable of safely performing these job demands. Thus, injury risk of two or more job designs could be compared. (3) The LI can also be used to prioritize ergonomic redesign efforts. Jobs can be ranked by LI, and a control strategy can be implemented based on a priority order of the jobs or individual lifting tasks.

The Frequency-Independent Recommended Weight Limit (FIRWL) and the Frequency-Independent Lifting Index (FILI) are additional outputs of the NIOSH lifting calculator. The FIRWL is calculated by using a frequency multiplier (FI) of 1.0 along with the other task variable multipliers. This effectively removes frequency as a variable, reflecting a weight limit for a single repetition of that task, and it allows equal comparison to other single repetition tasks. The Frequency-Independent Lifting Index (FILI) is calculated by dividing the weight lifted by the FIRWL. The FILI can help identify problems with infrequent lifting tasks, if it exceeds the value of 1.0.

How to Use the NIOSH Lifting Equation

Step 1: Measure and Record Task Variables

The first step is to gather the needed information and measurements for lifting task variables and record the data to be used later to calculate the RWL and LI for the tasks being evaluated. The evaluator prepares by interviewing and observing workers to gain a complete understanding of all required lifting tasks. Selection of the lifting tasks to be evaluated is based on the most significant and demanding manual material handling tasks. If the job requires a wide variety of lifting tasks, a multi-task evaluation can be performed using a composite of all single-task lifting assessments performed, discussed later in the chapter, but first the focus is on single-task assessments.

For each lifting task analyzed, the evaluator will need to determine the task variables as outlined above. The following worksheet can assist with data collection:

Department: ERGONOMICS PLUS	Job: NIOSH Lifting Variables								
Lifting Task	H Horizontal Location (10-25")	V Vertical Location (0-70")	D Travel Distance (10-70")	A Angle of Asymmetry (0°-135°)	C Coupling (1=good, 2=fair, 3=poor)	F Frequency (0.2-15 lifts/min)	L Ave. Load Lifted (lbs.)	L Max. Load Lifted (lbs.)	Dur Duration (1, 2, 8 hours)

The following task variables are evaluated to calculate the multipliers used in the NIOSH equation to determine the RWL. Here are some quick explanations and guidelines to gather the needed measurements:

1. *Horizontal Location of the Hands (H)*—Measure and record the horizontal location of the hands at both the start (origin) and end (destination) of the lifting task. The horizontal location is measured as the distance (inches) between the employee's ankles to a point projected on the floor directly below the mid-point of the hands grasping the object as pictured below:

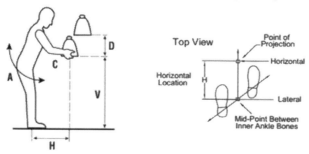

2. *Vertical Location of the Hands (V)*—Measure and record the vertical location of the hands above the floor at the start (origin) and end (destination) of the lifting task. The vertical location is measured from the floor to the vertical mid-point between the two hands as shown below. The middle knuckle can be used to define the mid-point.

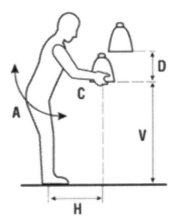

3. *Vertical Travel Distance (D)*—The vertical travel distance of a lift is determined by subtracting the vertical location (V) at the start of the lift from the vertical location (V) at the end of the lift. For a lowering task, the V location at the end is subtracted from the V location at the start.

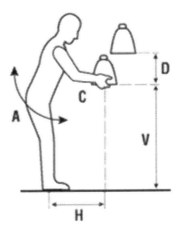

4. *Asymmetric Angle (A)*—The degree to which the body is required to twist or turn during the lifting task is measured. The asymmetric angle is the amount (in degrees) of trunk and shoulder rotation required by the lifting

task. Note: Sometimes the twisting is not caused by the physical aspects of the job design, but rather by the employee using poor body mechanics. If this is the case, no twisting (0°) is required by the job. If twisting is required by the design of the job, the number of degrees the back and body trunk must twist or rotate to accomplish the lift (i.e., 90° as pictured below) is determined.

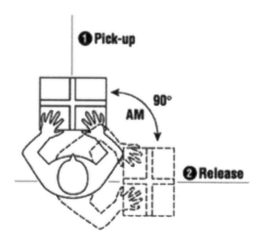

5. *Coupling (C)*—The classification of the quality of the coupling between the worker's hands and the object as good, fair, or poor (1, 2, or 3) is determined. A good coupling will reduce the maximum grasp forces required and increase the acceptable weight for lifting, while a poor coupling will generally require higher maximum grasp forces and decrease the acceptable weight for lifting.

   - **1 = Good**—Optimal design containers with handles of optimal design, or irregular objects where the hand can be easily wrapped around the object.
   - **2 = Fair**—Optimal design containers with handles of less than optimal design, optimal design containers with no handles or cut-outs, or irregular objects where the hand can be flexed about 90°.
   - **3 = Poor**—Less than optimal design container with no handles or cut-outs, or irregular objects that are hard to handle and/or bulky (e.g., bags that sag in the middle).

6. *Frequency (F)*—The appropriate lifting frequency of lifting tasks by using the average number of lifts per minute during an average 15 minute sampling period is determined. For example, the total number of lifts in a typical 15-minute period is counted and that number is divided by 15.

- Minimum = 0.2 lifts/minute
- Maximum is 15 lifts/minute.

7. *Load (L)*—Determine the weight of the object lifted. If necessary, a scale is used to get the exact weight. If the weight of the load varies from lift to lift, the average and maximum weights lifted are recorded.
8. *Duration (Dur)*—The lifting duration as classified into one of three categories is determined: Enter 1 for short duration, 2 for moderate duration, and 8 for long duration as follows:

   - **1 = Short**—lifting ≤ 1 hour with recovery time ≥ 1.2 X work time
   - **2 = Moderate**—lifting between 1 and 2 hours with recovery time ≥ 0.3 X lifting time
   - **8 = Long**—lifting between 2 and 8 hours with standard industrial rest allowances

Step 2: Enter Data/Calculate RWL and LI

In step 1, it the lifting task variables in the worksheet were determined and recorded. The following is an example of a completed worksheet:

Department: 250
Job: Unload product boxes from conveyor and place onto cart

Lifting Task	H Horizontal Location (10-25")	V Vertical Location (0-70")	D Travel Distance (10-70")	A Angle of Asymmetry (0°-135°)	C Coupling (1=good, 2=fair, 3=poor)	F Frequency (0.2-15 lifts/min)	L Ave. Load Lifted (lbs.)	L Max. Load Lifted (lbs.)	Dur Duration (1, 2, 8 hours)
Origin - Lift product box from conveyor	15	38	6	0	1	2	24	24	8
Destination - Place product box onto cart	20	32	6	30	1	2	24	24	8

Once this is completed the data can be input into the formula to determine the RWL and LI.

NIOSH Lifting Equation Example for Single-task Lifting Analysis

Task Description—This job task (pictured below) consists of a worker lifting compact containers full of copper component parts from the bottom shelf of storage rack with both hands directly in front of the body and then placing on a cart for transport to the assembly line. For this analysis, it is assumed that significant control of the object is required at the destination. The containers are of optimal design with handholds.

Origin and Destination Calculations:

Analyst	Task
Mark Middlesworth	Department 250

	Origin	Destination
Horizontal Location (min. 10", max. 25")	15	20
Vertical Location (min. 0", max. 70")	38	32
Travel Distance (min. 10", max. 70")	6	6
Angle of Asymmetry (min. 0, max. 135)	0	30
Coupling (1=good | 2=fair | 3=poor)	1	1
Frequency (min. 0.2 lifts/min.)	2	2
Avg. Load	24	24
Max Load	24	24
Duration (enter 1, 2 or 8)	8	8

		Origin		Destination
		51		51
Horizontal Multiplier (HM)	x	0.67	x	0.50
Vertical Multiplier (VM)	x	0.94	x	0.99
Distance Multiplier (DM)	x	1.00	x	1.00
Frequency Multiplier (FM)	x	0.65	x	0.65
Angle Multiplier (AM)	x	1.00	x	0.90
Coupling Multiplier (CM)	x	1.00	x	1.00
Recommended Weight Limit (RWL)		20.77		14.76
Frequency Independent RWL (FIRWL)		31.96		22.71
Lifting Index (LI)		1.16		1.63
Frequency Independent LI (FILI)		0.75		1.06

Ergonomics

Origin   Destination

Step 1—Measure and Record Lifting Task Variables

The horizontal distance at the origin of the lift is 15 inches and the horizontal distance at the destination of the lift is 12 inches. The height of the lift origin (bottom rack shelf) is 11 inches and the height of the lift at the destination (cart) is 40 inches. The travel distance between the origin and the destination is 29 inches. Ten degrees of asymmetric lifting is involved at the origin, and no asymmetry is involved at the destination. The container is of optimal design with handholds; therefore, coupling is defined as "good." The average frequency of lifting in this manner is 2 lifts/minute over a duration of 1–2 hours per day. The average load lifted is 12.5 lb., and the maximum load lifted is 26 lb. Control of the load is required at the destination of the lift. Therefore, the RWL is computed at both the origin and the destination of the lift.

Variable Summary:

H = 15" at the origin and 12" at the destination
V = 11" at the origin and 40" at the destination
D = 29"
A = 10° at the origin and 0° at the destination
C = 1 (good—container is of optimal design with handhold cut-outs)
F = 2 lifts/minute
L = 12.5 lb. average load and 26 lb. maximum load
Dur = 2 (task takes 1–2 hours per day with recovery time)

Record the task variables onto worksheet:

Department: 253 — Job: Unload parts from storage rack and place on cart

ERGONOMICS PLUS

Lifting Task	H Horizontal Location (10-25")	V Vertical Location (0-70")	D Travel Distance (10-70")	A Angle of Asymmetry (0°-135°)	C Coupling (1=good, 2=fair, 3=poor)	F Frequency (0.2-15 lifts/min)	L Ave. Load Lifted (lbs.)	L Max. Load Lifted (lbs.)	Dur Duration (1, 2, 8 hours)
Origin - Lift product container from rack	15	11	29	10	1	1	12.5	26	2
Destination - Place container on cart	20	32	6	0	1	1	12.5	26	2

Step 2—Conduct Risk Assessment using NIOSH Lifting Equation Calculator

Origin and Destination Calculations:

Analyst	Task
Mark Middlesworth	Department 253 - Assembly Line Utility

	Origin	Destination
Horizontal Location (min. 10", max. 25")	15	12
Vertical Location (min. 0", max. 70")	11	40
Travel Distance (min. 10", max. 70")	29	29
Angle of Asymmetry (min. 0, max. 135)	10	0
Coupling (1=good | 2=fair | 3=poor)	1	1
Frequency (min. 0.2 lifts/min.)	2	2
Avg. Load	12.5	12.5
Max Load	26	26
Duration (enter 1, 2 or 8)	2	2
	51	51
Horizontal Multiplier (HM)	x 0.67	x 0.83
Vertical Multiplier (VM)	x 0.8575	x 0.93
Distance Multiplier (DM)	x 0.88	x 0.88
Frequency Multiplier (FM)	x 0.84	x 0.84
Angle Multiplier (AM)	x 0.97	x 1.00
Coupling Multiplier (CM)	x 1.00	x 1.00
Recommended Weight Limit (RWL)	20.91	29.13
Frequency Independent RWL (FIRWL)	24.89	34.68
Lifting Index (LI)	0.60	0.43
Frequency Independent LI (FILI)	1.04	0.75

Origin Summary: The average weight to be lifted (12.5 lb.) is less than the RWL at the origin (20.9 lb.); however, the maximum load to be lifted (26 lbs.) is greater than the RWL and FIRWL. The LI is .60 and the FILI is slightly above 1.0 at 1.04, indicating a nominal overall risk to healthy employees and a slight risk when lifting the maximum load of 26 lb. from the origin.

Destination Summary: The average weight to be lifted (12.5 lb.) is less than the RWL at the destination (29.1 lbs.), and the maximum load to be lifted (26 lbs.) is less than the RWL and FIRWL. The LI is .43 and the FILI is .75, indicating a nominal risk to healthy employees at the destination.

http://ergo-plus.com/niosh-lifting-equation-single-task/

http://www.cdc.gov/NIOSH/DOCS/94-110/DEFAULT.HTML

DISCOMFORT SURVEYS

Discomfort surveys and special tests offer a means for detecting problems that may be missed in more general medical exams and reports. Workers completing a symptom survey form similar to the one shown below can identify parts of their bodies experiencing increased levels of discomfort as a result of poor job design. Although this survey is fairly easy to administer, the following procedures should be followed for best results:

- No names should be required on the forms, and the collection process should ensure anonymity.
- Survey participation should be voluntary in nature.
- Workers should fill out the form on their own (but, if needed, the surveys should be administered to groups by a trained person offering explanations).
- The survey should be conducted on work time.

Unless the company is prepared to act on the results of a symptom survey, it should not be conducted. Analysis of the information from a symptom survey is complex. One of the major difficulties is deciding what responses on the questionnaire indicate a problem that may need further evaluation. One approach for scoring results from a survey of this type is to rank-order the number and severity of complaints by body part from the highest to the lowest in frequency and severity. Those jobs linked with the body part showing the most complaints or the highest severity ratings would become the primary candidates for follow-up efforts at analyzing job risk factors and determining needs for risk-reduction measures.

A second survey, using the same form, completed after ergonomic changes have been made to correct problem jobs, can indicate whether the intended benefits have been achieved. Comparisons of the worker survey data gathered before and after ergonomic changes can furnish this information. One caution here is to allow sufficient time after the intervention to permit the workers to become accustomed to the job change and allow other novelty effects to subside. The second survey should be made no less than two weeks (and preferably one month) after the changes and should be made at the same time and day of the week as the initial survey. Comparisons of Monday morning results with those obtained on Friday afternoon may give faulty results because of differences in employee motivation.

ERGONOMIC COMFORT SURVEY

Name: _____ **Occupation:** _____ **Date:** _____
Location: _____ **Phone:** _____ **Rm:** _____ **Supv.** _____

1. Hours worked: Per day: _____ **Hrs**   Per week: _____ **Hrs**
2. Time working in present position: _____ **Years** _____ **Months**
3. Total time working in your occupation: _____ **Years** _____ **Months**
4. Average amount of time that you work at a computer: ___ **hrs per day**
5. Average amount of time you work while seated in a chair:___ **hrs per day**
6. Do you take stretch breaks during the day? ___ **Yes** ___ **No** ___ **Seldom Occasionally Frequently**
7. How stressful do you consider your job to be on a daily basis:
   None or Minimal Level _____ **Moderate Level**_____
   High Level _____ **Severe Level** _____
8. Have you had any discomfort during the last year? **Yes** _____ **No** ____
   (If No, stop here)
   If yes, carefully shade in area(s) of discomfort on the drawings.

9. Other jobs you have done in the last year (for more than 2 weeks):
   (If more than 2 jobs, include those you worked on the most)
   _____
   _____

10. Check body area(s) where you experience or have experienced discomfort in past year: (<u>Complete question #16 for each of the discomfort areas you check off below</u>)

Neck Shoulder Elbow Forearm Hand Wrist Fingers
Upper back Middle back Lower back Thigh
Knee Leg (below knee) Ankle Foot

11. Please describe your discomfort in the area checked:
Aching (soreness) Numbness Pins and needles (tingling)
Burning Pain Cramping Swelling Stiffness
Other: _____
12. When did you first notice discomfort? _____ Month _____ Year
13. When was the most recent occurrence of your discomfort? Now Today
Several days ago
Weeks ago   Months ago   Years ago
14. How long does each episode last? _____
15. How many separate episodes of discomfort did you have in the last year?
_____
16. What do you think caused or causes the discomfort?
_____
_____
17. Have you had medical treatment for your discomfort? _ Yes___ No
18. What are your hobbies, involvement in sports, or home activities?
_____
_____
_____
_____
_____

19. Do any of your hobbies, activities, or sports cause the same discomfort?
Yes _____ No _____
If yes, which one(s)
_____
_____
_____
_____

20. Please comment on what you think would ease or eliminate your discomfort:

Employee Signature:

CONCLUSION

Ergonomic considerations are an important part of the SMS process. A careful and methodical approach to addressing injuries and illnesses related to ergonomic problems can result in substantial improvements to incidence rates and reduce overall expenditures in the safety and health program, as well as by the organization.

REFERENCES

Badley, E.M., Rasooly, I. and Webster, G.K. (1994). Relative Importance of Musculoskeletal Disorders as a Cause of Chronic Health Problems, Disability, and Health Care Utilization: Findings from the 1990 Ontario Health Survey, *Journal of Rheumatology*, 21, 505–14.

Ergonomics Program Management Guidelines for Meatpacking Plants. (1993). *U.S. Department of Labor*, OSHA, Publications Number 3123.

Lawrence, R.C., Helmick, C.G., Arnett, F.C., Deyo, R.A., Felson, D.T., Giannini, E.H., Heyse, S.P., Hirsch, R., Hochberg, M.C., Hunder, G.G., Liang, M.H., Pillemer, S.R., Steen, V.D. and Wolfe, F. (1998). Estimates of the Prevalence of Arthritis and Selected Musculoskeletal Disorders in the United States, *Arthritis and Rheumatism*, 41(5): 778–99.

Marras, W.S., Allread, W.G. (n.d.). How to Develop and Manage an Ergonomics Process, Institute for Ergonomics, The Ohio State University, Columbus, OH.

Middlesworth, M., Rapid Upper Limb Assessment (RULA), *A Step-by-Step Guide*, Ergonomics Plus Inc., www.ergo-plus.com.

Middlesworth, M., Snook Tables. *A Step-by-Step Guide*, Ergonomics Plus Inc., http://ergo-plus.com/wp-content/uploads/SNOOK-Tables-A-Step-by-Step-Guide.pdf. U.S. Bureau of Labor Statistics, 2004, U.S. Department of Labor, Washington, D.C.

Praemer, A., Furner, S. and Rice, D.P. (1999). *Musculoskeletal Conditions in the United States*, Rosemont, IL: American Academy of Orthopaedic Surgeons.

Snook, S.H. (1978). The Design of Manual Lifting Tasks, *Ergonomics*, Vol. 21, No. 12, pp. 963–85.

U.S. Department of Labor, Occupational Safety and Health Administration. (1990). *Ergonomics Program Management Guidelines for Meatpacking Plants*, OSHA 3123, Government Printing Office.

U.S. General Accounting Office. (1997). *Worker Protection: Private Sector Ergonomics Programs Yield Positive Results*, GAO/HEHS Publication No. 97-163, Washington, D.C.

Van Overbeek, Thomas T. (March 15, 1991). *CRT Display Quality and Productivity*, Cornerstone Technology.

Water, T.R. Putz-Anderson, V., Harg, A. (January, 1994). *Applications Manual for the Revised NIOSH Lifting Equation*, U.S. Department of Health and Human Services.

Yelin, E.H., Trupin, L.S. and Sebesta, D.S. (1999). Transitions in Employment, Morbidity, and Disability among Persons Aged 51 to 61 with Musculoskeletal and Non-musculoskeletal Conditions in the U.S., 1992–1994, *Arthritis and Rheumatism*, 42, 769–79.

WEBSITE RESOURCES

A Primer Based Workplace Evaluations of Musculoskeletal Disorders (select the Table of Contents): http://www.cdc.gov/niosh/ephome2.html.
American Apparel & Footwear Association: *www.apparelandfootwear.org*.
American Conference of Governmental Industrial Hygienists: *http://www.acgih.org/*.
American Furniture Manufacturers Association: *http://www.Afma4u.org*.
American Industrial Hygiene Association: *http://www.aiha.org/*.
Board of Certified Professional Ergonomists (to find certified ergonomics practioners): *www.bcpe.org/*.
Center for Disease Control Ergonomics: *http://www.cdc.gov/od/ohs/INDHYG/ergolead.htm*.
Cornell University (office ergonomics): *www.ergo.human.cornell.edu*.
Elsevier (publisher of scientific, technical and health information): *http://www.elsevier.com/*.
ErgoWeb (current news): *www.ergoweb.com*.
Ergoworld (ergonomics and human factors meta site): www.interfaceanalysis.*com/ergoworld/*.
Federal OSHA index page for the Meatpacking Industry: *http://www.osha.gov/SLTC/meatpacking/index.html*.
Federal OSHA Ergonomics Guidelines for the Meatpacking Industry: *http://www.osha.gov/Publications/osha3123.pdf*.
Liberty Mutual Lifting calculations: https://libertymmhtables.libertymutual.com/CM_LMTablesWeb/taskSelection.do?action=initTaskSelection.
National Institute for Occupational Safety and Health: *http://www.cdc.gov/niosh/*.
North Carolina State University – Department of Industrial Engineering: *http://www/ie.ncsu.edu/*.
OSHA (federal developments): www.osha.gov.
PubMed (research academic journal articles): *www.ncbi.nlm.nih.gov*.
Taylor and Francis Publishing (academic journals): *http://www.tandf.co.uk/journals/*.
Washington State Department of Labor and Industry, Ergonomics: *http://www.Ini.wa.gov/wisha/ergo/*.

Chapter 10

Process Safety Management
Mark D. Hansen

CASE

RJ Long is safety director for a small company that recently began using chemicals in some of their manufacturing processes. Most of the chemicals have been in small quantities, and the potential adverse consequences appear to be relatively small. As the company shifts away from metal to composite use in many of their products, they will also be introducing new chemicals into the work processes. RJ is concerned about what steps should be taken to ensure safety of the employees and the facilities as these new chemicals are introduced.

INTRODUCTION

The Occupational Safety & Health Administration (OSHA) 29 CFR 1910.119 "Process Safety of Highly Hazardous Materials" was phased in through 1997. Since this time, many other international standards have emerged. What many companies have done is to implement comprehensive management systems to comply with 1910.119. This also includes safety considerations in the design of systems. Many companies are endeavoring to engineer systems to eliminate the hazards, by design. If hazards cannot be designed out, they are controlled using safety mechanisms. Lastly, procedures and training must be implemented. The Process Safety Management (PSM) program elements will be discussed in detail.

The systems approach to PSM revolves around identifying potential hazard severity and probability to determine a risk assessment. The methods for tailoring this approach to particular applications will be discussed in detail. Several example risk assessment matrices will be provided to illustrate the different approaches to eliminating and controlling hazards.

The benefits of implementing PSM's fourteen program elements will be discussed and some inherent requirements for success will be identified. This will include what the standard (29 CFR 1910.119) requires employers to do to successfully implement each element. Methods of implementing rigid structures to guarantee the process safety function included in all areas required by the standard will be discussed. Also, methods for ensuring that these methods are flexible and responsive to change via the management of change function will be discussed.

HISTORICAL BACKGROUND

A number of serious incidents occurred over the last fifteen years, resulting in increased attention to companies that use, manufacture, store, or handle highly hazardous materials. These incidents caused considerable human casualties and property losses. Table 10.1 provides a global perspective of the top twenty loses since 1974. PSM operations are inherently complex. As a result, the complexity of the operations rely on multiple systems to help prevent losses. These systems or barriers are a combination of

- *Hardware:* Physical systems that may help to control the exposure;
- *Management systems:* Management and procedural steps that can be taken to help mitigate the risk; and
- *Emergency controls:* Systems that can minimize the fire, explosion, or other emergency consequences of the risk.

Accidents that result in major losses generally occur because of the failure of a number of these systems or barriers within the process safety management system, often occurring at the same time. Interestingly, none of these losses are the result of the failure of a single barrier or safety mechanism.

The 29 CFR 1910.119 and Clean Air Act Amendment (CAAA) paragraph 301(r)/304 are the result of industry incidents involving fires, explosions, and/or toxic releases since 1988. The regulations demand more stringent management controls, including more thorough and effective training, review, documentation, and auditing requirements to ensure that all personnel involved in an operation understand and accept their role in process safety.

Table 10.1 The 20 Largest Losses (Marsh, 23rd Edition)

Date	Plant Type	Event Type	Location	Country	Property Loss US$ (Millions[1])
07/07/1988	Upstream	Explosion/fire	Piper Alpha, North Sea	UK	1,810
10/23/1989	Petrochemical	Vapor cloud explosion	Pasadena, Texas	USA	1,400
01/19/2004	Gas processing	Explosion/fire	Skikda	Algeria	940[2]
06/04/2009	Upstream	Collision	Norwegian Sector	North Sea	840
03/19/1989	Upstream	Explosion/fire	Gulf of Mexico	USA	830
06/25/2000	Refinery	Explosion/fire	Mina Al-Ahmadi	Kuwait	820[2]
05/15/2001	Upstream	Explosion/fire/ sinking	Campos Basin	Brazil	790
09/25/1998	Gas processing	Explosion	Longford, Victoria	Australia	750
04/24/1988	Upstream	Blowout	Enchova, Campos Basin	Brazil	700
09/21/2001	Petrochemical	Explosion	Toulouse	France	680
05/04/1988	Petrochemical	Explosion	Henderson, Nevada	USA	640
05/05/1988	Refinery	Vapor cloud explosion	Norco, Louisiana	USA	610
03/11/2011	Refinery	Earthquake	Sendai	Japan	600[3]
04/21/2010	Upstream	Blowout/ explosion/ fire	Gulf of Mexico	USA	600
09/12/2008	Refinery	Hurricane	Texas	USA	550
06/13/2013	Petrochemical	Explosion/fire	Geismar, Louisiana	USA	510[4]
04/02/2013	Refinery	Flooding/fire	La Plata, Ensenada	Argentina	500[4,5]
12/25/1997	Gas processing	Explosion/fire	Bintulu, Sarawak	Malaysia	490[2]
07/27/2005	Upstream	Collision/fire	Mumbai High North Field	India	480
11/14/1987	Petrochemical	Vapor cloud explosion	Pampa, Texas	USA	480

[1] Inflated to December 2013 values. Values are ground-up, property damage only.
[2] New, higher value for the property damage for this loss supplied from the insurance market.
[3] New data received of the value of loss at the refinery, following the Tohoku earthquake.
[4] New loss since publication of 22nd edition of The 100 Largest Losses.
[5] It is understood that this value is still subject to resolution.

These regulations require employers to initiate considerable management changes and improvements to their industrial process operations. Implementation of a sound PSM system will ensure that process hazards are properly managed, and that formalized engineering and administrative controls become an integral part of the design, construction, operation, and maintenance, including startups, shutdowns, turnarounds, and other phases of a facility's life.

In essence, CAAA Paragraphs 301(r)/304 require hazard assessments for certain chemicals to determine potential effects from accidental releases of both toxic and/or flammable chemicals, including downwind effects and potential exposures affecting populations. Risk management plans are also required and are intended to help prevent or minimize accidental releases and provide prompt emergency response following such releases. Industry associations in the United States have developed their own process safety standards by publishing the following programs:

1. American Petroleum Institute (API), "Recommended Practice RP750, Management of Process Hazards."
2. American Institute of Chemical Engineers (AIChE), Center for Chemical Process Safety (CCPS), "Guidelines for Technical Management of Chemical Process Safety."
3. Chemical Manufacturer's Association (CMA), "Responsible Care Program."

These laws and standards have made an unprecedented impact on industry's ability to more safely manage its operations.

DEFINITION OF PSM

PSM is the application of management systems to policies, procedures, and practices for the identification, understanding, and control of process hazards. The goal is to prevent episodic processrelated injuries and incidents. Process safety, as defined by OSHA, involves the operation of facilities that handle, use, process, or store hazardous materials/chemicals in a manner free from episodic or catastrophic incidents. Process safety applies to chemical and refinery processes and to other processes, such as mechanical and electrical processes, in the manufacturing, aerospace, nuclear, transportation, and distribution industries. Many of the procedures outlined are also useful in other applications of system safety.

PROCESS SAFETY, SYSTEM SAFETY, AND INDUSTRIAL SAFETY

System-safety engineering programs that began in the U.S. military in the 1960s were the first successful efforts that addressed process safety management. These longstanding programs have matured over the past decade and have become highly successful in fulfilling integrated process safety requirements in the aircraft, missile, spacecraft, and nuclear industries. Although these programs concentrated on major military hardware systems and their design, operation, and maintenance, the principles they use for identifying, evaluating, and reducing process-related hazards also apply to the chemical and petroleum industries.

Furthermore, process safety differs from traditional personnel or industrial safety activities, which have been in place in industry for many years. Traditional industrial losses have resulted in personnel injuries from falls, slips and trips, vehicle accidents, and property damage from material handling mishaps, utility failures, etc. Process-related losses involve operating equipment and processes, whether chemical, mechanical, or electrical systems and/or subsystems. Process safety is primarily concerned with the use, manufacture, handling, and storage of hazardous materials/chemicals and with the design, construction, operation, and maintenance of these facilities.

FACILITY MANAGEMENT COMMITMENT

Facility management plays a vital role in the success of a PSM system. Top management commitment and support combined with similar commitments by a facility's line management and staff personnel helps ensure that PSM programs can be successfully implemented. Management's role should be to

- *Demonstrate Commitment.* Promote interest and instill commitment among employees to develop effective PSM controls to reduce and/or minimize catastrophic failures. Establish corporate policy to ensure fulfillment of commitment.
- *Allocate Resources.* Delegate personnel to implement PSM system elements through assigned duties and responsibilities. Assign accountability to meet overall objectives.
- *Set Standards.* Ensure that adequate procedures, drawings, and documentation control systems are established and maintained to implement

the program; ensure that training programs are developed and updated as needed; perform audits periodically.
- *Define Priorities.* Establish priorities to meet specific objectives relative to personnel, process technology, facility, and equipment integrity for the reduction of process hazards.
- *Measure and Assess Performance.* Develop benchmarks for performance and assess needs for continuing improvement of program effectiveness.

ELEMENTS OF PSM

An effective PSM program consists of various integrated management elements that can result in a systematic and formalized program to reduce and/or eliminate workplace process hazards, thus meeting industry standards and regulatory requirements.

A list of the conventional fourteen PSM elements is shown in figure 10.1. Collectively, these elements, when implemented satisfactorily, can achieve an optimum level of process safety. They have been developed from various sources primarily OSHA, EPA, API, CCPS, and CMA and also meet the basic requirements of the United Kingdom's "The Control of Industrial Major Accident Hazards (CIMAH) Regulations of 1990."

PROCESS SAFETY MANAGEMENT

Elements

1. **PROCESS SAFETY INFORMATION**
2. **PROCESS HAZARD/RISK ANALYSIS**
3. **MANAGEMENT OF CHANGE**
4. **MECHANICAL INTEGRITY**
5. **OPERATING PROCEDURES**
6. **SAFE WORK PRACTICES**
7. **PROCESS SAFETY REVIEWS**
8. **TRAINING**
9. **CONTRACTORS**
10. **EMERGENCY RESPONSE**
11. **INCIDENT INVESTIGATION**
12. **AUDITS**
13. **EMPLOYEE PARTICIPATION**
14. **TRADE SECRETS**

Figure 10.1 Process Safety Management Elements

Each of these elements is summarized below.

ELEMENT 1: Process Safety Information. A compilation of written safety records and documentation that provides the foundation of process safety information. It establishes the means by which the hazards of a chemical process or mechanical and electrical operation are identified and understood. The element consists of hazards of materials used, process, and mechanical design of equipment/systems, and documentation (i.e., engineering drawings, equipment specifications, process materials, and relief system design).

ELEMENT 2: Process Hazard Analysis. The heart of the process safety management system that consists of the systematic identification, evaluation and elimination, and/or control of processrelated hazards to prevent catastrophic incidents. Various hazard analysis methodologies are suggested to evaluate and assess process hazards, and determine the consequences of catastrophic failures.

ELEMENT 3: Management of Change. Written procedures to manage changes in technology, facilities, and equipment including personnel. Changes, although inevitable, must be properly managed to avoid serious process incidents. Basis for change, safety and process operating impacts, consequences of deviations, engineering, and administrative control of facility changes must be considered.

ELEMENT 4: Mechanical Integrity. An ongoing program of written documentation and activities to maintain the integrity of process equipment in an asbuilt condition. Effective preventive and predictive maintenance programs, testing, inspection, and quality control system for process equipment including mechanical and electrical systems and subsystems are necessary to ensure safe operation of the facility and prevent premature failures or emergencies (i.e., reliability engineering, quality assurance, preventive maintenance, materials of construction, and fabrication and inspection procedures).

ELEMENT 5: Operating Procedures. Written procedures providing clear instructions for conducting activities safely in a facility for each operating phase and procedure, operating limits, and safety/health considerations, including steps taken to correct and/or avoid deviations.

ELEMENT 6: Safe Work Practices/Hot Work Permits. Written procedures to ensure the safe conduct of process equipment operations, maintenance, and modification activities involving lock out/tag out of energy sources, confined space entry, line breaking and entering, hot work/welding operations, and use of heavy equipment (i.e., cranes).

ELEMENT 7: Process Safety Reviews. Formal programs to identify, evaluate, and reduce hazards by intensive reviews at various stages of a project such as design, construction, installation/assembly, commissioning, startup, operation, maintenance, and demolition. These reviews are divided into capital project, prestartup, and process hazard analysis reviews. Process hazard analysis reviews involve a review team using one or more of the analytical techniques in Element 2 to evaluate and control process hazards.

ELEMENT 8: Training. An essential ingredient to effectively keep operating personnel knowledgeable in maintaining process equipment and machinery operating safely and be informed of the hazards of the processes and chemicals they are using. Specific hazards and safe practices applicable to job tasks should be stressed.

ELEMENT 9: Contractors. Provisions for instructions, including safety rules to contractors working on or near company process equipment and/or machinery to ensure they can safely perform their tasks. This element includes informing contractors regarding potential emergency actions.

ELEMENT 10: Emergency Response. The establishment and maintenance of a comprehensive emergency response program involving emergency plans, defined accountabilities, exercises/drills, and periodic updates. Emergency equipment and procedures should reflect credible catastrophic scenarios that might occur in the facility.

ELEMENT 11: Incident Investigation. The formation of an investigative team of knowledgeable persons to evaluate incidents that could have or did result in a major accident in the workplace.

ELEMENT 12: Audits. Periodic, detailed examinations of process facilities by a qualified team to determine compliance with process safety elements and provide corrective feedback and recommendations where improvements are needed.

ELEMENT 13: Employee Participation. Employers must have a written plan outlining employee participation in the conduct and development of Process Hazard Analyses (PHAs) and in the development of other elements of PSM. Employees and their representatives must have access to the information developed for PSM activities.

Process Safety Management 173

Table 10.2 Process Safety Management Comparison of Basic Program Elements with Similar Management Systems

Title	Guidelines for Management of Process Hazards API RP750 Section	Process Safety Management of Highly Hazardous Chemicals OSHA 1910.119 Paragraph	Guidelines for Technology Management of Chemical Process Safety AIChE, CCPS Chapter	Process Safety Code of Management Practices CMA Code for Responsible Care Tabs	1990 Clean Air Act Amendment OSHA-CAAA S 1630 Air Toxics Section
GENERAL	1	a,b,c	1,2,3	1,2,3	301,304
Process Safety Information	2	D	4	5.5	301 Regs, 304 std
Process Hazard Analysis	3	E	4,6	5.7	301 Regs, 304 std
Management of Change	4	L	7	5.10.8	304 Standard
Operating Procedures	5	F	4		301 Regs, 304 std
Safe Work Practices	6	K	4	7.4	301 Regs, 304 std
Training	7	G	10	7.7	301 Regs, 304 std
Assuring Quality and Mechanical Integrity of Critical Equipment	8	J	8	6.10	301 Regs, 304 std
Pre-Start-up Safety Review	9	I	5		304 Standard
Emergency Response and Control	10	N		6.16	301 Regs, 304 std
Investigation of Process Related Incidents	11	M	11	4.10	301 Regs, 304 std
Audit of Process Hazards Mgt Systems	12	O	13	6.7	301 Regs, 304 std
Contractors		H		7.13	304 Standard
Capital Projects			5		
Human Factors			9	7.9	
Standards, Codes and Regulations			12	6.4	304 Standard
Waste Management				2.5-2.7	
References	13	Appendices A,B,D	14	9	

Table 10.3 Comparison of Consensus Management Systems Elements with OSHA's PSM

Title	OSHA PSM	ANSI Z10-2010	OHSAS 18001
Hazard Identification, Risk Assessment and Controls	✓	✓	✓
Employee Participation	✓	✓	✓
Management of Change	✓	✓	✓
Education and Training	✓	✓	✓
Incident Investigation	✓	✓	✓
Audits	✓	✓	✓
Contractor Safety	✓	✓	✓
Emergency Preparedness	✓	✓	✓
Management Review		✓	✓
Document Control	✓	✓	✓
Monitoring and Measurement		✓	✓
Non-Conformity and Corrective Actions	✓	✓	✓
Work Procedures and Permits	✓		✓
Operating Procedures		✓	✓
Feedback and Planning Process		✓	
Design Review	✓	✓	✓
Mechanical Integrity	✓		✓

Source: The regulatory and consensus standards reflected in this comparison.

ELEMENT 14: Trade Secrets. Employers must make information necessary for compliance with the regulation available to those persons (1) compiling process safety information; (2) developing PHAs and operating procedures; and (3) involved in accident investigations, emergency planning and response, and compliance audits, without regard to the possible trade secret status of such information. Employers may require confidentiality agreements with personnel receiving trade secret information. Employees and their designated representatives shall have access to trade secret information contained in PHAs and other required documents.

Table 10.2 provides a comparison of PSM elements as they apply to each standard and U.S. regulatory requirements for OSHA and EPA. Table 10.3 illustrates the comparison between the consensus standards and OSHA's PSM.

ENFORCEMENT TRENDS

Since its promulgation, enforcement of the PSM standard by OSHA has become Enforcement data shows key trends and areas where both EPA and OSHA focus their efforts. The top ten most cited PSM elements are shown in table 10.4.

Table 10.4 Top Ten OSHA PSM Citations Sorted by Frequency (February 26, 2014–XXXXXXXX)

Cited Element 1910.119(X)(X)(X)(X)	Description	Frequency
j2	Written Mechanical Integrity procedures	1231
e1	Perform an initial process hazard analysis	913
l1	Written procedures to manage changes	848
d3ii	Document that equipment complies with Recognized and Generally Accepted Good Engineering Practices (RAGAGEPs)	719
c1	Written plan of action regarding employee participation	673
f1	Develop and implement written operating procedures	671
j5	Correct deficiencies in equipment that are outside acceptable limits	620
g1i	Trained in an overview of the process and in the operating procedures	598
e5	Establish a system to promptly address the PHA team's findings and recommendations	588
N	Emergency planning and response (.38 and .120)	527

Table 10.5 PSM Inspection History Summary (February 26, 2014–XXXXXXXXXX)

Item	Value
Total Number of PSM Inspections with at least one violation	20,100
Total Number of Violations	3,721
Total Initial Penalty	$93 Million
Total Number of CHEM NEP Inspection (Opened November 2011 to February 26, 2014)	890
Total Number of Refinery NEP Inspections (opened as of April 4, 2013)	74

Data Source: OSHA Office of Statistics.

Permit to Work System for Contractors

Contractors tend to have a safety disadvantage due to their ever-changing working environment. One week they are working in a chemical plant and the next day they are working in a refinery. As a result, the requirements may change several times a week, perhaps even in a day. Both of these facilities have different safety and emergency response procedures with different phone numbers, different evacuation areas, different alarms, etc. Even within the same facility, different units will have different alarms, different chemical hazards, different PPE requirements, and numerous other safe work practices. This is a lot to keep straight from facility to facility and unit to unit on a daily basis. Keep in mind that almost all businesses train their contractors on an annual basis. A contractor work permit provides daily refresher training for these workers, as well as provides a hazard assessment based on the current working conditions and the work the contractor will be conducting. Like other permit processes, the daily contractor work permit must be kept at the site of the work being performed. This permit reviews required safety equipment, emergency procedures, other required permits, and safe work practices, such as lockout. The work permit process also fulfills the safe work practices of controlling the entrance, presence, and exit of contract employers and contract employees in covered process areas (1910.119[h][2][iv]).

Car Seal Program

This is very important in PSM. In fact, in some circumstances, it is even required by ASME Section VIII, Appendix M. The program is essential for control devices, such as valves. Safety professionals need to ensure there is no deviation from their "Safe" position. "Safe" could mean that the device is either required to stay open (car seal open) or stay closed (car seal closed) depending on the application. This program can be applied to any type of control device, such as valves preceding pressure relief valves, valves at site glasses and level sensors that control material flow, and valves for cooling loops on equipment.

Mistakes made in the implementation and management of car seal programs can lead to compliance issues with OSHA's 1910.147—*The control of hazardous energy (lockout/tag out)* standard. It is not unusual to find "tag out" tags being used in a car seal program, and this is a clear violation of 1910.147(c)(5)(ii).

Lockout devices and tag out devices shall be singularly identified; shall be the only devices(s) used for controlling energy; shall not be used for other purposes.

A car seal program can add a lot of safety to a PSM/RMP process, but only when it is developed, implemented, and managed properly. We will discuss the "do's and don'ts" of the program.

MANAGEMENT OF CHANGE FOR PERSONNEL CHANGES

Management of change (MOC) for personnel changes is a hotly debated topic in process safety circles; even OSHA and EPA are getting into the debate with their recent "Request(s) for Information" on their PSM Standard and RMP Rule.

Clarifying paragraph (l) of the PSM Standard with an Explicit Requirement that Employers Manage Organizational Changes. The existing standard does not explicitly state that employers must follow management of change procedures for organizational changes, such as changes in management structure, budget cuts, or personnel changes; however, as noted in a March 31, 2009, Memorandum for Regional Administrators from Richard Fairfax, it is OSHA's position that paragraph (l) covers organizational changes if the changes have the potential to affect process safety. Since the original promulgation of the PSM rule, it has become well established in the safety community that organizational changes can have a profound impact on worker safety and, therefore, employers should evaluate organizational change like any other change. Illustrating the significant hazards that organizational changes can produce, the CSB identified a lack of organizational management of change as a significant factor behind the 2005 BP Texas City Refinery accident that killed 15 workers and injured over 170 others (CSB Report No. 2005-04-ITX).

Require Owners and Operators To Manage Organizational Changes. In its Request For Information (RFI), OSHA notes that while the PSM standard requires employers to establish and implement written procedures to manage change, including all modifications to equipment, technology, procedures, raw materials, and processing conditions other than replacement in kind, the standard does not explicitly require employers to follow management of change procedures for organizational changes, such as changes in management structure, budget cuts, or personnel changes. However, OSHA highlights a policy interpretation indicating that it is OSHA's view that the PSM standard does cover organizational changes if the changes have the potential to affect process safety.

Additionally, OSHA notes the 2005 BP Texas City Refinery explosion, where the Chemical Safety and Health Investigation Board (CSHIB) identified a lack of organizational management of change as a significant causal factor in the accident. EPA's Risk Management Plan (RMP) rule contains management of change requirements for Program 3 processes (see § 68.75) that are virtually identical to the PSM standard. Therefore, EPA is also interested in receiving public comment on whether the RMP rule's management of change requirements should be expanded to include management of organizational changes.

There are distinct advantages to performing MOCs on personnel changes; here are two examples:

One of EPA's top-cited RMP deficiencies is not updating the emergency contact information in the RMP within 30 days of the change.

Both PSM and RMP require the Emergency Shutdown Procedures (ESD) procedures to assign shutdown responsibilities to qualified operators to ensure that emergency shutdown is executed in a safe and timely manner—is not uncommon for us to see four operators listed in the ESD procedures, but recent staffing changes took a shift down to three, and thus, the ESD procedures is now not accurate and in some cases would not have been able to be executed in a timely manner.

THE SYSTEMS APPROACH TO PROCESS SAFETY MANAGEMENT

Few programs or initiatives have had the potential of affecting so much of a facility's operation as PSM. Initial evaluations by many companies have determined that the requirements set forth by OSHA and EPA will have an enormous impact, and that it will take three to four years to fully implement the twelve PSM elements.

Several major oil companies have conservatively estimated that, for an average refinery producing 100,000 barrels of gasoline per day, compliance will cost between $1,015 million to implement and will require an additional 2,540 more employees and/or contractors to initially implement a PSM program for compliance. These are conservative estimates, but they indicate the significant impact the regulations will have on the industry. Similar costs are expected to be incurred by other industries affected (i.e., chemicals plant processes, explosive and pyrotechnic manufacturers, and gas plant operations).

Key Elements. A large proportion of the cost will occur in technical, operation, and maintenance activities to formalize necessary records, documentation, drawings, equipment specifications, operating procedures, inspection, and testing programs, etc. Over the years, engineering drawings (i.e., piping and instrumentation drawings, process flow diagrams, and process equipment specifications and descriptions) have typically become outdated. Additionally, operating procedures, including relief system designs, materials of construction, and quality assurance programs may need updating to reflect current design after changes to operations or facilities.

In addition, the consequences of process deviations to equipment design, operations, and their relative safety impact must be assessed. Safe operating limits must be defined and procedures developed to address how to correct and/or avoid dangerous process deviations.

Two important elements in the proposed regulations are directed at management of change and Process Hazards Analysis (PHA).

Management of change will affect all aspects of the facility's operation: technology, instrumentation, equipment, and personnel. Process changes often become necessary to improve a unit's design or operation, and in some operations may be as frequent as 50,100 changes per year. Moreover, changes may adversely impact a facility's design, operation, and/or maintainability. Comprehensive, yet flexible, management of change procedures must be developed and implemented to meet these concerns.

PHA is the analytical heart of the PSM system. When properly performed, PHA can result in improvements affecting design, operation, maintenance, and training.

PHAs can be accomplished by using different hazard identification and analytical techniques. Common examples are

1. Preliminary Hazard Analysis
2. What If or What If/Checklist
3. Hazard and Operability Study (HAZOP)
4. Failure Mode and Effects Analysis (FMEA)
5. Fault Tree Analysis (FTA)

Each technique has specific advantages and disadvantages in hazard identification, evaluation, and control.

Once process hazards have been identified and their potential causes evaluated, the risk (i.e., consequences and frequency) can be evaluated. In simpler terms, the following questions can be answered:

1. What can go wrong?
2. What are the causes?
3. What are the consequences?
4. How likely is it?
5. What are the risks?
6. Is the risk acceptable?

Risks can be evaluated qualitatively and/or quantitatively. Qualitative risk analysis is the most widely used method of evaluating episodic risk. A common approach to qualitatively analyzing hazards to determine potential risk is the use of a risk matrix method as illustrated in tables 10.6 and 10.7. These techniques, although based on intuitive reasoning and the analyst's experience, can be very effective in determining the relative risk of each incident scenario and in identifying risk-reduction methods. The objective of the risk matrix is to determine how the severity and/or frequency of the identified hazards can be reduced through some mitigating action. For example, water spray system installation, reduction of hazardous material inventory, or improvement in the flare system design can be used to reduce risk from high to moderate or moderate to low.

Table 10.6 An Example Risk Matrix based on Probability and Severity Using Numerical Values

Probability of Success / Category	A. Frequent (1) Event Likely to Occur Once or More Per Year	B. Probable (2) Event Likely to Occur Once Every Several Years	C. Occasional (3) Event Likely to Occur Once in Lifetime of Facility	D. Unlikely (4) Event Unikely, But Not Impossible
I. Catastrophic (1) Personnel-Life Threatening Environment-Large, Uncontrolled Release Equipment-Major Damage Resulting in Loss of Unit	■	■	■	4
II. Critical (2) Personnel-Severe Injury Environment-Moderate, Uncontrolled Release Equipment-Moderate Resulting in Unit Downtime	2	■	■	8
III. Marginal (3) Personnel-Lost Time Injury Environment-Small, Uncontrolled Release Equipment-Minor Damage Resulting in Unit Slowdown	■	■	9	12
IV. Negligible (4) Personnel-Minor Injury Environment-Small, Controlled Release Equipment-Negligible Damage	4	8	12	16

■ High: Requires Action   ■ Moderate: Further Study Required   Low: Investigate As Time Permits

Quantitative risk analyses are performed using engineering techniques and computer models to analyze incident scenarios to determine their impact.

More specifically, what damage may occur from a fire, explosion, and/or toxic release and how frequently will it occur? The potential impacts (e.g., fatality, injury, equipment loss, and/or environmental damage) are systematically estimated to determine the magnitude and severity of the incident.

A Management Function. Because PSM is comprehensive and addresses all aspects of a facility's operation, the implementation of a PSM program should be considered a line-management function responsibility. Organizationally, many companies have formed a process safety group reporting directly to the plant manager or "once removed" from the plant manager via the Safety, Health and Environmental (SH&E) manager. The process safety group is typically staffed with technical personnel who have operations experience, and is charged with the responsibility to coordinate and facilitate the PSM program implementation. This requires direct interaction among employees and departments responsible for implementation of PSM elements (e.g., technical, operations, maintenance, SH&E, and training.)

Table 10.7 An Example Risk Matrix based on Probability and Severity Using a Hybrid Approach

SEVERITY CATEGORY / FREQUENCY CATEGORY	A—FREQUENT Will occur twice or more in system lifecycle	B—PROBABLE Will occur at least once in system lifecycle	C—REMOTE Possible, but unlikely to occur in system lifecycle	D—IMPROBABLE A credible accident event cannot be established
I. CATASTROPHIC PERSONNEL: Death FACILITIES/EQUIPMENT/VEHICLES: System loss, repair impractical, requires salvage, or replacement	IA	IB	IC	ID
II. CRITICAL PERSONNEL: Severe injury/occupational illness. Requires admission to a health care facility. FACILITIES/EQUIPMENT/VEHICLES: Major system damage. Damage greater than $1,000,000 or which causes loss of primary mission capability	IIA	IIB	IIC	IID
III. MARGINAL PERSONNEL: Minor injury/occupational illness. Lost-time accident of more than one day that does not require admission to a health care facility. FACILITIES/EQUIPMENT/VEHICLES: Loss of any non-primary mission capability, or damage more than $200,000 but less than $1,000,000 with no direct impact on mission capability.	IIIA	IIIB	IIIC	IIID
IV. NEGLIGIBLE PERSONNEL: Less than minor injury/occupational illness. May or may not require first aid but lost time is less than one day. FACILITIES/EQUIPMENT/VEHICLES: Damage equal to or less than $200,000	IVA	IVB	IVC	VD

CONCLUSIONS

Recent worldwide experience has shown that a majority of the 100 largest property losses have occurred since 1977 as shown in Figure 10. As previously discussed, federal and some state governments have recently proposed extensive legislation for industry process safety and risk management programs. Companies affected by these regulations must accept the challenge to continue improving the safety of their operations.

Companies are continually striving to be better corporate citizens because it makes good business sense and because, "it is the right thing to do." A safe and healthful workplace is essential for employees, contractors, communities, and the environment. In addition, corporate management recognizes the importance of PSM programs in a global marketplace.

Recently, companies have been working toward improving process safety. Initial audits and reviews are now being conducted to evaluate the impact of OSHA and EPA PSM regulations and to identify areas where improvements are required. PHAs are being conducted by experienced teams to provide preliminary screening of process hazards in their facilities. These plans assess what is required to attain full compliance with the regulations.

Realistically, risks are an unavoidable, but controllable aspect of petroleum and petrochemical operations. Moreover, risk is an unavoidable part of our day-to-day activities. As individuals and as a society, people often engage in risky activities because they believe that the rewards and potential benefits outweigh any potential harm. Likewise, risk management decisions for "industry-imposed" risks involve value judgments that integrate technical, social, economic, and political concerns with scientific risk assessments.

Thus, the management of process hazards and the prevention of catastrophic incidents in the workplace requires a recommitment of manpower and financial resources. Assuming this PSM commitment is made, many positive benefits beyond SH&E performance will result, including improved quality and reliability, and enhanced worker and community safety.

The complexity, magnitude, and impact of the fourteen PSM program elements and the interrelationships among facility operating departments require the development of a formalized, comprehensive, and organized network to implement an effective program. Most importantly, management and employees must learn to apply these systems to insure that process safety becomes integrated into day-to-day operations.

REFERENCES

"100 Largest Losses, 23rd Edition," Thirteenth Edition, March, 2014.

Alderman, J.A. (March, 1990). *PHAA Screening Tool*, Presented at 1990 Health & Safety Symposium.
Bird, Frank E. and Germain, George E. (1990). *Practical Loss Control Leadership*, Atlanta, GA, Institute of Publishing.
Center for Chemical Process Safety (CCPS). (1989).
CCPS. (1991). *Plant Guidelines for Technical Management of Chemical Process Safety*. New York: AIChE.
CCPS. (1992). *Guidelines for Auditing Process Safety Management Systems*. New York: AIChE.
CCPS. (1994). *Guidelines for Implementing Process Safety Management Systems*. New York: Wiley-AIChE.
CCPS. (1995). *Guidelines for Safe Process Operations and Maintenance*. New York: Wiley-AIChE.
CCPS. (1996). *Guidelines for Integrating Process Safety Management, Environment, Safety, Health and Quality*. New York: Wiley-AIChE.
CCPS. (2000). *Fundamentals of Process Safety*. New York: AIChE.
Chemical Process Safety Management, Section 304, Environmental Protection Agency, 1990 Clean Air Act Amendment, Promulgated Nov., 1990, Federal Register.
Chemical Facility Safety and Security Working Group. (2014). *Executive Order 13650 Progress Report*.
Crowl, D.A. and Louvar, J.F. (2001). *Chemical Process Safety: Fundamentals with Applications* (2nd ed.). Upper Saddle River, NJ: Prentice-Hall.
DeHart, R.E. and Gremillion, E.J. (February, 1992). *Process Safety—A New Culture*, Occupational Safety & Health Summit.
Department of Defense Military Standard. (1993). *System Safety Program Requirements*, MILSTD882C, Washington, DC.
Eastman, M. and Sawers, J.R. (1998). *Documentation: The Key to MOC*. Chemical Processing.
Guidelines for Hazardous Evaluation Procedures. (1992). 2nd ed. New York: American Institute of Chemical Engineers.
Guidelines for Technical Management of Chemical Process Safety, Center for Chemical Process Safety (CCPS), 1989, *American Institute of Chemical Engineers*, 345 East 47th Street, New York, NY, 10007.
Haight, Joel, M., Yorio, Patrick, Willmer, Dana R. (2013). Health and Safety Management Systems—A Comparative Analysis of Content and Impact (Session 539); American Society of Safety Engineers Annual Professional Development Conference.
Hansen, M.D., Gammel, Geralkd, W. (October, 2008). Management of Change: A key to Safety-Not Just Process Safety. *Professional Safety*, pp. 41–50.
Hawks, J.L. and Mirian, J.L. (August, 1991). HGP, Inc. Houston, TX, "Create a Good PSM System," Hydrocarbon Processing.
Krembs, J.A. and Connolly, James, M. *Analysis of Large Property Losses in the Hydrocarbon & Chemical Industries*, 1990 NPRA Refinery and Petrochemical Plant Maintenance Conference, May 23–25, San Antonio, Texas.
Lees, F.P. (1980). *Loss Prevention in the Process Industries*. London: Butterworths.

Manuele, F. (February, 2003). Severe Injury Potential: Addressing an Often-Overlooked Safety Management Element. *Professional Safety*, 48(2): 26–31.

Manuele, F. (February, 2006). ANSI/AIHA Z10-2005: The New Benchmark for Safety Management Systems. *Professional Safety*, 51(2): 25–33.

Management of Process Hazards. (January, 1990). *API Recommended Practice (RP) 750*, 1st edition. American Petroleum Association, 1220 K Street, NW, Washington, DC 20005.

OSHA's Process Safety Management of Highly Hazardous Chemicals (29 CFR 1910.119) in Plain English, NUS Training Corporation, undated.

Ozog, H. and Stickles, R.P. (2003). *Management of Change or Change of Management? (White Paper)*. Salem, NH: ioMosiac.

Petersen, D. (1998). *Safety Management* (2nd ed.) Des Plaines, IL: ASSE.

Prevention of Accidental Releases. (November, 1990). Section 301, Paragraph 1, Environmental Protection Agency, 1990 Clean Air Act Amendment, Promulgated, Federal Register.

Process Risk Reviews Reference Manual, Safety and Environmental Resources Chemicals and Pigments Department, E. I. du Pont deNemours and Company, copyright 1990., Wilmington, DE 19898.

Process Safety Management of Highly Hazardous Chemicals, Occupational Safety & Health Administration (OSHA) 1910.119, Federal Register August 26, 1992.

Renshaw, F.M. (July, 1990). Rohm & Haas Co., Bristol, PA, "A Major Accident Prevention Program," Plant/Operations Progress.

Sherrod, R. M. and Sawyer, M. E. (July, 1991). Stone & Webster Engineering Corporation, Houston, TX, "OSHA 1910.119 Process Safety Management: It's Effect on Engineering Design," System Safety Conference.

Shrivastava, P. (1987). *Bhopal: Anatomy of a Crisis*. Cambridge, MA: Ballinger Publishing.

United States Chemical Safety and Hazard Investigation Board (CSB). (August 20, 2001). *Management of Change (Safety Bulletin No. 2001-04-SB)*. Washington, DC: Author.

United States Environmental Protection Agency. (2014). *Enforcement and Compliance History Online*.

United States Occupational Safety and Health Administration. (2014). *Office of Statistics*.

United States Occupational Safety and Health Administration. (2014). *Injury and Illness Prevention Programs*.

United States Occupational Safety and Health Administration. (2014). *Request for Information on Potential Revisions to Process Safety Management Standard*.

Zimmerman, Jonathan, and Haywood, Bryan. Process Safety Management Enforcement Trends and Best Practices (Session 721); American Society of Safety Engineers Annual Professional Development Conference, 2015.

Chapter 11

Systems Safety Engineering

Henry A. Walters

CASE

An operator is putting a pump online after a repair. A drain valve located downstream behind a wall is overlooked and left open, resulting in spilling acid to the ground. This is an environmental reportable incident and could have easily resulted in an injury. Even though there are tagging systems that should prevent this from happening, it could also be a design problem. Do they need the drain line at that point and/or is there a system problem with the tagging procedure? If either of these two items is driving the at-risk behavior of "failing to secure the drain line" then there is a system barrier that is a root cause of at-risk behavior (Brown and Hodges, 1997).

INTRODUCTION

The preceding example highlights the need for occupational safety professionals to become familiar with and incorporate the principles of systems safety engineering as they relate to safety management systems. In the truest sense, systems safety engineering is a technical field, requiring specialized skills and training that pertain primarily to the design of safe equipment. Unfortunately, most engineers receive little if any formal education in systems safety engineering so it is not difficult to understand why occupational safety professionals often hesitate to embrace and practice its concepts and are afraid to routinely use the many valuable tools available through its practice. The release of the Process Safety Management Standard in 1992 forced many industries, such as the chemical industry, to incorporate systems safety principles (Roughton, 1993; Goyal, 1996; Roughton and Buchalter, 1997).

Another area closely related to principles of systems safety is total quality management (Esposito, 1993; Weinstein, 1996; Manzella, 1997). In fact, if one examines the basics of systems safety engineering, it becomes obvious that several of the principles are often practiced, but that companies often fall short of formalizing the process. It is past time that all organizations begin formally adopting the principles of systems safety.

One of the primary goals of systems safety management and engineering is to eliminate risk, thus reducing the likelihood of an injury or loss of property. A large part of this depends on being proactive rather than reactive—designing out the chance of risk rather than attempting to minimize the loss (Brown, 1993). This is accomplished by critically examining all aspects of a system in order to identify and control the possible hazards before the worker is exposed to them (Ragan, 1994). In other words systems safety management attempts to address all possible hazards before they become an incident. To accomplish this goal, one must understand that in the life cycle of any system there are possibly different hazards introduced at each stage and these hazards may react differently based on the system interface.

The purpose of this chapter is to identify some of the systems present in the workplace that must be evaluated and provide an introduction of many systems safety concepts and tools available for use in occupational safety. It is impossible to discuss in a single chapter all the techniques of systems safety engineering. For additional information on the topic, consult the references at the end of this chapter.

System-Safety Tools

The old saying that "people don't plan to fail, they simply fail to plan" is true in the safety field. Therefore, one of the most critical elements in any SMS effort is the System-Safety Program Plan (SSPP).

System-Safety Program Plan

The SSPP is the road map of the entire system-safety effort. The success of the system safety effort is dependent on a well-written SSPP. A good SSPP does the following at a minimum:

- Identifies all activities and steps to be completed as part of the system-safety effort.
- Establishes which analyses will be conducted at each stage of the system life cycle.
- Delineates management review process and controls.

An important part of the SSPP is to establish the methods of risk assessment and determine how to test the system to ensure acceptable risk has been obtained. The training program is also identified in the SSPP.

As a final check of the effort, a method of auditing is established. Auditing is important to ensure that

- The elements of the SSPP are being followed and there are no problems preventing completion of the plan.
- The planned analyses and other activities are sufficient.
- The SSPP is still viable.

When properly developed, the SSPP provides a checklist that can be used to measure progress toward real system safety. Once the SSPP is developed, the various hazard analyses specified by it become the major focus of the system-safety effort.

In developing the SSPP, it is important to recognize what the various systems needing analysis are and how the interface among these systems changes over time. The following paragraphs discuss some of these systems and how the life cycle of the system as a whole comes into play while developing the SSPP. The systems discussed are by no means all-inclusive, as every plant or workplace is unique and requires its own system analysis.

Occupational Safety Systems

In its most complex form, the overall system in occupational safety is the entire workplace (Fitzgerald, 1997). The workplace consists of various parts or subsystems. The four major subsystems might be addressed as follows:

1. Physical plant
2. Tools and equipment
3. People
4. Managerial systems that link the other subsystems together

Each of these, in turn, consists of subsystems. The major subsystems are examined here in more detail. Obviously, the physical plant can be examined in a myriad of ways. One way is to think of the plant as consisting of the following eight subsystems (Ferry, 1990):

1. Basic structures and facilities
2. Internal storage and transport
3. Heating and cooling systems
4. Electrical power

5. Cleaning and waste
6. Utilities
7. Fire protection
8. Security and exclusion from the area

Each of these subsystems can be further subdivided. For example, fire protection consists of the fire-retardant material used in construction, the sprinkler system, fire doors, fire extinguishers, and so forth. How one determines the level of safety desired in each is addressed later.

Tools and equipment comprise a system consisting of the physical items employees use to do their job. The system includes subsystems and elements of each subsystem, such as these:

1. Workstation (seating, tables, lighting, conveyor belts, etc.)
2. Tools used to do the job (hammers, drills, wrenches, keyboards, etc.)
3. Equipment used in processes (fork trucks, presses, punches, extruders, copiers, etc.)

The workforce consists of subsystems and elements of each subsystem, such as the following:

1. Management (CEO, sales manager, production manager, front-line manager, etc.)
2. Foremen
3. Hourly employees
4. Teams or cells (logistics, production, quality control, etc.)
5. Committees (safety, union, benefits, etc.)
6. Social systems (male, female, ethnic, etc.)

Managerial systems, and elements of each subsystem, include the following (Ferry, 1993):

1. Written procedures (work standards, safety rules, disciplinary procedures, etc.)
2. Visual and audible warnings (placards, signs, notification systems, alarms, etc.)
3. Training programs
4. Supervisory personnel to oversee enforcement of procedures

It is not as important to correctly categorize one subsystem as a part of a specific system as to insure that all critical subsystems are identified. Once the system and subsystems of interest are identified it is most important to realize how each of them works with each other. One area of interest is when they impact each other. This leads to the need of understanding life cycles.

Life Cycle

From a system-safety management perspective, it is critical to consider the fact that all systems have a life cycle. This means a system goes through a specific series of steps on its way from inception to the time it no longer serves a useful purpose. Life cycles can be described in various ways. For example, the life cycle of hardware is discussed in the chapter "Staffing and Job Process Control." In the sense of occupational safety, referring to the workplace as the system, the phases of the life cycle may be described as follows:

1. Concept
2. Definition
3. Construction
4. Operation
5. Termination

Concept Phase

The concept phase consists of the initial plans for a new company facility. System-safety considerations in this phase may include initial hazard analyses for:

- Industrial safety (working surfaces, area access/egress, material handling)
- Fire protection
- Industrial hygiene (lighting, ventilation, noise, sanitation, heat, toxic agents)
- Human factors
- Vehicular safety (loading zones, parking, traffic flow)
- Emergency preparedness (alarms, evacuation procedures, abatement)
- Radiation protection
- Environmental protection

Someone must use the principles of systems safety in this phase to begin identifying potential hazards before plans proceed too far. It is much easier and less costly to implement safety during this phase than to wait and redesign a facility. Because it is difficult at this stage to identify how various subsystems may interface, the safety professional should conduct preliminary hazard analyses (PHAs). These analyses identify the obvious hazards that are encountered in almost every facility.

Systems safety engineering is proactive by identifying and attempting to eliminate all hazards before anyone is exposed to them. This means ideally all hazards could and would be identified in the concept phase. However, this is rarely possible as the interaction of subsystems often generates hazards that might not be identifiable in the analysis of the independent subsystem. For example, in the design of an aircraft's landing gear, the hydraulic system may be designed to an acceptable level of risk. If there is a minor leak of hydraulic fluid, it is determined that there is no safety hazard involved within the hydraulic system. The brake system has also been designed to an acceptable risk. However, when the two subsystems are integrated, if the leaking hydraulic fluid can drip onto the brake system then the heat generated by the brake system might ignite the leaking hydraulic fluid, creating a safety hazard. The same concept applies within the occupational system. Therefore, the principles of systems safety must continue throughout the life cycle.

Definition Phase

The definition phase begins when the initial concept has been solidified and it is time to begin intricate design of the facility. It is at this point that the interaction of subsystems may become obvious and more detailed. As the concept becomes better defined, safety professionals should conduct systems and subsystems hazard analyses (SHAs and SSHAs). Additional analyses include fault hazard analysis (FHA) and failure mode and effects analysis (FMEA). For example, it is important to analyze such things as the role of personnel in fire protection, or what happens to the fire protection system if the electrical system shuts down. The definition phase should be completed before construction of a new installation begins, which means that safety professionals should make a safety related "go/no go" decision prior to the beginning of construction.

Construction Phase

The construction phase is when the facility is being built. In product or hardware development, this corresponds to the fabrication and/or installation and/or occupancy-use phases(s). During this phase, the safety professional's primary role is monitoring and quality control—ensuring the plans developed in the previous phases are being followed. Training of the workforce often begins during this phase. Because of the new interaction between person and workplace, their introduction to the overall system requires verification of previous analyses and the initiation of operating and support hazard analyses (O&SHAs).

Operation Phase

By the time the life cycle reaches the operation phase, safety professionals will have worked themselves out of a job, except for monitoring compliance. All potential hazards have been previously identified and either eliminated or their effects controlled. However, during the actual operation of the subsystems, new and previously unidentified hazards may be discovered. For example, management control systems are not usually fully introduced until this phase, and, therefore, new hazards may become apparent. It is also in this phase that safety professionals can gather data, identifying how the systems are actually working together; therefore, new analyses may be required in the operation phase. Since the objective of systems safety engineering is to identify and remove hazards before workers are exposed to them, it is during the operation phase that such innovative approaches as performance-based safety (PBS) can play a vital role.

PBS allows a systematic method of checks and balances for various subsystems. In 85 percent to 95 percent of all incidents, the last common factor before an injury is a behavior. This does not mean the behavior caused the incident, but that had a different behavior occurred, the injury may have been completely prevented or at least lessened. For example, an employee is injured in a vehicular accident. An accident investigation discovers that the employee was crossing an intersection with a green light when a truck ran a stop light and broadsided the vehicle and the employee was thrown from the vehicle and suffered a broken back. The employee stated that he or she had not seen the truck and was not wearing his or her seat belt. Was the employee at fault for the accident? Probably not. However, the employee may have prevented the accident had he or she looked before entering the intersection, seen the truck, and taken evasive action. Regardless, had the employee been wearing his or her seatbelt, he or she would not have been thrown from the vehicle and the injuries may have been lessened.

Since employees are not usually performing at-risk behaviors intentionally, PBS provides for the identification of behaviors involved in accidents, methods of collecting data to determine the actions that are leading to the injuries, feedback to reinforce safe behaviors and identify barriers leading to at-risk behavior, and methods of problem solving to remove the barriers. Thus, PBS is another proactive tool to be used in systems safety engineering during the operation phase of the life cycle. The example given in the introduction of this chapter highlights the importance of considering behavior in systems safety approaches.

The operation phase usually also sees a greater use of some of the other systems safety engineering tools to be discussed later.

Termination Phase

As stated previously, the goal of systems safety engineering is to identify all hazards before anyone is exposed; therefore, the safety professional should consider the residual hazards after a facility is vacated (termination phase), even as the system is being developed and operated. This is an area that has historically been overlooked. The EPA and specific acts such as the Clean Water Act, the Toxic Substances Control Act, and the Resource Conservation and Recovery Act, among others, play a key role in determining factors that should be considered during the termination phase. Otherwise, companies may be held liable for injuries and illnesses occurring years after the property has been vacated. As with the other phases, planning for the termination phase should be conducted as early as possible during the previous phases. Responsible companies begin considering the termination phase during concept formulation.

Numerous types of hazard analysis have already been mentioned in each phase of the life cycle. Tools to be used in these analyses will now be addressed.

Hazard Analysis

A *hazard* can be defined simply as the potential to do harm (Manuele, 1997). Since the ultimate goal of all system-safety efforts is to create a workplace with a minimal acceptable level of exposure to harm, hazard analysis is the heart of systems safety (Roland and Moriarty, 1990). Each of the hazard analyses described previously contains at least three elements. These are a description of hazard severity, hazard likelihood, and a plan for hazard control.

Hazard Severity. *Hazard severity* can be defined in many ways. In its simplest form, hazard severity could be based on the extent of the injury and use standard categories such as fatality, lost-time injury, medical treatment, and first aid. Table 11.1 is a system of classifying hazard severity as described in MIL-STD-882B .

The problem is that the definition of each category often leaves much to be desired. For example, it is not unusual for companies to lower their lost-time injury rate by implementing limited duty (reduced responsibilities) for the injured person. Does this make the working environment less hazardous? Not really. Therefore, it is probably best for a company to develop its own definitions for severity.

Hazard likelihood can also be defined in many ways. If possible, one should use an actual probability based on *a priori* (before the fact) or *a posteriori* (result of test or experience) probabilities. Methods to determine these are presented subsequently. If actual probabilities cannot be established, simpler methods to determine categories of hazard likelihood have been developed. They include the MIL-STD-882B example found in Table 11.2.

By combining Tables 11.1 and 11.2, it is possible to develop a hazard risk matrix that allows a semi-quantitative basis of resource allocation. Table 11.3 is such a combination and is similar to one used in marine safety (Wang and Ruxton, 1997).

Hazard Severity Classification

Category	Name	Characteristics
I	Catastrophic	Death
II	Critical	Severe injury
III	Marginal	Minor injury
IV	Negligible	No injury

Classification of Hazard Likelihood

Description	Level	Specific definition
Frequent	A	Likely to occur frequently
Probable	B	Will occur several times in life of exposure
Occasional	C	Likely to occur sometime in life of exposure
Remote	D	Unlikely but possible to occur
Improbable	E	So unlikely assumed may not be experienced

Hazard Risk Matrix

Frequency of Occurrence	Hazard Category			
	I	II	III	IV
A. Frequent	1A	2A	3A	4A
B. Probable	1B	2B	3B	4B
C. Occasional	1C	2C	3C	4C
D. Remote	1D	2D	3D	4D
E. Improbable	1E	2E	3E	4E

The resulting hazard risk index (HRI) is Table 11.4 (Vincoli, 1993). The HRI might then be used so that management would require that all HRI I items receive immediate attention, all HRI II items be scheduled to be addressed, all HRI III items be monitored, and HRI IV items be reviewed infrequently. A more precise method to allow prioritization within an HRI is a formal risk analysis.

Hazard Risk Index (HRI)

HRI	Classifications	Suggested Criteria
I	1A, 1B, 1C, 2A, 2B, 3A	Unacceptable
II	1D, 2C, 2D, 3B, 3C	Undesirable
III	1E, 2E, 3E, 4A, 4B	Acceptable with review
IV	4C, 4D, 4E	Acceptable without review

Risk Analysis. Risk analysis uses a factual probability and a sophisticated estimate of the loss as the criteria for a quantifiable risk assessment. *Risk* is simply the probability of an event occurring, times the most reliable cost estimate of loss. Thus, an occurrence with a probability of .001 and a resulting loss of $1,000 (risk of 1) is a greater risk than an occurrence with a probability of .000001 but an estimated loss of $500,000 (risk of .5). The problem with quantifying loss is how does one account for intangibles such as personal anguish and lost production due to morale after the event? It is therefore advisable that the use of quantifiable methods might be a basis of decision-making, but it should not be the sole factor considered.

Hazard Control. The third part of all hazard analyses is the development of a method of hazard control. The development of hazard control begins with the hazard reduction precedence. The best solution to any hazard is to eliminate it totally through engineering. For example, if workers are exposed to a toxic substance in the workplace, is it possible to use a nontoxic substance in place of the toxic substance or use a method to avoid toxicity, such as removing the need for that substance completely?

If not, go to the second step, which is to initiate controls for the hazard through engineering. Following the example, design a ventilation system that removes the toxic fumes before the worker is exposed to them. The need to conduct FMEA becomes apparent in this case. The ventilation system may sufficiently reduce the risk of injury, but what happens if the system fails?

If the hazard is still not lowered to an acceptable risk, the third step is to provide safety devices. These can take the form or such things as emergency shut-off buttons or personal protective equipment such as an appropriate respirator.

The fourth step in hazard reduction is to provide warning devices. Thus, the worker would be warned that the toxic substance is present or that the respirator is not functioning properly. Warning systems are as simple as signs and posters or more elaborate such as alarms and flashing lights.

Should the risk still not be acceptable, the fifth step is to provide special procedures or training. In this example, training on how and when to evacuate the area may lower the risk of exposure to the hazard.

Finally, if the system is still not at an acceptable level of risk, it is time to determine whether to accept the risk or dispose of (eliminate) the system altogether. It may become necessary to develop a different means of accomplishing the need without that system.

With these three parts—hazard severity, hazard likelihood, and hazard control-in mind—the safety professional must closely examine some of the types of hazard analyses. Although there are numerous types of hazard analyses that should be conducted, the following ones are samples of those most often used in occupational systems safety efforts.

Preliminary Hazard Analysis

The preliminary hazard analysis (PHA) is one of the first activities described in an SSPP. This analysis is initiated during the concept phase and may be very cursory because it can prove difficult to identify many of the hazards or their actual risks. During a PHA, the safety professional compares the original concept for the system to existing systems and uses historical data to determine what hazards may be encountered in the new system. Specific items considered in the PHA include the following:

- Examination of basic energy systems such as fuels, power sources, and pressure systems.
- Identification of all legal regulations pertaining to the system.
- Identification of initial subsystem interface hazards that may be generated.
- Identification of environmental hazards such as possible natural disasters and radon exposure.

In simple terms, the PHA is a basic evaluation of existing, known hazards that may be present in the system.

System Hazard Analysis

The system hazard analysis (SHA) is an in-depth investigation of the entire system to determine its level of safety. It consists of numerous parts. A detailed description of how the system operates identifies the hazards present and provides information on the hazard severity and likelihood. Each hazard is then investigated to determine the causes. Since the hazard may be a result of the interface of subsystems, it is necessary to conduct an extensive review of the impact of the interface on the various subsystems. An evaluation of the acceptability of the risk is followed by the development of methods to control the hazard.

Subsystem Hazard Analysis

An subsystem hazard analysis (SSHA) focuses on subsystems, or elements within the system, that may be systems in and of themselves (Utley, 1994). Thus, an SSHA is basically an SHA conducted on a subcomponent of the system. The SSHA identifies all the hazards generated by the operation of the subsystem's components. It identifies the hazards that the subsystem generates in both its normal and fault occurrence modes. SSHAs should be conducted on every major identifiable subsystem. In essence every major subsystem of the total system needs to be identified as soon as possible. Then each major subsystem of these subsystems should be examined to determine if it is possible to identify hazards that may be generated by this subsystem itself. The depth to which subsystems should be further divided depends primarily on the size of the system itself and the level of risk generated by each lower-level subsystem.

Fault Tree Analysis

Fault tree analysis (FTA) and failure modes and effects analysis (FMEA) are valuable tools to determine ways to reduce the risks (Stephenson, 1991). FTA is a method that allows an investigator to go from the generic fault to the specific root causes for its occurrence. FTA is effective because it allows one to utilize qualitative analysis using deductive logic or quantitative analysis using reliability or failure data. By working one's way backward through what has led to each level of failure, one can identify and then develop solutions to the most powerful root causes of a hazard. An example of an FTA is provided later.

Failure Modes Effect Analysis

Safety professionals are often aware of the dangers of a system in normal operation. However, what happens when the system fails may not be identified as easily. Failure modes effect analysis (FMEA) is borrowed from the science of reliability. It is usually performed by reliability personnel to determine how the reliability of equipment affects the overall safety. The FMEA is generally an important part of the definition phase. When structured toward safety investigations, FMEA focuses on the single events that pose a state of hazard in the system. An FMEA identifies each subsystem of the system that may have failures and then determines what hazards may result from the failure. For example, a fire alarm system requires the tripping of a switch to activate it. What happens if that switch fails? At the risk of oversimplification, an FMEA can be used to identify the faults to investigate using FTA.

Risk Reduction

Once hazards have been identified, there are a variety of techniques to develop acceptable solutions to address the hazards. The hazard reduction precedence should be followed as it is best, whenever possible, to remove the dependence on personal behavior from the hazard. In fact, an engineering solution often removes a barrier that leads to at-risk behavior. For example, it is better to design a machine guard that prevents someone from placing his or her hand in a pinch point than to depend on training employees not to place their hands near pinch points. Whichever method of problem solving is used to address hazards, it is important to follow some structured process, as it is too easy to jump to wrong or less-effective solutions when rigorous procedures of analyses are not followed.

Operating and Support Hazard Analysis

This is very similar to the SHA except that it is performed continually during the operating phase and the final stages of the construction phase, as it is here that one can get actual data to develop the analyses. It is necessary to perform this analysis as the initial SHA may not identify all the hazards generated by the actual operation of the system due primarily to interface synergism. Also, this analysis investigates issues surrounding support systems that may be technically outside of the original system. For example, a building on a large military complex once experienced a major flood due to the successful operation of its sprinkler system. Somehow the signal that was supposed to be received by an outside fire department failed, and because of a sensor within the building declaring a fire, the sprinkler system activated and remained discharging for an entire weekend until someone opened the building on Monday morning. The steps required to conduct the operating and support hazard analysis (OaSHA) are similar to the SHA in that the primary purpose is to still identify all hazards created by the operation and support of the system and mitigate the risk.

Fault Hazard Analysis

Fault hazard analyses (FHAs) are conducted as a subset of the SSHA, SHA, or OaSHA. An FHA is a limited analysis in that it inductively investigates the effect of a single fault rather than the effects of multiple faults occurring at one time. It is similar to the FMEA. The final effect of a fault is traced through the subsystems and interfaces.

The following is a brief description of some of the statistical tools that may be used to determine actual probabilities of hazard exposure or system failure.

Statistical Tools

There are basically two ways of determining a probability: *a priori* and *a posteriori*. *A priori* (before-the-fact) probabilities are based on the inherent nature of the events. For example, before a coin is ever tossed, it is known that the probability of the coin landing with the "heads side up" on any given toss is 50 percent. Unfortunately, *a priori* probabilities do not often occur in occupational safety situations. An example of where an *a priori* probability may occur is with a switch position. Disregarding human behaviors that may play a role, the probability that a three-position switch will randomly be in any given position is one in three (33.33 percent).

An *a posteriori* (after-the-fact) probability is based on past experience. This means one must have factual data on which to base the probability. An *a posteriori* probability is calculated by dividing the number of times something has occurred in the past by the exposures (times it could have occurred). Mathematically this appears as

Probability of occurrence = occurrences/exposures

Since the laws of statistics allow one to expect the likelihood of something occurring in the future to be the same as has happened in the past, an a posteriori probability means that on any single future exposure one can predict the probability of a specific outcome. However, some issues must be considered. One is that there must be enough historical data on which to base the probability. Otherwise, the calculated probability may not be accurate, because there have not been enough statistically significant occurrences. For example, using limited data to calculate the probability of having an automobile accident, one accident in ten trips would lead to a probability of .1 of an accident happening on any given future trip. However, if additional data were available, there may have really been two accidents in 100 trips, or a probability of .02.

The space shuttle program is an example of predicting probabilities on limited data. Prior to the first accident on mission twenty-five, was it logical to assume a 0.0 probability of an accident? After that accident was the real probability of an accident .040 (1/25). Accident probability lowered with the successful completion of each subsequent mission. However, after mission one hundred thirteen, based on a posteriori probability, the probability of a space shuttle crashing became .018 (2/113). By the final launch the probability of a crash was .015 (2/135). A table outlining these relationships will be provided in chapter 13. A general rule of thumb is to have at least thirty independent data points before even considering using a posteriori probabilities.

A second issue even more important from an occupational safety perspective is that one does not want to use historical data that is too old to be reliable. Processes or equipment may have changed and that can have had an impact on the previous experience. For example, automobiles are much safer today than sixty years ago and driving practices have changed, so it would not be valid to use all of the trips driven in the last sixty years as the pool of exposure to determine the probability of an accident. One should try to use as much historical data as are still valid.

Once one understands how to calculate the probability of an occurrence on any single, given trial, it is possible to determine the probability of a certain number of successes or failures occurring in the future dependent on the type of distribution of the exposure. The actual formulae used to do these calculations are beyond the scope of this chapter, but many of the more useful ones in occupational safety applications may be found in several of the references. One concept that needs to be understood is that the total probability of everything that can possibly happen is 1. Therefore, in any system, the probability of success plus the probability of failure must equal 1. This means that if one knows the probability of success or failure, one can compute the other by subtracting the known probability from 1 (Complimentary Law). Mathematically this appears as:

$$P_{(success)} + P_{(failure)} = 1$$

or

$$P_{(success)} = 1 - P_{(failure)}$$

or

$$P_{(failure)} = 1 - P_{(success)}$$

Since there are so many factors to consider as to whether or not trusting an *a posteriori* probability, how confident can one be that the probability one uses is accurate? The solution is to generate a confidence limit when basing designs on desired probabilities of success.

Confidence Intervals

Based on an experienced failure rate, one can determine a more likely probability based on the degree of confidence one wants. A 95 percent confidence interval means that there is a 5 percent or less chance that the true probability falls outside of the calculated interval. For example, if one had experienced 10 automobile accidents in 1,000 trips, would one want to use the a posteriori probability of .010 in predicting failure rates of a system? If one decides that he or she wants to be 95 percent confident of her/his predicted probability, it would be better to develop a one-tailed, upper-bound confidence test. This would yield a probability of .017 (Walters, 1995), so in designing the system a .017 probability of failure is expected rather than the tested .010. This is a more conservative estimate of expectations. The degree of confidence has an effect on the value used. A 90 percent test of the same parameter yields a probability of .015. Therefore, in most cases, safety predictions should use higher confidence intervals in order to err on the side of safety. The development of confidence intervals is a fairly simple technique found in most books on elementary statistics.

Event Systems

As stated previously, most systems consist of numerous subsystems. To determine the probability of success or failure of a system, it is often easier to determine the probabilities of success or failure of the subsystems and then to use one of various methods to generate a probability of success for the entire system. Event system analysis allows this.

Systems are usually arranged in such a manner as to be considered series or parallel systems. However, problems arise when a system does not fall into a true series or parallel arrangement. It is important to determine when systems should be designed in series and when they should be designed in parallel.

Series Systems. A series system is one in which all the subsystems are in a linear relationship. For example, a fire can be modeled as a series system consisting of subsystems of oxygen, fuel, ignition, and reaction (Figure 11.1). All four subsystems have to be present and working in order for a fire to occur—no fuel, no fire. Assume it has been determined that the probabilities of success (occurrence of a fire) of the subsystems are .99, .3, .1, and .01, respectively. The probability of success of a series system is the product of the probabilities of success of all of the subsystems. Mathematically, this is

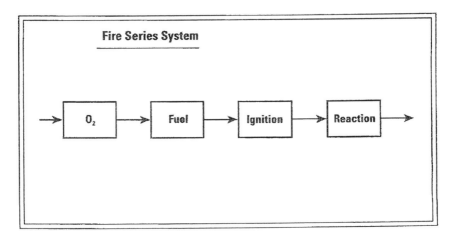

$P_{(Fire)} = P_{(Oxygen)} \times P_{(Fuel)} \times P_{(ignition)} \times P_{(Reaction)} = .99 \times .3 \times .1 \times .01 = .000297$

The probability of failure of a series system is 1 minus the probability of success. In the previous example, the probability of failure (not having a fire) is 1 minus .000297, or .999703.

The success of the fire example is an undesired outcome—a fire. Therefore, it can be called an accident system. There are also series safety systems. A safety system is one in which the successful operation of the system is a desired outcome. For example, assume there is a fire extinguishing system consisting of a water source, a water distribution system, and sprinkler heads. Once more, all three must work in order for the fire to be extinguished. The probabilities of the individual subsystems working are .98, .9, and .85, respectively. The probability of success of the system then is:

$P_{(extinguish)} = .98 \times .9 \times .85 = .7497$

and the probability of failure of the system is

$1 - P_{(extinguish)} = 1 - .7497 = .2503$

Note the probability of success of a series system is less than the individual probabilities of success of each of the subsystems. Therefore, one wants to design accident systems in series because the more subsystems are in series, the greater the decrease in probability of the accident occurring. Conversely, it is best to refrain from designing safety (success) systems in series, since the overall probability of success of the system is less than the probability of success of any single subsystem. Success systems should be designed as parallel systems.

Parallel Systems. Parallel systems are those allowing a system to work if any of the subsystems works by itself. For example, the space shuttle had three auxiliary power units (APUs) on board, arranged in such a way that any one of the three could supply sufficient power for the shuttle to operate safely (Figure 11.2). Just as in series systems, parallel systems can be either accident systems or success systems.

An example of an accident system in parallel is provided by the numerous independent ways that an accident might occur. There are numerous ways a person might fall off a ladder. The ladder might blow over, or it may have been improperly leaned against a building and slides, or the person may just reach too far and lose his or her balance. Consequently, there are numerous ways the person could fall off the ladder. The probability of falling off the ladder could be determined by calculating all of the various ways it could occur and then adding them together, but this would prove somewhat tedious. However, there is only one way the person will not fall off the ladder—if all three of the possible ways to fall fail to occur. Using the Complimentary Law, it is easy to compute the probability of falling off the ladder by first determining the probability of not falling and subtracting that from 1. Mathematically, the success of any parallel system is

$$P_{(success)} = 1 - P_{(failure)}$$

Assume the probabilities of each of the ways of falling off the ladder are .001, .010, and .020, respectively. Using the complimentary law, the probabilities of not falling off the ladder for each are .999 (1–.001), .990 (1–.01), and .980 (1–.02), respectively. Again, in order not to fall off the ladder, all three ways to fall must fail. Therefore, the probability of system failure is

$$P_{(failure)} = .999 \times .990 \times .980 = .969$$

Since the probability of success of the parallel system is 1 minus the probability of failure, the probability of success of the ladder accident system is

$$P_{(success)} = 1 - P_{(failure)} = 1 - .969 = .031$$

Note the probability of success of the system is higher than the probability of success of the individual subsystems. Accident systems should not be designed in parallel. If any one of the ways the person could fall of the ladder could be eliminated, there would be a decrease in the overall probability of the person falling.

Success, or reliability, systems should be designed in parallel. The shuttle provides a good example. Assume the probabilities of success of the

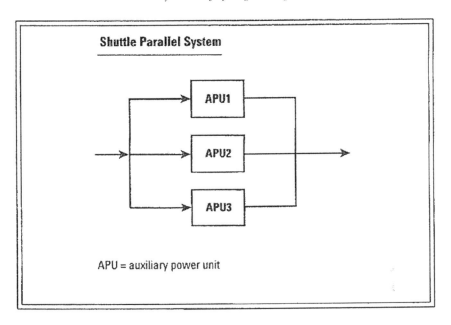

individual APUs are .900 (Figure 11.2). Since they are identical, one will assume identical probabilities of success. Since any one of the APUs will power the system, the only way for the system to fail is for all three to fail at once. The individual probability of failure is .100 (1–.900) so the probability of system failure is

$$P_{(failure)} = .100 \times .100 \times .100 = .00100$$

The probability of success is

$$P_{(success)} = 1 - P_{(failure)} = 1 - .00100 = .999$$

Note that the probability of success of the system is considerably higher than the success of the individual APUs. One way to look at it is that if only one APU was on board, the probability of success of the shuttle would be 9 of 10 trips, or 900 of 1,000 trips. With the three APUs acting in parallel, the probability of success is 999 in 1,000 trips. Safety (success) systems should be designed in parallel.

Cut Sets

Sometimes a system cannot be designed as purely a series or parallel system. Figure 11.3 illustrates such an irregular system. Note that Block 2 leads to

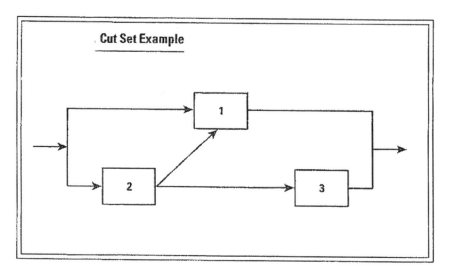

both Blocks 1 and 3. This means that there is no true series or parallel relationship among the subsystems. Therefore, it is impossible to use the formulae for series and parallel systems to determine success of the system.

To determine system success, one must first determine all the individual ways the systems can fail. This is called determining the cut sets. By determining the minimum cut sets, it is then possible to calculate a probability of system success or failure. It is beyond the scope of this chapter to discuss this method, and the reader is referred to the references at the end of this chapter.

Fault Tree Analysis

As discussed previously, fault tree analysis (FTA) is a method allowing both quantitative and qualitative analysis of a system. The qualitative analysis allows for the determination of the root causes of a system failure. This is accomplished by identifying the ultimate fault and working backward to identify each step that led to the fault at that level. The analyst works through each level until he or she can no longer identify a cause for that level. At that point, it is determined that a basic fault has been reached. Once the basic faults are identified, the safety professional can then analyze the steps needed to prevent or reduce the probability of that fault occurring.

Through the use of Boolean Logic, a system can also be quantitatively analyzed in order to determine the probability of a fault occurring. The following diagram is an example of a fault tree.

In fault trees, the symbols have specific meanings. A rectangle indicates a fault requiring analysis. The "or" gate means that any of the faults going through it will by itself cause system failure (analogous to series systems).

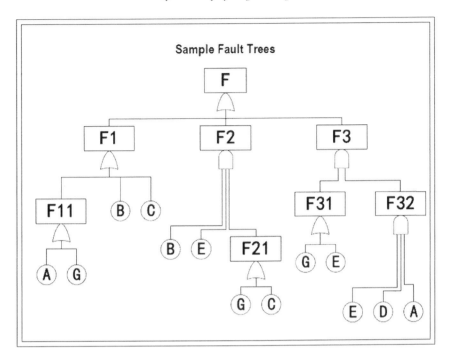

The "and" gate means all the faults going through it must occur before there is system failure (analogous to parallel systems). The "or" gate has a curved bottom and the "and" gate is straight. The development of a fault tree continues until there is nothing but circles at the lowest level of every branch. A circle indicates a basic fault, which requires no further investigation. Safety professionals should be able to determine a fairly reliable probability for the occurrence of basic faults. Additional symbols may be used, but are not discussed here.

Boolean Algebra allows for a quantitative analysis of the fault tree which simplifies the original tree and identifies the root causes of system failure. Using Boolean Logic, mathematically the fault tree depicted can be reduced to its Boolean equivalent in the following manner.

Assume the following probabilities are the probabilities of the individual faults:

$P_{(A)} = .001$ $P_{(B)} = .003$ $P_{(C)} = .002$ $P_{(D)} = 0004$ $P_{(E)} = .006$ $P_{(G)} = .005$

Using the rules of Boolean Algebra, an "or" gate is the same as addition and an "and" gate equates to multiplication. Also, anything times itself is equal to the original value ($A * A = A$) and any value added to one equals one (1 +

A = 1). Once more providing a complete understanding of Boolean Algebra is beyond the scope of this text, but these simple rules should allow one to follow the discussion here. See the references for more detail (Walters, 1996).

To reduce the figure, simply perform the math as follows:

$F = F1 + F2 + F3$
$F1 = F11 + B + C$
$F1 = (A + G) + B + C$
$F2 = (B * E) * F21$
$F2 = (B * E) * (G + C)$
$F3 = F31 * F32$
$F3 = (G + E) * (E * D * A)$
$F = (A + G) + B + C + BE(G + C) + (G + E)EDA$
$F = A + G + B + C + BEG + BEC + GEDA + EEDA$

Factoring the final equation using the simplest number of faults first yields:

$F = A(1 + GED + ED) + G(1 + BE) + B(1 + C) + C$

which is then reduced to:

$F = A + G + B + C$

Using the given the probabilities of the basic faults in the sample fault tree, the probability of a fault occurring could then be calculated as follows:

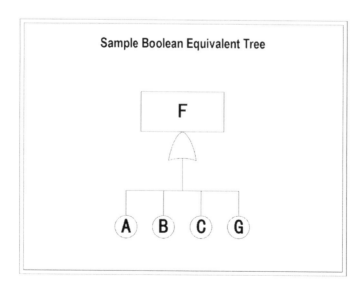

F = .001 + .005 + .003 + .002 = .011

The following figure is the Boolean equivalent of the fault tree depicted. It is easy to see the value of this simplification as one can clearly see which areas need addressing in order to prevent the fault. For additional information on FTA, refer to the references at the end of the chapter.

Conclusion

Systems safety engineering is no longer a concept that is practiced by safety design engineers alone. It is readily incorporated into the overall SMS process. The principles allowing for the proactive, safe design of equipment must be adopted by occupational safety professionals. When examined realistically, the principles are not difficult to adapt to that role.

This chapter has noted some of the systems in place in the workplace and highlighted the general concepts such as the need to be proactive and yet perform different analyses at each stage of a system's life cycle. The failure to conduct these analyses may result in the introduction of unexpected hazards leading to inury. Each analysis serves a different purpose. The reasons for the various analyses and how to perform them were noted with suggestions on how to incorporate the basics to meet individual workplace needs.

The discussion of statistical tools points to the availability and importance of quantitative methods of analyses. The reader is encouraged to take advantage of the numerous references offered below.

REFERENCES AND SUGGESTED FURTHER READING

Brown, B.A. (February, 1993). Simplified System Safety, *Professional Safety*, p. 24.

Brown, B. and S. Hodges. (August, 1997). "A Behavioral Strategy for HAZMAT Safety, Cuts, Injuries, and Incidents," *Industrial Safety and Hygiene News*.

Esposito, P. (December, 1993). "Applying Statistical Process Control to Safety," *Professional Safety*, p. 18.

Ferry, T. (1990). *Safety and Health Management Planning*, New York: Van Nostrand Reinhold.

Fitzgerald, R.E. (June, 1997). "Call to Action: We Need a New Safety Engineering Discipline," *Professional Safety*, p. 41.

Goyal, R.K. (May, 1996). "Cost-Effective Implementation of Process Safety Management in a Refinery," *Professional Safety*, p. 33.

Manuele, F.A (July, 1997). "Principles for the Practice of Safety," *Professional Safety*, p. 27.

Manzella, J.C. (May, 1997). "Achieving Safety Performance Excellence through Total Quality Management," *Professional Safety*, p. 26.

Ragan, P.T. and B. Carder. (June, 1994). "Systems Theory," *Professional Safety*, p. 22.

Roland, H.E. and B. Moriarty. (1990). *System Safety Engineering and Management*, Second Edition. New York: John Wiley & Sons, Inc.

Roughton, J. (August, 1993). "Process Safety Management: An Implementation Overview," *Professional Safety*, p. 28.

Roughton, J. and D.S. Buchalter. (January, 1997). "OSHA's Process Safety Management Standard vs. EPA's Risk Management Plan: A Comparison of Requirements," *Professional Safety*, p. 36.

Rsiwadkar, A.V. (April, 1995). "ISO 9000: A Global Standard for Quality," *Professional Safety*, p. 30.

Stephenson, J. (1991). *System Safety 2000*, New York: Van Nostrand Reinhold.

Sutton, I.S. (1992). *Process Reliability and Risk Management*, New York: Van Nostrand Reinhold.

U.S. Department of Defense. (1984). *MIL-STD-882B 1984: System Safety Program Requirements*. Washington, DC: U.S. Government Printing Office.

Utley, P.T. (September, 1994). "System Safety Hazard Analysis: A Tool for Determining Confined Space Entry Requirements," *Professional Safety*, p. 25.

Vincoli, J.W. (1993). *Basic Guide to System Safety*. New York: Van Nostrand Reinhold.

Walters, H.A. (1995). *Statistical Tools of Safety Management*, New York: Van Nostrand Reinhold.

Wang, J. and T. Ruxton. (January, 1997). "Design for Safety: U.K. Marine and Offshore Applications," *Professional Safety*, p. 24.

Weinstein, M.B. (July, 1996). "Total Quality Approach to Safety Management," *Professional Safety*, p. 18.

Chapter 12

Accident Investigation

Mark A. Friend

CASE

Rhonda Belot is manager of safety at a small company that manufactures cases for computers. Recently a worker was injured while working at her station when a forklift meandered out of the aisle and knocked her off her work stool. The worker was immediately transported to the hospital and sustained non-life-threatening injuries. Rhonda is concerned about this accident and wants to assure it does not happen again. Instead of simply going out and investigating the accident herself, as she as often done in the past, Rhonda decides to enlist the help of various stakeholders in this particular process. Her thinking is that supervisors of forklift drivers and machine operators, based on their experience and backgrounds, can offer insights of which she may be unaware. Rhonda determines that she'll use a representative from each of those two groups, but is a little unsure as to where to proceed from here.

MANAGEMENT'S STAKE IN ACCIDENT INVESTIGATION

In some organizations accident investigation is seen merely as an afterthought, a mere sequel to an accident, mishap, or near-miss. Properly understood, however, accident investigation can become an integral part of an effective and profitable safety management system. The keys are as follows:

1. Accident investigation can be used as an effective *accident prevention technique.*

2. The accident investigations can be used as a *diagnostic tool* to identify faults in the management system throughout the organization.
3. Of all the parties who have a stake in accident investigation, the *senior manager is actually the greatest stakeholder*. It is in his or her own best interest to conduct thorough accident investigations.
4. Knowing what to investigate, when to investigate, and where to locate the investigative function in the organization is critical.

Nearly all actions in any business or organizational operation will be subject to risk, and the possibility of an accident is always present. Even in the best-managed operations or systems, accidents can occur. Management can reduce the risk to as low as reasonably achievable (ALARA), but some level of risk always exists. In June 2015, a tire fell from a plane in the sky and crashed through a roof in Montreal, Canada (CTV News, June, 2015). Was that risk accounted for when the structure was built? Was the roof capable of withstanding the impact of a tire falling from hundreds of feet above? Was it expected and were actions taken ahead of time to prevent damage in the event such an occurrence happened? Of course, the answer is, "No." It was an extremely unlikely occurrence, so no preventive steps had taken place. Sometimes accidents happen regardless of predications or preparations. In this particular instance, no one was killed or injured. Risk cannot ever be completely eliminated. The key is to attempt to determine the areas of greatest risk and reduce those to an ALARA. With a properly conducted accident investigation, the manager can unearth the *causes* of accidents within the system and take remedial action that will not only benefit safety and health concerns, but can actually strengthen the overall profitability of the organization and the effectiveness of the SMS. The point of the investigation is to identify the fault in the safety management system and correct it.

THE IMPACT OF ACCIDENTS

There are few times when management needs help more than it does when there has been a significant (and perhaps costly) mishap on the premises. When an accident has taken place it may seem as if everyone and everything puts pressure on the principals of the company. A few of these pressures are as follows:

- A disruptive event calling for some type of decisive action
- Chaos accompanying the accident
- Work processes disturbed or shut down
- Profit margins shrunk or vanished

- Expensive machinery damaged or lost
- Dollar losses far beyond insurance coverage
- Valuable members of the organization taken out of action
- Bad press coverage damaged the corporate image
- Public confidence in the organization shaken
- Families of the injured need attention
- Stockholders and board members demand explanations
- Government investigations resulted with attendant finger-pointing and time-consuming demands for attention
- Violations of regulations surface
- Doubt cast on other operations in the organization
- Pressure for operational, supervisory, and management changes
- Insurance and workers' compensation costs likely to rise as a direct result of the mishap
- On-the-fence customers and clients take their business to the competition
- Top and middle management have to spend long, valuable hours seeing the affair through investigation and corrective action
- The morale of the whole workforce adversely affected

These are only a few of the management concerns when a mishap occurs. They may not all be present, but they are all *potential* after an accident, and they point to management's need for involvement. A serious consideration of the process of accident investigation, *before the accident occurs*, will ease many of these pressures that can later arise from an accident.

The Cost

Estimates of indirect costs range from two to 200 times as much as direct costs. A conservative ratio is 3:1; that is, the indirect costs run an average of three times the direct costs. In other words, if an accident involved hospital bills and lost wages of $20,000, then one should consider at least $60,000 in indirect costs as an accurate estimate—for a total of $80,000. Indirect costs do not actually come out of company pockets, but if the accident total includes lost production, cost of a temporary replacement, paperwork, damaged property to replace, and lost-sales or services, it costs quite a bit more! Moreover, accidents cost over 130,000 Americans their lives every year (Centers for Disease Control [CDC], 2015). Some businesses have actually been forced to close because of accidents. These closures, in turn, have cost thousands of workers their jobs. It is estimated that roughly 1.1 million days of work are lost due to industrial accidents each year (Bureau of Labor Statistics [BLS], 2014). The fact is that business cannot afford the cost of accidents.

Management Inefficiencies

Allowing accidents to occur is a costly way of doing business and reflects poorly on the SMS. There is little justification or reason, whether on moral or just sound business grounds, for not preventing mishaps to every extent possible. Mishaps signal there is something wrong with the way their organization operates, and accidents are evidence of management malfunction at some level of the operation. This strongly infers that management's inefficiencies are powerful contributing factors to accidents.

It may take many such inefficiencies, even hundreds or thousands, before a recordable mishap occurs. In fact, management inefficiencies may *never* result in a recordable accident, but these inefficiencies are costing the company money day in and day out. Without an accident to bring them to management's attention, they can go on draining company assets and diminishing profits for years. Mishaps can serve as a sign of inefficient operation, bad business, and poor operating practices, giving management a sharp focal point for examination. A solid accident investigation will uncover, among other things, management inefficiencies which, if corrected, will not only lessen the opportunity for similar mishaps, but will result in more efficient operations.

Morale

After an accident at work, people generally stand around and talk for a while, returning to work only slowly and reluctantly, subdued and very much sobered by the event. Their work pace slows down, and they are wrapped up in their own thoughts and concerns. In some situations, the business may close early, or people will go home before quitting time. Morale is shot, and employees don't feel like going ahead with the job. Both their interest and productivity drop. The effect often results in a chain reaction, spreading to other departments or activities. When the activity does continue, another mishap may quickly occur, since people are distracted and don't have their minds on the tasks at hand. Even if no second mishap occurs, production and activity drop noticeably. If the mishap also calls attention to a bad or hazardous work situation, there may be far-reaching results, even beyond lost work time or lowered production. Low morale may involve worker complaints or refusal to work. Teamwork may fall to a low point, as each individual is preoccupied with the accident. While lowered morale resulting from mishaps cannot easily be measured, its presence is definitely felt, both on the job and in the profit picture. Good mishap investigation can be a vital part of management's efforts to restore morale. Where workers are aware of management taking immediate steps to track an accident to its sources, many of the less obvious effects of lowered morale are eased. Rumors can be quickly dispelled, the record can be set straight, worker fear or concern over hazardous conditions can be relieved—even the process of answering specific questions about the accident can help the worker re-focus attention and dismiss the accident from the employee's mind, confident that "something is being done" about it. Properly conducted, an immediate accident investigation demonstrates management's interest, concern, and good intentions. Even more importantly, it can prevent a similar mishap in the future that could further damage an already-demoralized workforce.

Public Concern

Every few months some spectacular accident hits the headlines—an aircraft crash, a school-bus collision, a fire in a senior citizens' home, a mishap in a mine, or a broken dam. These events have repercussions far beyond the effects or people immediately involved. They generate a high degree of public interest and concern, and often bring adverse public opinion to bear on the organizations and operations involved. There may be cancelled aircraft sales, demands for a new school-bus design, public demand to bring criminal charges, replacement of government officials, calls for indemnity payments, or demands for higher regulatory standards, all as a result of public concern. Whether this concern is appropriate or totally inappropriate, it does get action—usually costly action that might well have been avoided.

On a smaller scale, thousands of times each year, there are local mishaps that generate the same kind of action, stemming from a loss of public confidence. The public may become antagonistic toward an organization. Once adverse public reaction builds due to a mishap, it may take years to overcome, and it has been known to put some companies out of business entirely. Public concern is a strong weapon against those whom it considers responsible for an accident, particularly if they represent an establishment already subject to public doubt or suspicion.

WHO ARE THE STAKEHOLDERS?

Who has a stake in accidents? The most concerned stakeholder of all, naturally, is the victim, whether individual or corporate. On the corporate level, the prime stakeholder is the accountable executive, who must somehow cope with the consequences an accident may have on the organization as a whole. Often this person has the most direct stake in uncovering the causes of an accident, because the executive must review the investigation and apply the findings in some way. In the case of an industrial concern, the stakeholders may be those who carry out investigations, such as the supervisor/foreman, line manager, safety professional, staff manager, special committee, or outside consultant.

Supervisor/Foreman

The supervisor/foreman is the closest to the action. Since the accident takes place in the supervisor's domain, he or she is often involved in the investigation of the mishap. In large corporations the supervisor or foreman may conduct only those investigations involving minor injuries or small losses of resources. Whether or not the supervisor actually conducts the investigation, he or she should be involved in furnishing data and information. The supervisor may also prepare all or some of the reporting forms. The supervisor/foreman seldom has specialized investigative training, but if is the supervisor's job to investigate, he or she should be prepared.

It is reasonable to expect the supervisor to investigate mishaps occurring in his or her department. The supervisor's personnel, equipment, and operations were involved, and departmental operations were being carried out when the mishap occurred. The supervisor likely knows more about the operations than anyone else and may be considered responsible for the mishap in this area. He or she may, in fact, face reprimand or demotion if mishaps occur too frequently. The supervisor definitely has a stake in determining and correcting the causes of accidents. From another perspective, the very reasons that make the supervisor/foreman the logical person to conduct the investigation are also reasons why he or she should *not* be involved. The supervisor's reputation is on the line. The causes uncovered may reflect adversely on the effectiveness or method of operation. The closeness to the situation may preclude an open and unbiased approach. The more thorough the investigation, the more likely he or she is to be implicated as contributing to the event. While it should never be the objective of an investigation, some stigma may invariably be involved. For that reason, it requires all the integrity and objectivity a supervisor can muster to carry out an internal investigation.

Regardless of who actually performs the investigation, the supervisor/foreman should be informed of the findings. To keep the event from being repeated, the investigator must turn in a thorough report with firm findings and recommendations for specific correction. The supervisor must carry out most of the recommendations and can do more to prevent a recurrence than anyone else, since he or she controls the day-to-day operations. The supervisor can also communicate directly with department personnel to assure corrective action or compliance with rules.

Finally, the supervisor/foreman is a major stakeholder in the smooth and efficient operation of the department. Correcting hazardous conditions and operational errors uncovered by a thorough investigation is essential to the welfare of immediate operations. It assures increased production and reduced operating costs, and also demonstrates control of operations. In short, the supervisor is in a better position than anyone else to prevent a recurrence.

Line Manager

The *line manager* in this section refers to the person in the management chain of command who has authority at least one level above the front-line supervisor. The line manager deals with the supervisor/foreman, rather than the workers, and can be considered middle management. Since this line manager's personnel and operation are involved, he or she has overall authority to take corrective actions on deficiencies uncovered by an investigation. It is to the line manager's advantage to ensure mishaps are promptly investigated and corrective actions quickly implemented.

The line manager may merely review the investigations carried out by supervisors or foremen but must be able to recognize when a thorough job has been done, as well as to tell when findings are credible. The line manager, as much as the supervisor, has a record riding on a good investigation, so the same drawbacks apply here as to the supervisor. The deficiencies appearing as the cause of an accident may be the line manager's, rather than the supervisor's.

Safety Professional

In an organization large enough to have a full-time safety professional, this person should be considered the organizational expert on accident investigations. While the supervisor may actually conduct most of the investigations, the safety professional has a major role in reviewing accident reports for adequacy, thoroughness, and—most importantly—assurance the appropriate corrective actions are suggested and carried out. The safety professional's role in evaluating investigations includes maintaining an advisory relationship with top management. As desirable it may be to keep the safety manager in the role of evaluator, there are appropriate times for this individual to conduct the investigation.

The safety professional is more likely than anyone else to have developed investigative skills, particularly from special training as a part of professional education. If the supervisor receives special training in investigation, the safety professional will often be the instructor. When the supervisor does not have adequate time for investigative duty, the safety professional may have to do it. There are also instances in which it would be preferable to have the safety professional, rather than the supervisor or line manager, perform the investigation. The safety professional should make the investigation when

- There are implications of high public interest.
- There have been a series of similar mishaps.
- A complex case calls for investigative skills beyond those of the supervisor.
- There is a considerable potential for litigation.
- There has been a loss of life or large resources.
- There is likely to be an investigation by OSHA.

Another reason for having the safety professional investigate accidents is that he or she is not normally a stakeholder; that is, there is no personal, vested interest in the mishap or the findings. The safety professional can often assess facts and information free from bias, prejudice, and the pressures of responsibility for operation. This advantage may be reduced when workers or supervisors fear that the safety professional is too management-oriented or when the findings of an investigation may reflect on his or her own competence.

Staff Managers

Staff managers are at the middle management level and serve in an advisory capacity. Examples are the personnel director, medical director, labor relations manager, or the supply/logistics manager. These persons are seldom involved in line operations, but their functions are often involved in the causal factors of accidents, so they should be familiar with the investigative process. For example, if mishaps are occurring because workers lack the physical ability to handle the task, it might be necessary to review the worker selection process with the personnel and medical directors. Should it turn out that faulty materials or equipment are involved, the purchasing manager may be a stakeholder. Since they are far removed from the front line of first-level supervision, there may be a tendency to overlook staff managers. They can play a critical role in resolving the causes of mishaps and may serve as valuable experts on whose knowledge the investigator can draw.

COMMITTEE INVESTIGATIONS

Investigating committees typically fall into two categories:

1. General or standing safety committees investigating mishaps as one of their normal functions.
2. Special investigative committees appointed for a particular investigation.

The organization's general safety committee is often charged with investigating accidents when there is no one else trained or available to do the job. In a smaller organization, mishap investigation falls naturally within their scope. This can work to the organization's advantage where committee members serve for extended periods. It allows members to develop wide experience and knowledge of plant-wide activity and in applying investigative methods. In many organizations where the safety committee performs the investigative functions, special training is provided in investigative techniques and strategies.

Since they operate on an organization-wide basis, members will conduct more investigations than if they were investigating only their own departments, further sharpening their investigative skills and experience. In addition, the committee approach has the advantage of bringing people with diverse backgrounds and knowledge together to examine a specific problem associated with a mishap. Since they do not represent just the department involved, they have no vested interest and can be more objective and unbiased. Committee findings also tend to be better received than those coming from representatives of the management/staff function, particularly where committee members are drawn from the supervisory and worker levels.

Even in organizations where investigations are routinely handled by other persons, special investigative committees are sometimes appointed after unusual or very serious mishaps. The very nature of a particular incident may call for special expertise or knowledge (as in the case of fires, mishaps involving radiation or explosives), in order for an in-depth investigation to be made. This is particularly true where public concern, liability, workers' compensation, or regulatory violations may be an issue. Where consequences of a mishap are particularly grave, only a special committee may be considered objective enough to perform a thorough and unsparing investigation.

As with general safety committees, a special committee's findings may be better received, particularly when fellow workers or supervisors appear to be involved as causal factors in the accident. Committee findings may also receive a better reception when the investigation results are made public or reported to the workforce at large.

Care should be taken in choosing the members of a special committee, for several reasons. First, it is essential to secure members who actually have the appropriate expertise or knowledge. Second, they should be generally respected by workforce and management and should not be regarded as representing one viewpoint exclusively (i.e., a committee comprised only of members of the top management, supervisory or worker levels).

Committee action also has certain disadvantages. It takes people away from their normal routines to do what some consider the supervisor's job.

Occasionally, committee members will resent the system, feeling it makes them perform the unpleasant task of finding fault with their peers. There is also the danger of unwise compromise to reach consensus, and this may simply obscure essential points.

OTHER INVESTIGATIONS

It may not be necessary for the organization to perform all its own investigations. An insurance carrier, using its own investigators, may investigate certain mishaps, as in the case of incidents involving boilers, vehicles, or fires. In some instances OSHA or other government agencies will conduct an investigation. In either of these cases, facts needed for other purposes can be taken from their reports. The manager truly interested in using an investigation as an accident prevention technique may find that the insurance company or government agency investigation is not adequate, or that it concentrates on issues other than management inefficiencies or control systems. An in-house investigation may still be desired, if only to supplement the outside work.

PRIORITIES FOR INVESTIGATION

There is no question companies should investigate matters of great public concern, serious injury, and high property damage or fatalities. Where organizations fail in this regard is in making a sufficiently detailed investigation to aid in preventing similar incidents, or in applying the findings to correct the underlying causes.

The term *mishap* indicates all of those unexpected, undesired events (many of which are referred to as *accidents*), that cause a loss of resources. There are various definitions of *accident*. For example, state workers' compensation laws may define it quite narrowly as an event both sudden and unpredictable. The firm's auto insurance carrier may operate on a broader definition. Government regulations often refer to the dollar loss involved (i.e., the dollar loss determines whether or not the event is considered a reportable and investigatable accident). Many firms, in fact, conduct investigations only if required by law. This is short-sighted and may end up costing the organization enormous amounts in the future. Say, for instance, the regulations define an event as a reportable accident causing $500 worth of loss or damage. If the loss is only $499, the company reasons an investigation is not really required. This reasoning contradicts the whole purpose of the investigation. All of the same causal factors present might later cause a $100,000 accident. Had it not been so intent on that one-dollar difference, management might have conducted an investigation that would have prevented the second, costlier event.

A dollar value may not be the proper basis for determining if an investigation is required, but by the same token, neither is an injury. For each recorded injury, there may be hundreds of similar mishaps that caused no injury at all. The same causal factors are still present. The same mistakes of management and supervision are there, as are the same operator errors and design deficiencies. The ideal approach is to investigate all undesired, out-of-the-ordinary events that prevent doing the job efficiently. They deserve the same careful scrutiny that a loss of life or a serious injury would bring. In fact, investigating these apparently minor mishaps is an efficient and profitable way to prevent the very costly accidents that may eventually result from the identical causes.

James Reason (18 March 2000) discusses how defenses, barriers, and safeguards often work to prevent accidents, but there may be gaps in each. In an ideal world, the each defensive layer always works, but each is like a piece of Swiss cheese. It may work most of the time, but given enough exposures, occasionally the holes align, and an accident occurs. Anytime there is a gap in a defense, unless there is a 100 percent fool-proof, backup system, an accident can occur. Those holes must be identified, and the faults in the system corrected, before they align again.

Organizations should, therefore, investigate any mishap that may reveal causal factors possibly leading to reportable events. Where resources do not permit an investigation of all mishaps, selective sampling may be useful, as might the investigation of mishaps with the greatest potential for damage.

Senior managers sometimes question the wisdom of investing resources in the prevention of insignificant events, rather than concentrate on costly or catastrophic events. Since investigation is a preventive technique, they reason, inquiring into minor mishaps is not a profitable use of resources. If even the minor mishaps require thorough inquiry, the manager may be doubtful as to what constitutes a "thorough" or "proper" inquiry. When should there be a detailed investigation? Consider the following:

1. If the operating environment involves high hazards, the prevention efforts (investigation included) deserve a corresponding high investment of resources. High-hazard operations involve the same causal factors as low-hazard operations; only the size of the probable loss or the likelihood of mishap is different. Even the low-hazard operations can involve very costly resources and deserve careful attention.
2. Related to that is when the operating environment is one of potentially high-severity accidents. Even though the probability may be low, every

effort must be made to prevent high-severity accidents from occurring. A good example is in the field of aviation. Although the probability of failure on a given flight may be low, failure and accidents are not acceptable. Even minor incidents must be investigated to prevent future problems.
3. What seems to involve minimum loss potential is often deceptive, ultimately involving very high losses. For example, the relatively common back injury or eye injury is a case in point. This may be given scant attention because of its very simplicity and the fact that extended absence from the job is either rare or is fully covered by workers' compensation. In reality, these are two of the most expensive injuries and spell great hidden expenses for management. The total costs of lost production, increased insurance premiums, training replacements, worker rehabilitation, and other benefits for a single worker disabled by a back injury may be equal to that of several deaths. Careful consideration of all reported incidents must be given. Many organizations establish parameters to determine whether a given incident warrants investigation. Parameters are sometimes based on actual dollars expended, but that can be misleading. Consideration should be given to potential costs associated with a type of mishap.

Sophisticated investigators and managers realize that even common, minor events seemingly causing only inconvenience, irritation, or delays are symptoms of inefficient management and operations. They will consider the investigation of each as simple good business. Not only can the elimination of minor problems bring major safety and health benefits, but also become a means whereby management can achieve greater overall efficiency and profitability.

The astute manager will recognize all these factors, and will frame investigative to account for them. It must be recognized that a senior executive is a top stakeholder in accident investigation, and that it is in his or her own—and the company's—best interests to make accident investigation a key part of the overall safety management program.

HOW TO ENSURE A QUALITY INVESTIGATION

Management's main concern is ensuring every accident investigation is a good one, thoroughly and properly conducted. There are three keys to successful investigations:

1. Being thoroughly prepared for investigations *before* the accident.
2. Knowing how to gather and analyze the facts surrounding an accident.

3. Having a good accident report that can serve as a basis both for investigation and for corrective action.

One of the perennial management problems is that there never seems to be enough time, personnel and expertise to perform really strong accident investigations. Regardless of available resources, the following twelve steps, if carefully considered, can ensure a quality effort:

1. Understand the need for investigation.
2. Prepare for the investigation.
3. Gather the facts about the accident.
4. Analyze the facts.
5. Develop conclusions.
6. Analyze conclusions.
7. Make a report.
8. Make appropriate recommendations.
9. Correct the situation.
10. Follow through on recommendations.
11. Critique the investigation.
12. Double check the corrective action.

Only steps 3 through 6, respectively, are properly in the domain of the accident investigator. Steps 1, 2, and 8 through 12, are strictly management's responsibility. Note that Step 7 (reporting) is a gray area. Even when an expert investigator is available, he or she may simply be using opinions and facts placed in a report by someone else. In fact, the supervisor or foreman generally fills out the accident report form, a task that does not often call for great skill or depth, and may be performed in the most routine and perfunctory manner. By breaking down the investigative process into these twelve steps, the senior manager can tell when the organization is making the best use of the available investigative resources. Also, these steps allow the manager to evaluate the investigative work of subordinates; make valid observations; and take effective, corrective actions. *The thoroughness and effectiveness of any accident investigation is far more dependent on management actions than on an accident investigator gathering factual data and arriving at conclusions.* While the manager cannot do without the investigator, the efforts of the investigator will be wasted if management is not a part of the total process, from advance preparation to follow through.

In actual practice, the manager may discover the twelve steps are not clear-cut or readily separated. When an accident occurs, the steps may run together, overlap, and be carried out simultaneously. The need for preparation naturally comes first, with advance preparation for an accident investigation made before the fact. After the accident occurs, the fact-gathering process begins. At the same time, the reporting—and part of the analyzing—are already underway. In fact, sometimes four or more steps may be under way simultaneously. By understanding what is involved in each of the separate steps, the manager can still assure that each will be properly undertaken, carried out, and completed.

Step 1. Understand the Need

What exactly is the perceived problem or need? Is it to prevent future accidents? To comply with reporting requirements? To support the company in the event of litigation? To find a scapegoat? The purpose should be to determine faults in the system that permitted or even encouraged the mishap to occur. The nature of the findings will actually be determined largely by the need, so it is critical to understand it. It is natural that different findings will emerge for different purposes. For example, if investigating the faulty performance of a company product, one may pay closer attention to what has taken place in engineering or quality control than if investigating an operator error.

Accidents are a result of circumstances, sometimes repeated again and again, with no evidence of consequences. The most important reason for investigation is that accidents indicate something is wrong with the management system. Uncovering the cause(s), and remedying the defects can strengthen the system and eliminate future accidents triggered by a particular fault. The accident investigation becomes management's tool for discovering and analyzing faults and inefficiencies in the entire management system.

A critical error in accident investigations is blaming the perpetrator. This is referred to as the *pilot-error syndrome*. The pilot may be blamed, and in the case of airplane accidents, the pilot may be dead and cannot defend him or herself. As soon as the investigator blames the pilot, management and the system are off the hook and there is no need to continue the investigation. The investigator must maintain the mindset that the employee(s) always does the rational thing. It may not seem rational to the investigator, but for whatever reason from the employee perspective, it seemed like the thing to do at the time. Whether the act was a result of horseplay, taking a shortcut, or simply breaking a rule, what in the system permitted the employee to think this behavior could be condoned or would ultimately be hidden from management? Why did it seem like the thing to do at the time? Did the employee believe getting the job done quickly was more important than safety? What was the employee thinking and why did the employee believe this thinking and the related actions would be tolerated? These questions must be answered and every attempt made to correct a system that allows or even encourages employees to take shortcuts or break safety rules.

Step 2. Prepare for the Investigation

After the unexpected happens, it is too late for preparation; a plan (sometimes called a pre-accident plan, or preplan) is needed. This preplan can range from the very simple to the enormously complex; however, it should never be so complicated that it becomes a hindrance to prompt action. The briefer, the better. There are four essential elements in the preplan:

1. An alarm system to let certain parties know there has been an accident. Normally, one phone call to a person who will notify the critical parties activates the plan. Based on the incident reported on the phone, the person receiving the call will work through a list to contact all pertinent individuals.
2. A plan to have needed fire and rescue people on the scene as soon as possible. This often requires coordination with local agencies in advance. They should be aware in advance of particular hazards they are likely to encounter when entering the scene.
3. Procedures to preserve evidence and property at the scene.
4. Immediate notification of the investigator, so investigation can get underway promptly. The sooner the investigation starts, the better—while evidence is still intact and memories of the event are relatively fresh.

Any pre-accident plan will have some holes in it, and dry runs or tests are useful. Test of the preplan will ensure that all parties know what to do and when and how to do it. It will also determine whether the plan itself might expose someone to injury, or whether any trouble spots or breakdowns are inherent in it. Test whenever there have been changes in the plan itself, or when several key personnel have been changed. Also, if the plan has not been used for a while, a test can serve as a refresher course, as well as a checkup.

A critical part of the plan is recognizing that those who respond may encounter hazards, and to take steps to counter those hazards. For the investigator, this includes hazards likely to be encountered at the accident scene: broken machinery, fumes, leaks or spills, possible fire or explosion debris, or damage. Far too many people are hurt at the accident scene because there were no plans to counter the hazards.

Too often, the investigative personnel, who may be on the scene immediately after an incident, are poorly prepared. Plans, then, should provide for safety at the accident site. Not only do rescue personnel or investigators get hurt in responding quickly to a call, but helpers and onlookers can also get hurt. A quick analysis of the hazards at the site should be part of the plan. All responders must be appropriately trained and all regulations must be obeyed. For example, any responder to a hazardous material spill or leak must have training under 29 CFR 1910.120. Depending on the actions the responders are required to take, the training may be relatively extensive. In some cases the local fire department may call an outside team to handle hazardous materials incidents. The fire department may also require the company to handle hazardous materials spills internally or through an outside contractors. Anticipated incidents of this type should be coordinated with local agencies, so everyone is aware of and prepared for problems of this nature.

Someone should be in definite charge at the accident site. This may be difficult, particularly within the confines of a plant. The problem is usually one of overlapping jurisdictions, since different agencies may be operating out of their normal territory or situation. The investigative effort crosses all lines, so plans need to be made for this situation, clearly spelling out who is to take charge and making certain all involved personnel are aware of the chain of command in response to an accident.

The preplan should also contain arrangements for funding or support services such as lab work, transportation, witnesses, chemical analyses, etc. The availability of independent consultants should be established in advance, particularly where the firm is involved with exotic substances or processes that may require expert investigation. There should be pre-established procedures for reporting mishaps to corporate headquarters, families of victims and, where advisable, to the news media. All media interactions should occur through an authorized, organizational representative who is well prepared to handle the public-relations aspects of an accident. Considerations relating to the release of information to OSHA or any governmental authorities should be spelled out. The preplan should also specify reporting methods used, both with internal management and any external interests.

Step 3. Gather the Facts

While Steps 1 and 2 should be completed well in advance of any actual accident, Step 3 begins only when an accident has actually taken place. It marks the beginning of the actual investigation. First and foremost in investigation, the facts must be discovered—the facts about what happened and what allowed it to happen. Facts can be gathered more quickly and efficiently if the investigator is organized and prepared for the investigative action, as with an effective preplan. He or she must know where the investigator is going, what the investigator is going to do and how to do it, regardless of the specific situation to be found at the accident site.

In gathering facts, the investigator should first take care of transient evidence—that is, evidence likely to be removed, rained out, snowed on, or obliterated by crowds. The investigator should get the names and numbers of witnesses likely to move on, so they can be called later. Photos of marks, debris, and evidence not likely to be visible or in place later should be taken. Sketches and measurements should be made, so essential information will be a matter of record. Most permanent evidence can be addressed later. This includes witness statements, examination of records, laboratory tests, detailed examination of evidence, more exact measurements, and evaluation of procedures and operations.

The details involved in gathering facts are simple if a systematic, clear procedure has been established, detailing what kinds of facts should be collected and how they should be compiled. These simple procedures may have to be repeated for each new relevant fact or phase of investigation. If the organization of this phase is well done, even complicated accident investigations are far easier. The best approach is having a good plan, knowing the on-site priorities, and knowing where to start. As the facts are gathered, the investigation then moves naturally into the next phase. Investigations typically seek to discover the who, what, when, where, why, and how of the accident. A series of questions involving each of those variables is answered, either through interviews or discovery of evidence.

Step 4. Analyze the Facts

This is an ongoing process beginning when the first facts are gathered and statements of witnesses are considered. What is their credibility? How much accuracy or inaccuracy can be expected from them? How do their stories compare with other evidence, on-site or laboratory findings, and with each other? What do the facts indicate about management and supervisory processes, housekeeping, work environment, and so on? This analysis merges with the gathering of facts, and may suggest new questions and offer new direction for further fact-gathering.

Organization and direction can be given to this step if everything is carefully documented and all actions are made a matter of record. While senior management is not typically involved in this step, these documents and records can enable the manager to assess how thoroughly the fact finding was done, and thereby assess the overall quality of the investigation.

Step 5. Develop Conclusions

As facts are gathered and analyzed, the investigator can begin to draw conclusions about what happened and what caused it to happen. There are numerous approaches to determining the cause(s) of an accident. The 5-Why technique asks a series of *Why* questions to determine the primary causal factor in an attempt to ultimately correct it (Six Sigma, n.d.a). This is done within the context of the other W questions and how. As each question is asked, the correct response is, *Because* . . . Following an injury, the question may be asked, "Why did the operator back over an employee's foot?" The response may be, "Because the operator wasn't paying attention." So, "Why wasn't the operator paying attention?" Because, "He was distracted by an alarm sounding in the vehicle." So, "Why was the alarm sounding in the vehicle?" "Because. . ." This process continues through the asking of the 5 *Whys* and the appropriate *Because* responses in an effort to determine the root cause(s) (Six Sigma, n.d.a).

Some investigative teams utilize the *fishbone diagram* to make a final determination of causes. Factors considered include the people, material, job or task, environment, and machine or equipment (Six Sigma, n.d.b). Each of these variables should be scrutinized before any final conclusions are drawn.

A determination of what happened and what caused it to happen should be formally presented, with statements of conclusions and the relevant facts on which they are based, as a matter of verifiable record. A formal and systematic presentation will permit the investigator to see any gaps in knowledge or reasoning, and may point to areas where more facts and analysis are needed. It will usually direct the investigator back to either the fact-finding or analysis process to repeat these steps until adequate and reasonable conclusions can be formed. The ultimate goal in this step is to find whatever fault in the overall system permitted or even encouraged the accident to occur. If no fault is found in the system, or if blame is simply placed on an employee, no overall changes are likely to occur. The investigation was a waste of time and resources. If human behavior by an employee or a member of the management team is to "blame," the questions must be asked, "Why was this behavior deemed appropriate or acceptable by the individual?" "What can be done to ensure this behavior does not occur again?" "What changes will occur in the overall management system or safety management system?" These questions are necessary because accidents must always be considered a fault in the system.

Step 6. Analyze Conclusions

This analysis may refer either to tentative or partial conclusions, while still gathering facts or to final conclusions after all facts are known, weighed, and analyzed. This is a continuous step that never really stops but may be repeated several times during the investigative process. As time was taken to formally develop tentative conclusions, they can now be examined and analyzed. This, too, may send the investigator back to earlier steps to gather more facts or to review them from a different perspective. Eventually, tentative conclusions can either be made firm or discarded altogether.

Step 7. Formulate a Report

The accident report should bring all the material together—facts, analyses, and conclusions. The information has been gathered on the people involved, the situation, or specific incident. The facts have been reviewed and analyzed and conclusions have been advanced, reviewed, and revised. Now all the information should be formalized in a report. Everything included in the report should be supported by facts and evidence; unsubstantiated statements or mere speculations do not belong here. The report narrative should begin with a short synopsis, one or two lines that tell what happened, without detail or causes. A good example would be, "On the afternoon of November 5th, in Building Number 3 of the Fairmont plant, a grinding wheel exploded. The operator was seriously injured and the foreman who was watching the operator received minor cuts on the face."

That is an adequate synopsis. It should be followed by a detailed and more complete account. In selecting detail the report should stick to essentials and leave out anything not leading to an understanding of the incident or conclusions. A good report should be clear and concise, free of extraneous material. The reader may know nothing of the process or location involved and yet must understand clearly what has happened and why. The reader is not interested in all of the work the investigator has done or questions asked, but rather in what significant was found, as supported by the evidence. Diagrams, charts, and appendices should be avoided unless they are essential to understanding the event and conclusions. Omit photographs showing the accident site but revealing nothing at all about what took place there. Some investigators include complete texts of regulations, complete laboratory reports, or charts, even when they show no causal factors. This may make the report seem more impressive, but it does not add to its validity. Before the final report is submitted to management, the investigator must consider the possibility that it could be subpoenaed by a government agency or by the court in the event of litigation. All conclusions must be based purely on factual evidence.

Senior management should not have to waste time reading material that does not contribute to an understanding of the event. Worse, an overly inflated or complex report may be only lightly scanned, thus obscuring the real and vital conclusions. The one or two relevant facts may be so buried in a welter of minor details that even a careful reading might not help. The report may also be used as a source document for corporate-wide data gathering, so pertinent facts and conclusions should be easily found and retrieved.

Step 8. Make Appropriate Recommendations

Recommendations are probably the most important part of the report. The best report in the world has failed if it merely states facts and draws conclusions. Corrective actions are needed, and the report should recommend them. It should start with single, specific recommendations. That is, each recommendation should cover just one item, spelling out precisely what should be done to correct the situation. Thus, the report may present a list of several recommendations, each one separately stated, along with its specific required actions. This allows management to take specific corrective actions. From this list of recommendations, the manager can say to someone, "This is your responsibility. This is what you need to do. Do it." In this way, one person can be held responsible and accountable for the correction of each specific causal record.

In most investigation, numerous causal factors will surface—each subsequently paired with a recommendation for corrective action. A thorough job, using solid techniques, may discover even more. In one thorough investigation of what seemed to be a simple event, thirty-seven specific recommendations for corrective action were developed. If only one or two recommendations for corrective actions are made, there is a strong probability the investigator was not thorough enough.

Step 9. Correct the Situation

A good investigation should result in action, correcting the causal factors that allowed or encouraged the accident to take place. Mere recommendations in the report are not enough. Someone in top management should insist the corrective actions be taken as recommended and personnel will be held accountable for those actions. A record should be made of who is assigned to perform corrective actions for later follow-up.

Step 10. Follow Through on Recommendations

Management cannot rest after recommendations for corrective actions have been made, or even after responsibility for taking those actions has been assigned. Someone has to check on the corrective action to make sure it has been taken, and it does, in fact, match the recommended or desired action. Has it been done properly and completely? Have all recommendations been acted upon? Were undesired changes made in the process? Follow through should include data collection so all accident reports can be filed, analyzed, and interpreted for later management decision-making.

Step 11. Critique the Investigation

Oversights and omissions may occur during an investigation, but the pressures of completing the job often preclude taking action at that time. There may be flaws in advance planning or omissions in gathering facts or analyzing them. There may be a variety of errors, particularly in emergency situations. While these problems are still fresh in everyone's minds, before the next accident happens, the investigative process itself should be critiqued, taking action to keep these slip-ups from reoccurring. This may be the task of the safety professional, but implementation will require full management support.

The little things that caused the big problems to occur must be identified. For example, the wrong number was posted for the fire department, too many people were allowed to remain at the accident scene, the supervisor did not know she was supposed to call a member of the safety committee who would begin the investigation, or the digital camera used for on-site photos had been borrowed by the personnel department. These are the type of things a critique will bring out. Parties to the investigation should be brought together for this step, since the safety professional will not know all of the problems encountered.

Step 12. Double Check the Corrective Action

A double check ensures every item was thoroughly addressed and recommended change is continued and enforced. If supervisors and workers revert to the old systems out of habit, a one-time performance is not enough. The double check ensures everything possible has been done to make every future investigation successful and profitable.

MANAGEMENT OVERVIEW

Organizations cannot always afford the expenditure of resources for a large, full-scale investigation. They can, however, make any investigation as effective as possible by always including the twelve steps. Resources can and should be tailored to the size, nature, and type of investigation, but the twelve steps should always be retained. A large-scale accident involving a fatality, high costs, or high public interest will naturally warrant the greatest investigative investment. Not only will it require a large expenditure of resources, but it should be the most carefully followed up, particularly where corrective actions are concerned. Reports of such investigations should go directly to top management. Where serious injuries have taken place or valuable property has been damaged, a full-scale investigation might be warranted, but not quite as extensive as that involving fatalities or the public. The aim is to develop a report indicating further actions to be taken. When injuries or damage are slight, a small investment of investigative resources may be adequate by providing information and taking appropriate corrective action.

Finally, there is the routine investigation that does not justify the use of many resources, but still must be done, as a matter of record. It may consist of a simple accident report filed by the supervisor, with the entire investigation performed in the process of filling out the report or questionnaire. To some extent, even one of these should still include all twelve of the steps.

Too often, senior management considers accident investigation a no-win situation, a case where good money is spent looking into an event already costing too much. Management may even consider accidents as "a cost of doing business," expected and unavoidable. The most vital role in accident investigation is *preventing costly future accidents.* By pinpointing weaknesses and inefficiencies, malfunctions and unsound practices, a good investigation can save funds. Recommendations for corrective actions may give management valuable suggestions for tightening overall management control and efficiency. When accidents occur, they are always considered a fault in the management system and changes must be made to that system to ensure they do not occur again.

REFERENCES

Accident/Incident Investigation Manual. (1976). Washington: National Technical Information Service (NTIS).

Benner, Ludwig E., Jr. (1979). *Four Accident Investigation Games*, Oakton, VA: Lufred Industries.

Brassard, M. and Ritter, D. (2013). *Safety Management Systems Memory Jogger 2.* Salem, NH: GOAL/QPC.

Bureau of Labor Statistics (BLS). (2014). *Nonfatal Occupational Injuries and Illnesses Requiring Days Away from Work, 2013.* Retrieved from: http://www.bls.gov/news.release/pdf/osh2.pdf.

Centers for Disease Control (CDC). (2015). *Accidents or Unintentional Injuries.* Retrieved from: http://www.cdc.gov/nchs/fastats/accidental-injury.htm.

CTV News. (June, 2015). *Jet Tire Crashes through Kitchen Ceiling of Montreal Home.* Retrieved from: http://www.ctvnews.ca/canada/jet-tire-crashes-through-kitchen-ceiling-of-montreal-home-1.2438323.

Ferry, Ted S. (January, 1981). "Accident Investigation and Analysis: A Dozen Steps for the Safety Professional." *Professional Safety*, Park Ridge, IL: American Society of Safety Engineers.

Elements of Accident Investigation. (1978). Springfield, IL: Charles Thomas.

Jefferies, J.W. "Accident-Injury Loss Sequence," from *the Journal of the National Safety Management Society.*

Lederer, J.C. (April, 1980). Personal Correspondence.

Modern Accident Investigation: An Executive Guide. (1981). New York: Wiley.

Nertney, R.J. and Fielding, J.R. (1976). *A Contractor Guide to Advance Preparation for Accident Investigation.* Washington: National Technical Information Service (NTIS).

Pope, W.C. (1975). *Manualizing Your Safety Function.* Alexandria, VA: Safety Management Information Systems.

Reason, J. (18 March, 2000). *Human Error: Models and Management. British Medical Journal.* Retrieved from: http://www.ncbi.nlm.nih.gov/pmc/articles/PMC1117770/.

Six Sigma. (n.d.a). *Determine the Root Cause: 5 Whys.* Retrieved from: http://www.isixsigma.com/tools-templates/cause-effect/determine-root-cause-5-whys/.

Six Sigma. (n.d.b). *Fishbone Diagram.* Retrieved from: https://www.moresteam.com/toolbox/fishbone-diagram.cfm.

Chapter 13

Financial Aspects for Safety and Health Professionals

Mark D. Hansen

CASE

Jeremy Butler was recently hired as the safety director for ABC Weaving, a textile manufacturer. He isn't on the job long when he notices ongoing sound levels near many of the workers were in the 95–100 dB range much of the day. After considering the problem, he learns that he can permanently fix it for the whole facility with an initial investment of $1.5 million plus annual expenditures of $100,000. When he presents this to his boss, she immediately denies his request for funds. Her explanation is that marketing has come up with a great new idea that will cost the company a one-time fee of $2 million and return $200,000 per year. Jeremy is stumped and doesn't know how to proceed.

INTRODUCTION

Safety and health professionals today are pressed with understanding the financial concepts to determine the impact on their companies. These demands have grown more prominent over the recent years. As safety and health professionals attempt to integrate into the business function, understanding these concepts is vital to gaining acceptance at the senior management level. Many of the concepts used are also applicable to problems related to systems safety engineering, as pointed out in the previous chapter.

Safety and health is attempting to migrate from compliance to best practices; however, compliance is still a significant driver in many companies today. As a result, safety and health professionals find themselves having to justify expenditures, and they can use financial principles to do just that.

Safety and health professionals can establish the economic value of expenditures, including the risk averted or incurred by a safety and health action, as well as the monetary outlay associated with that action. Safety and health professionals often choose one of two basic strategies concerning adherence to occupational safety and health regulations: compliance or intentional avoidance. Within each of these basic strategies, there are specific alternatives ranging *from full compliance* by fully implementing a program, to *extreme avoidance* by doing nothing (and risking OSHA fines as well as losses associated with workers compensation).

Although compliance is intended to protect the safety and health of workers, economically struggling companies may not accept compliance and may instead opt for cost-reduction activities, even if such action requires avoiding a risk-taking strategy. Regardless of the strategy selected, some health and safety-related costs will occur, and therefore impact a company's (tangible) costs, as well as possible product quality, as measured by production rejects and customer feedback. In addition to these tangible factors of cost and quality, there are "vague" or difficult-to-measure factors (intangible) such as production flexibility, public image, and employee morale. Because most health and safety-related decisions involve both tangible and intangible factors, a method of analyzing risks is vital for objectively calculating decisions to be made regarding the identification of optimal safety and health program actions.

Poor decisions concerning compliance can be costly for an employer, so it is important to analyze costs and related risks factors to arrive at an optimal decision. There are numerous methods to address expenditures for safety and health. Several will be addressed in this chapter, as follows:

- risk mapping;
- the time value of money;
- cost-benefit analysis;
- a basic overview of budgeting; and
- understanding financial statements.

USING RISK MAPPING FOR INVESTMENT DECISIONS

Risk mapping is a tool used to manage risk, optimize resource allocations, and adjust project schedules based on cost and risk information. It combines an order-of-magnitude, integrated, risk-analysis approach with cost data and importance measures. Thus, risk mapping extends safety and health risk-analysis efforts to have true business value to an organization.

Deriving value from safety and health expenditures has become universally important as a core competency of modern enterprises. However, deriving value from safety and health expenditures is no longer as simple as meeting customer requirements or improving product performance. Value is always multifaceted, often subjective, and occasionally bewildering. Fundamental activities in the process of delivering value include the following:

1. Determining the dimensions of value
2. Systematically identifying and balancing the technological, financial, environmental, and societal risks
3. Establishing measures for value
4. Establishing an organization suited for value creation
5. Identifying and acquiring new technologies
6. Managing the deployment of technologies

Risk mapping was developed with value creation at its core. It is a tool used to establish dimensions and measures of value and to provide a balance among the various aspects of value. Companies have used it for identification of value creation opportunities in a risk management format. That is, by identifying and evaluating an integrated risk profile of safety and health issues, companies have determined where to invest to achieve the highest value of return on their safety and health expenditures.

Overview of Risk Mapping

Value from safety and health expenditures most often comes in the form of cost or risk avoidance but can result in increased productivity, which translates directly to the corporate bottom line. A true value determination must account for both costs and risks. The difference between costs and risk can be summarized as follows:

- Costs are expected expenditures that can be included in a budget of financial forecast for an economic time frame of interest.
- Risks represent expenditures or liabilities that are potential but not expected within the same economic time frame; hence, they are not generally included in a budget or financial forecast. A probability exists that an expenditure or liability will actually be incurred within each time frame of interest. Thus, the expense will be zero if the loss incident does not occur. The expense or liability can be very high if it does occur and can have a significant impact on a business.

To combine costs and risks, they must both be in the same units of measure. Since costs are generally in monetary units and decisions are generally made on an economic basis, it follows that risks must also be converted to monetary values.

As mentioned previously, *Risk* is a combination of the probability of occurrence and the severity of consequences of unexpected loss incidents. To combine risk with costs, the risks are put into units of dollars per year. The "dollars per year" risk measure is thus an annualized liability or loss rate. Eliminating that liability adds value to an organization.

Risk mapping provides risk management and optimized resource allocations based on cost and risk information. It combines an order-of-magnitude, integrated, risk-analysis approach with cost data and importance measures. Thus, risk mapping extends safety and health risk management efforts to an organization's true business value. By using the risk-mapping tool, decisions can be made in a cost-effective manner based on cost and risk information.

Defining the Scope of Risk Mapping

Risk-mapping methodology may be used to address a wide range of objectives at varying levels of detail. It is important at the outset, however, to define the goal clearly and, therefore, limit the scope as necessary. Examples of applications range from a site-wide risk prioritization, which may include not only performance risk but also the risk of delaying or eliminating a project to a top-level strategic issue prioritization. See figure 13.1.

To develop an understanding of risk requires addressing three specific questions: What are the hazards? What are the possible undesired outcomes of hazards? How likely, in terms of probability, are these outcomes to occur? To understand risk it is essential to view an accident as a sequence of events (see figure 13.1). A *hazard* is the presence of a material or condition that has the potential for causing loss or harm. An accident scenario begins with an unplanned initiating event or deviation involving a process hazard. The

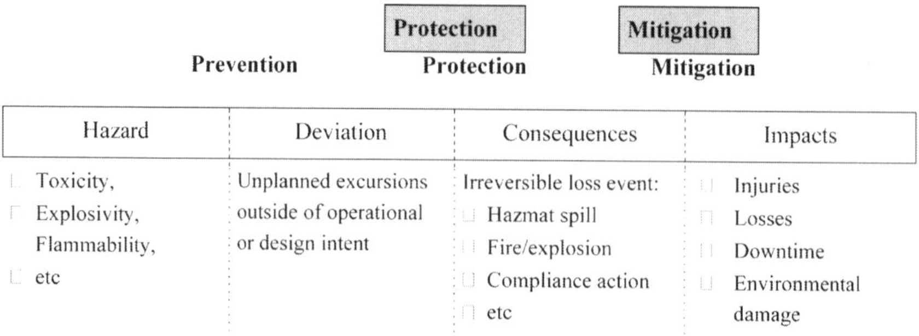

		Protection	**Mitigation**	
	Prevention	Protection	Mitigation	
Hazard	Deviation	Consequences		Impacts
Toxicity, Explosivity, Flammability, etc	Unplanned excursions outside of operational or design intent	Irreversible loss event: ☐ Hazmat spill ☐ Fire/explosion ☐ Compliance action ☐ etc		☐ Injuries ☐ Losses ☐ Downtime ☐ Environmental damage

Figure 13.1 Anatomy of an Accident

effects of the deviation are undesired outcomes or consequences and potential harmful impacts. Given that a deviation occurs, *preventions* reduce the likelihood of the deviation occurring, while, *protections* reduce the probability of the consequences occurring.

Order-of-Magnitude Methodology

As was indicated in chapter 11, estimating the risks of safety and health issues involves determining the likelihood of an undesired outcome and the impact of that outcome, should it occur. To simplify the risk-analysis portion of risk mapping, cost and risk parameters are based on an order-of-magnitude basis. Furthermore, to simplify the display and combination of cost and risk parameters, only the exponents of the magnitudes are used. Using logarithms and exponents are common methods to measure order-of-magnitude impact, both based on the power of 10. For example, a risk of 100 times per year, is recorded as a 2, and only the exponent "2" is used.

The likelihood of occurrence of each undesired outcome, or scenario frequency, is based on the estimated frequency of the initiating event and the effectiveness of the preventive and protective features. An order-of-magnitude scale, as shown in table 13.1, can be used for capturing the likelihood of occurrence of each undesired outcome. When using the same number of times per year the table below can be used to solve any risk-mapping problem.

If an initiating event is expected to occur once every ten years, risk mapping assigns a value of –1 to the initiating event. If there is a 10 percent chance that a particular protection will *fail* to minimize the outcome of that event, risk mapping also assigns a value of –1 to the protection. The undesired event occurrence frequency is the frequency of the initiating event times the probability the protection(s) will fail.

Table 13.1 Likelihood Magnitudes

Magnitude	Times per Year	Alternate Description
+2	100	Twice a week
+1	10	Once a month
0	1	Once a year
–1	0.1	Once every 10 years, or 10% chance per year of operation
–2	0.01	Not expected to occur during facility, but may occur; 1% chance per year of operation
–3	0.001	Would be very surprising if occurred during facility life; 1 chance in 1000 per year of operation
–4	0.0001	Extremely unlikely, or not expected to be possible

Initiating Event   Protection Effectiveness
1/10 years × 0.1 =
0.1 × 0.1 = 0.01 or 1 in 100 years.

Since risk mapping deals with orders of magnitude, the result can be achieved by calculating the sum of the initiating event frequency and the protection effectiveness, as follows:

Initiating Event   Protection Effectiveness
−1 + −1 = −2

Using −2 as the exponent, based on the power of 10 equals 10^{-2} or .01.
This equates to an event frequency of 1 in 100 years.

Evaluating the impacts of undesired events includes assessment of the types of impact to be considered and the severity of each. The risk-mapping approach provides a framework for capturing the wide range of potential impacts that a given scenario might impose, such as worker and public safety, business impact, and social impacts. Table 13.2 gives an example scale for measuring the severity of consequences of undesired outcomes related to facilities handling hazardous materials.

Since impacts are additive rather than multiplicative (as is frequency), combining impacts from various impact types in risk mapping is not simple. Impacts must be added and combined in an absolute manner. Thus, if an event had outcomes of medical treatment for workers (severity magnitude of 3), exposure above limits for off-site populations (severity magnitude of 4), and localized, short-term environmental effects (severity magnitude of 4), the event impact calculation would be as follows:

| Impact Magnitude 3 = $1,000 Cost, Loss Liability | Impact Magnitude 4 + $10,000 Cost, Loss Liability | Impact Magnitude 4 + $10,000 Cost, Loss Liability | = | $21,000 |

The magnitudes can be expressed as shown in table 13.2.

Risk Determination

Because *risk* is defined as a combination of the likelihood of occurrence and the severity of impacts of unexpected loss incidents, order-of-magnitude approach in risk mapping allows a calculation similar to the frequency determination. In risk mapping, risk is the sum of the frequency and total impact

Table 13.2 Order of Magnitudes for Cost, Loss or liability for Worker, Public, and Environmental Effects

Magnitude	Cost, Loss, or Liability	Worker Effects	Public Effects	Environmental Effects
7	$10 M	Multiple fatalities or multiple permanent health effects	Fatality or permanent health effect	Widespread and long term or permanent
6	$1 M	Fatalities or multiple health effects	Severe or multiple injuries	Widespread and short term or localized and long term
5	$100,000	Severe or multiple injuries	Injury or hospitalization	Localized and long term
4	$10,000	Lost workday(s)	Exposure above limits	Localized and short term
3	$1,000	Medical treatment	Exposure below limits	Reportable spill
2	$100	First-aid case	Odor/noise concern	Variation from permit

magnitudes. For those familiar with the use of a log in calculations, using the previous equations, the risk would be determined as follows:

Frequency + Impact = Risk:
$\text{Log}_{10}(\$21,000) = 4.3$
$-2 + 4.3 = 2.3$ (Composite Risk), then taking the anti-log of 2.3 is

$10^{2.3} = \$199.5$ or rounding up to $200/year

An Example Risk-Mapping Application

A good example is a hypothetical corporation using risk mapping to identify key strategic safety and health issues. The primary objective is to develop and apply a systematic risk identification process in a cost-effective manner. It can be used by management in an ongoing basis to assist with risk management decisions. The process involves identifying key strategic issues from a set of high-level *potential* accidents. The strategy includes six steps:

1. Identify the issues
2. Determine impact categories

3. Develop accident scenarios associated with each issue
4. Obtain cost and probability information resulting from each scenario
5. Determine risk magnitudes based on #4
6. Establish risk tolerability criteria

Each facility in the corporation is asked to submit five key safety and health issues based on risk assessments and worst-case scenarios. These issues are compiled and combined into a single list now representing the key corporate issues. This composite list is used as a starting point to identify strategic environmental, health, and safety concerns.

Frequency and effectiveness categories are then established for risk mapping. Categories of this nature can be used in any study but are unique to each. Based on the information provided by the facilities, it is determined that the following impact categories are important to this corporation:

- Worker safety and health
- Public safety and health
- Capital assets
- Operational continuity
- Compliance
- Product and service liability
- Ecology
- Society

Based on expected levels of impact in each category, qualitative descriptors are applied, as in the last three columns in the top row of table 13.2, along with quantitative levels similar to those in the second column of the table. The qualitative descriptors are developed by safety, and health professionals collaborating with operations, engineering and senior management. Once established, the quantitative levels are then estimated, based on the expected quantitative impacts of each.

The third step in the process includes developing a sequence of events for each of the twenty-one issues. That is, for each issue, initiating events, prevention, protection, and expected impacts are postulated. For each scenario, the likelihood of the initiating event, based on the order-of-magnitude method, is estimated.

The scenario risk is then calculated based on the estimated scenario frequency and relative impact in terms of cost. Next, these risk estimates are combined to reflect the total risk for each key issue.

The final task is to establish criteria for making risk management decisions. These criteria are used to sort key issues into three categories:

1. Issues with risk high enough to require action regardless of economic return.
2. Issues with high risk; however, any risk-reduction effort must show an economic return.
3. Issues with low risk that warrant no risk-reduction efforts.

These risk tolerability criteria represent the corporate risk aversion. Because senior management team makes the decisions affecting the amount of risk to which the company is exposed, risk aversion estimates of members of this team are used to determine corporate risk aversion. This is calculated through a series of individual interviews among team members used to measure risk aversion.

The interview may begin with, "What level of annual economic loss from a single type of event at any facility would you consider to be a part of normal operations?" Estimate are recorded on a diagram as follows (see figure 13.2):

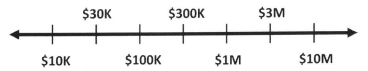

Figure 13.2   Example Range of Economic Loss Considered Part of Normal Operations

The next question may be, "What is the largest human loss you can conceive, resulting from a single event, over a single plant's lifetime?" Estimates are then recorded in figure 13.3:

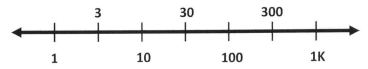

Figure 13.3   Example Range of Largest Human Loss for an Event

Based on the results of the risk mapping and the risk tolerability questionnaire, the risks associated with strategic safety and health issues are plotted, forming a risk matrix, as shown in figure 13.4. The matrix plots risk on a log-log scale with indices of frequency and impact, as shown above. The frequency and impact numbers are the order-of-magnitude values used to calculate risk. Levels of constant risk in the matrix go along the diagonal from the upper left to the lower right with the highest risk in the upper right-hand corner and the lowest risk in the lower left-hand corner.

The lower diagonal line plotted represents the level of tolerable risk. Any event with a risk that falls to the left and below that line warrants no action. The upper diagonal line plotted represents the level of intolerable risk. Events above and to the right of that line are characterized as high risk. Events above the intolerable risk line warrant risk-reduction actions regardless of whether there is a positive economic return associated with the risk-reduction action. The area between the lines represents the area where risk-reduction actions must have positive economic return equal to or greater than other corporate investments. The ranges are established based on collaboration between safety and health professionals, operations, engineering, and senior management.

In this example, safety and health professionals, operations, engineering, and senior management collaborated to identify twenty-one key issues to be considered. Of the twenty-one key issues considered in this analysis, fourteen fell into the high-risk area of the plot. These issues must be addressed and the risk associated with them must be reduced, even though there may not be a net positive economic return from the investment. Seven of the issues fall into the area between the criteria. These seven issues should be addressed only if risk-reduction measures offer a return equal to or greater than other corporate investments.

Determining value from safety and health issues is a process that sets the framework for value creation and continuously works within that framework

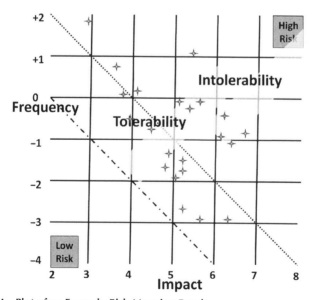

Figure 13.4 Plot of an Example Risk-Mapping Results

to identify the potential of value. Risk-mapping technology is a flexible, cost-effective tool that has been used as an input to a company's investment, decision-making process. It provides management with an established risk characterization method to help identify strategic issues, quantify risk, and confirm issues that warrant risk-reduction actions.

In the example application, risk mapping is used to establish a framework for the determination of key issues that have the potential for adding value. Risk mapping is also used to identify criteria for determining when the strategic issue of adding value applies and when other decisions apply to reducing risk. This framework satisfies the initial steps in the value creation process: determining the dimensions of value, identifying and balancing risks, and establishing measures for value.

TIME VALUE OF MONEY

Another approach to demonstrating the value of safety and health expenditures is by using calculations that explain the best use of a revenue based on the time value of money. This enables safety and health professionals to demonstrate the financial benefit of spending revenue to improve working conditions. It is often purported that the most valuable assessment of money is the present value. Being able to justify safety and health expenditures by utilizing principles to calculate the time value of money can demonstrate the value of safety and health expenditures. These methods include present worth, future worth, depreciation, rate of return, replacement analysis, and cost-benefit analysis. Once completed, safety and health professionals can tie these values to incident rate reductions, sales, and profit margins. This links the safety and health activity to the financial aspects, providing a demonstration of added value to the company's bottom line.

Present Worth

The *present worth* of an asset is the sum of all discounted, expected, future-cash inflows minus the sum of all cash outflows and discounted, expected, future-cash outflows. Restated in simple terms, this means that the value of a machine, or an improvement in a process, etc., is simply the sum total of all the money expected to be made or saved over the life of the asset minus present and future costs associated with that asset. All dollar amounts are adjusted for inflation and current interest or discount rates so that current monetary value is accurately reflected. Present worth is an easy check for the feasibility of a project. Projects with negative net present worths can be easily identified and eliminated in favor of those projects with better return potentials.

Present worth is a function of:

- Initial cost or investment
- Cash outflows, such as maintenance costs, time payments and the cost of:
  - Safety and health program implementation (e.g., confined space and lockout/tagout)
  - Equipment (e.g., fire extinguishers and fire protection)
  - Training
  - New employees (e.g., safety and health professionals and industrial hygienists)
  - Cash inflows such as cost savings or payments for goods and services and expected:
  - Workers compensation and insurance savings
    - Decrease in overtime, productivity, turnover, training costs, etc.
    - Decrease in legal liabilities, legal fees, settlements
    - Decrease in OSHA citations
    - Increase in good will (e.g., company reputation and union negotiating)
- Salvage value of the asset
- Adjustments for inflation (discounting) or interest rates

Calculations can also be performed to determine how much will be required to accumulate a given amount in the future. As an example: In three years, $400,000 will be required for safety and health modifications to a facility. How much money should be invested at 10 percent to have the required amount when needed? Using the formula:

$$PV = FV(1+i)^n$$

PV = Sum of money at the present time **(the unknown variable)**
i = Interest rate for a given interest period **(0.10)**
FV = A payment or receipts at the end of an interest period in a series of n equal payments or receipts **($400,000)**
n = number of years **(3)**

$$PV = \$300,526$$

That is, to have $400,000 in three years with a current interest rate of 10 percent compounded annually, the company must be able to invest $300,526 now to be able to make the allocated expenditure for safety and health modifications to the facility.

This is a benefit to illustrate to management that the current investment required is about $100,000 less than the planned target expenditure. This makes good fiduciary sense, especially if it is know that these modifications are going to be required.

Future Value

The future value of an asset is the current value of the asset plus the compound interest thereon. This value is also a good check for the feasibility of a project. It can be discounted to present worth in order to compare the value of a product or project with the investment necessary to create it. If the value is less than the investment required, the project should be terminated in favor of more profitable projects. Future value is a function of:

- Initial value of investment
- Interest rates (compounding), which are composed of:
  - True cost of borrowing money (currently 2–5%)
  - Risk of investment/project (junk bonds verses AAA bonds)
  - Rate of inflation (currently 3–6% annually depending on source)

As an example: A company may need to install pollution abatement equipment in three years to bring the facility up to EPA specifications. After speaking with vendors and accountants, it is estimated that at that time $2 million will be needed to make the necessary improvements. Interest rates are at 10 percent and the company has $750,000 to set aside and invest in anticipation of spending needs three years from now. In three years, what will be the value of $750,000 set aside today, and how much will the company need to borrow to make up the difference? Using the formula:

where:
FV = Future worth of a present sum of money after n interest periods, or the future worth of a series of equal payments **(the unknown variable)**
PV = The sum of money at the present time **($750,000)**
i = Interest rate for a given interest period **(10%)**
n = Number of years **(3)**

$F = \$750,000 (1+0.10)^3$
$F = \$998,250$

The company will need $2,000,000 for the project. Investing $750,000 at 10% interest for three years will net $998,250. In three years, the company will need to borrow an additional $1,001,750.

Depreciation

Depreciation is the process of allocating, in a systematic and rational manner, the expense of an asset to each period benefited by the asset. This entails spreading the cost of the asset across its estimated life and charging it against the corresponding accounting periods. The process allows companies to charge the expenses associated with an asset against the profits it generates during the periods in which it is used. There are many methods of calculating depreciation such as, sum-of-year digits, declining-balance, group and composite depreciation, and straight line. For the purpose of the example used, straight line depreciation will be used. The formula for straight line depreciation is

where:
D = Annual depreciation
P = Initial cost of the asset
SV = Salvage values of the asset
n = Expected depreciable life of the asset

This is a function of

- Cost of the asset
- Estimated lifetime of the asset
- Salvage value (if any) of the asset
- Method of depreciation used (straight line, Accelerated Cost Recovery Standard (ACRS), etc.)

As an example of straight line depreciation: Assume $10,000 is spent on a piece of equipment. It is planned to keep the equipment for four years, and the salvage value is $5,000. Using the formula:

where:
D = Annual depreciation (**unknown variable**)
P = Initial cost of the asset (**$10,000**)
SV = Salvage values of the asset (**$5,000**)
n = Expected depreciable life of the asset (**4**)

$$D = \frac{\$10,000 - \$5,000}{4}$$

D = $1,250 per year

This is a benefit to illustrate to management that spending $10,000 on equipment now will permit a $1,250 write off every year for each of the four years, and the equipment will still have a salvage or residual value of $5,000 at the end of the four-year period.

Rate of Return

The rate of return is a measure that allows comparison between two alternatives. It is a function of the ratio of the present value of the net income generated over time by the asset divided by the cost of the asset, usually expressed as a percentage. In other words, the amount of money generated by two alternative projects is translated into something resembling an interest rate. In this manner, the company can chose which project will yield the highest rate of return on its money. Many companies also have a minimum attractive rate of return, which is the lowest rate of return acceptable before a project will even be considered.

The rate of return is a function of

- Initial cost of the asset(s)
- Expected cash outflows include the cost of:
  - Safety and health program implementation (e.g., confined space and lockout/tagout)
  - Equipment (e.g., fire extinguishers and fire protection)
  - Training
  - New employees (e.g., safety and health professionals and industrial hygienists)
  - Maintenance
  - Repairs
- Cash inflows, such as cost savings or payments for goods and services and expected:
  - Worker's compensation and insurance savings
  - Decrease in overtime, productivity, turnover, training costs, etc.
  - Decrease in OSHA citations
- Salvage value (if any) of the asset
- Discount rate used by the company, such as
  - Minimum acceptable rate of return or
  - Cost of capital (company's borrowing costs) or rate of inflation

The formula for rate of return is

where:
R = Rate of Return
P_i = Net present value of all expected inflows or revenue gains from the investment (Savings or expected dollar benefit in this case)
P_o = Net present value of all expected outflows or expenses from the investment

As an example: There is an option between two emergency response vehicles—each available on a one-year lease. Vehicle A will cost $100,000; it has a residual value at the end of the lease of $50,000 and will require about $10,000 in maintenance. The cost of money is 15 percent interest, so a rate of return of at least 15 percent is generally expected by management. By having our own emergency vehicle, overall savings generated are expected to be $75,000. The alternative is a $200,000 Vehicle B with a residual value of $100,000 that will require about $5000 in maintenance. Because of increased levels of emergency equipment, this will also result in a net savings of $135,000. Using the formula below for each option:

Vehicle A:
P_i = $75,000 **Savings or expected benefit**
P_o = ($100,000 − 50,000) + $10,000= $60,000 **Expenses**
R = $75,000 − 60,000/ $60,000 = .25 or 25% **Rate of Return**

Vehicle B:
P_i = $135,000 **Savings or expected benefit**
P_o = ($200,000 − 100,000) + $5,000 = $105,000 **Expenses**
R = $135,000 − 105,000/$105,000 = .28 or 28% **Rate of Return**

This is a benefit to illustrate to management that even though Vehicle A is cheaper than Vehicle B, it has a less favorable rate of return. In the case of the negative calculation, the value closest to zero is the best option. When considering cost inflows, salvage value and depreciation are the included values. When considering outflows, the cost of the vehicle and forecasted annual maintenance are the included values. Expected inflows that appear to wash out as equal values include: worker compensation savings, insurance savings, overtime, productivity, turnover, training cost, savings, and a decrease in OSHA citations. Expected outflows that appear to wash out as equals include only training. Maintenance and repairs do not wash out and are included in the calculation. The other variables are not applicable to this particular problem. Present values must also be considered for investments that occur over multiple years. In this case, a return of at least 15 percent is expected. If the return is lower than 15 percent, management may invest its funds elsewhere or choose not to invest in this project.

Replacement Analysis

Replacement analysis provides an economic comparison of two asset choices, a defender (the current assets) versus a challenger (asset being considered for purchase). It is usually used when determining whether or not to replace an existing asset with a new or more efficient one or when comparing different options in procuring equipment such as, buying or leasing. The costs and expenses associated with the assets are converted into an equivalent uniform annualized cost (EUAC). This determines how much expense will be associated with a given asset in a one-year period,

Financial Aspects for Safety and Health Professionals 249

thus providing a uniform benchmark for comparison. Once the EUAC has been determined, all the company needs to do is choose the lowest cost option. For example, to determine which is more feasible, buying or leasing a particular asset, the formulas below can be used—the defender formula to calculate the purchase option and the challenger formula for the lease option.

The formula for the defender is *(Purchase Option)*

$EUAC_d = P - SV + AOC$

where:
P = Purchase cost of the asset
SV = Salvage value of the asset
AOC = Annual operating cost

The formula for the challenger is *(Lease Option)*

$EUAC_c = L + AOC$

where:
L = Lease cost of the asset
AOC = Annual operating cost

This is a function of

- Initial cost of the asset
- Salvage value (if any) of the asset
- Annual operating cost
- Lease cost

As an example: Using a similar example as that shown above, there is an option between two pieces of equipment. Each has a life expectancy of four years. Piece A will cost $10,000; it has a salvage value of $5,000 and will require about $1,000 per year in maintenance. The alternative is to lease Piece B for $2,000. It will require about $500 per year in maintenance. Using the formula below for each option:

The formula for the defender is

$EUAC_d = P - SV + AOC$

where:
P   = $10,000
SV   = $5,000
AOC   = $1,000 × 4
$EUAC_d$ = $10,000–$5,000 + $4,000
        = $9,000

The formula for the challenger is

$EUAC_c = L + AOC$

where:
L   = $2,000 × 4 years
AOC   = $500 × 4
$EUAC_c$ = $8,000 + $2,000
        = $10,000

This is a benefit of purchasing over leasing. Also, it can be used to illustrate the relatively low cost to purchase this vehicle in the first place. Using this calculation illustrates that there are several options have been evaluated, rather than merely indicating that the company needs to be purchase an emergency response vehicle. Depending upon interest rates and availability of funds, additional consideration may be given to the present value of one dollar; that is, one dollar to be received in the future is worth less than one dollar to be received today. In calculations that are close, present values of future amounts must also be considered.

Cost-Benefit Analysis

Cost-benefit analysis is a method used to analyze the effects of making a change in a process—usually in two or more proposals (see figure 13.5). Typically, cash flows of present procedures are compared against predicted cash flows incurred under the change. The advantage of using the cost-benefit analysis is the ability to monetize costs of intangibles (e.g., goodwill, reputation of a company, the cost of a life, cost of future injuries, decreased turnover, and decreased training). The estimates used must be accompanied by realistic, conservative accounting assumptions. Without realistic assumptions to force the solution to the worst-case scenario, errors could occur and invalidate the estimate basis.

The following example is dependent upon reasonable estimates of

- Decrease in legal liabilities, legal fees, settlements
- Decrease in OSHA citations
- Increase in good will (e.g., company reputation and union negotiating)

When Ford Motor Company performed a cost-benefit analysis to determine the benefits and costs relating to the fuel leakage associated with static rollover test portions of the FMVSS 208 (Ford Pinto), Ford failed to make conservative accounting estimates of the worst-case scenario. In 1970, Ford used $200,000 as the cost of a life (provided by the National Highway Traffic Safety and health Administration [NHTSA]); the value was based almost entirely on deferred future earnings (DFE). At the time this decision was made, there were at least three different DFE-based figures ranging from $200,000 to $350,000 being used by as many different federal agencies. Willingness to pay (WTP) has since replaced DFE as the preferred method of assessing the value of life. Furthermore, research has shown that various WTP studies have revealed a higher median value than the one used by Ford.

The Ford cost-benefit analysis presented a $137 million cost and $49.5 million benefit. In the formula, the number of deaths, the cost per vehicle to

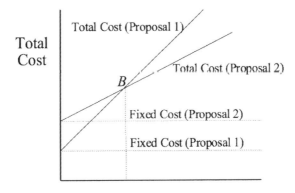

Variable Units

Figure 13.5 Cost-Benefit Analysis Between Two Proposals

make the design change, and the proportion of deaths to be attributed to small light vehicles were all subject to such dramatic changes that the approximate $2.75 cost to $1.00 benefit ratio achieved could have easily been changed so that the benefit exceeded the cost, even if the value of life used was accepted and left unchanged.

The formula Ford used was as follows:

Benefits:
Savings: 180 burn deaths, 180 serious burn injuries, 2100 burned vehicles
Unit Cost: $200,000 per death, $67,000 per injury, $700 per vehicle
Total Benefit: ((180 × $200,000) + (180 × $67,000) + (2,100 × $700)) = $49.5 million

Costs:
Sales: 11 million cars, 1.5 million light trucks
Unit Cost: $11 per car, $11 per truck
Total Cost: ((11,000,000 × $11) + (1,500,000 × $11)) = $137 million

The cost-benefit analysis performed by Ford on crash-induced fuel tank leakage and fires presents a startling example of the imprecision of this type of cost-benefit analysis, as well as, the strong possibility for manipulation of figures to achieve a desired result.

By using the high estimated death figure from the NHTSA, the so-called "benefit" total would have been $161.2 million, even if everything else would have held constant. By using the low estimated death figure of $5.08, the figure would have been $63 million. If only small cars had been used, rather than all automobiles and light trucks, the cost figures would have been lower still. Increasing the value of life would have further impacted the results. In retrospect, had a *conservative* accounting estimate rather than a *liberal* one been used, the design changes would have likely been made.

What is life worth today? Various federal regulatory agencies value life differently when determining the cost-benefit of a new rule, as follows:

Agency	Value in Millions
Consumer Product Safety and health Commission (CPSC)	$2M
Department of Transportation	$9.1M
Environmental Protection Agency (EPA)	$9.6M
Nuclear Regulatory Commission (NRC)	$3M
Occupational Safety and Health & Health Administration	$8.7M
Office of Management & Budget	$7–9M

The total cost to implement a new rule is divided by the number of lives expected to be saved. For example, if a new rule is estimated to cost $100 million to implement and it is expected to save 20 lives, at $5 million per life, the rule is too expensive for CPSC but acceptable for the others. It seems that the federal government values life more than some workers value their own lives. If employees are killed while they deliberately violate safety and health rules, the workers compensation payments are nowhere near these amounts.

When conducting a cost-benefit analysis, safety and health professionals may choose to use a particular agency's numbers depending what is being justified. For example, if they are justifying based on OSHA, they would want to use $8.7M; on the basis of EPA, they would want to use $9.6M for the cost of a life.

Tying the Cost to Lost-Time Accident Savings

Another way for safety and health professionals to justify the cost of their programs is to tie the reduction of Lost-Time Accidents (LTAs) to cost savings. The plant manager must be informed that none of this is free and cannot be done with minimal cost. This often takes a considerable financial investment. Assume the cost of an LTA is $40,000 and recent history indicates two LTAs/year. The annual costs for the company are $80,000.

As a general rule, to achieve an improvement in the incident rate from 5.0 to 2.5, safety and health professionals should double the cost it takes to achieve an improvement from 2.5 to 1.25. To achieve an improvement from 1.25 to 0.625, the figure should be doubled again. As the incident rate is cut in half, the cost to achieve that milestone is typically doubled. To be clear, see the example below:

Incident Rate	Cost to achieve
5.0–2.5	$40,000
2.5–1.25	$80,000
1.25–0.625	$160,000
0.625–0.00	$320,000

The cost increase is for equipment, staff, resources, and rewards for achieving the postulated goal. For example, early on, if the safety and health professional achieves one year without an LTA and a sustained incident rate of 2.5, the payback is 0.5 times the cost to achieve an improvement from 5.0 to 2.5. As 1.0 is approached, that multiplier increases to 0.75, 0.85, and may be as high as 1.0. Clearly, the payback can increase quite rapidly.

For more information, use the chart referenced in the next section regarding accident costs and the relationship to profit and profit margin.

Justifying Safety and Health Using Sales and Profit Margins

In times of keen competition and low profit margins, safety and health may contribute more to profits than an organization's best salesman. It is necessary for the salesman of a business to sell an additional $1,667,000 in products to pay the costs of $50,000 in annual losses from injury, illness, damage, or theft, assuming an average profit on sales of 3 percent. The amount of sales required to pay for losses will vary with the profit margin. Table 13.3 shows the dollars of sales required to pay for different amounts of costs for accident losses; that is, if an organization's profit margin is 5 percent, it would have to

Table 13.3 Safety and Health Department Payback for Achieving Safety and Health Metric Milestones

Goal		Cumulative Safety and Health Department Budget	LTA Averted Cost	Cumulative LTA Averted Cost	Payback to Safety and Health Department
IR	LTA				
2.5	One Year Without an LTA	$100,000	$80,000	$80,000	0.5 x $80,000 = $40,000
1.5	Two Years Without an LTA	$200,000	$80,000	$160,000	0.75 x $160,000 = $120,000
1.0	Three Years Without an LTA	$300,000	$80,000	$240,000	0.85 x $240,000 = $240,000
0.5	Four Years Without an LTA	$400,000	$80,000	$320,000	1.0 x $640,000 = $320,000

make sales of $500,000 to pay for $25,000 worth of losses. With a 1 percent margin, $10,000,000 of sales would be necessary to pay for $100,000 of the costs involved with accidents.

The formula looks like this:

Sales to Offset Sales = ($ losses × 100)/Profit Margin (%).

Table 13.4 Sales Required to Cover Losses

Yearly Incident Costs	Profit Margin				
	1%	2%	3%	4%	5%
$1,000	100,000	50,000	33,000	25,000	20,000
5,000	500,000	250,000	167,000	125,000	100,000
10,000	1,000,000	500,000	333,000	250,000	200,000
25,000	2,500,000	1,250,000	833,000	625,000	500,000
50,000	5,000,000	2,500,000	1,667,000	1,250,000	1,000,000
100,000	10,000,000	5,000,000	3,333,000	2,500,000	2,000,000
150.000	15,000,000	7,500,000	5,000,000	3,750,000	3,000,000
200,000	20,000,000	10,000,000	6,666,000	5,000,000	4,000,000

Table 13.4 shows the sales dollars required to cover losses at profit margins of 1–5%.

Understanding financial statements will assist safety and health professionals not only in sound decision-making but also understanding the companies in which they work. Safety and health professionals need not be certified public accountants, however, they do need to know the basics.

THE FINANCIAL STATEMENTS

Financial statements are reports to give investors and creditors additional information about a company's performance and financial standings. The purpose is to provide useful financial information to users outside of the company. Financial statements are designed to report the earnings and profitability, asset and debt levels, uses of cash, and total investments by the company for a specific time period.

A company's goal is to make a profit. One way to demonstrate this is to maintain strong financial statements. It is not unlike maintaining a company diary. Financial statements make up a portfolio that contains a balance sheet, income statement, statement of cash flow, and a statement of retained earnings. It shows the way business is conducted, where profit centers are, and where potential land mines are festering.

Lenders look at the balance sheet to see how much a company is worth or how "liquid" it is. The balance sheet shows what the company owns versus what it owes. Ultimately, it shows the net worth. Assets are described in terms of current and non-current items. Current items are those items that will be collected or are due within twelve months. Current assets include accounts receivable, inventory, and prepaid expenses. Non-current assets include property and equipment. Current liabilities include items such as payroll taxes, accounts payable, and deferred income. Non-current liabilities include multi-year loans and other debt commitments that stretch past a year.

Total assets minus total liabilities calculates a company's net equity or net worth. The goal is to keep the balance sheet positive rather than negative. Negative numbers mean a company owes more than it owns. This is not necessarily a death knell, but it should be a concern. Ideally, the debt-to-equity ratio should be no more than two-to-one.

The Income Statement

The income statement shows the company's bottom line as net income or net profit. A useful way to arrange income statements is in a year-to-year comparative format. The income statement lists annual net sales or gross sales minus returns and other allowances. For example, if there is $300,000 of income from sales, $160,000 can be deducted for the cost of goods leaving a gross profit of $140,000. Operating expenses are then subtracted from gross profit to yield the net operating income. Operating expenses consist of the sum of selling and administrative expenses, and they include such items as salaries, employee benefits, rent, payroll taxes, utilities, office supplies, and costs for marketing and advertising. For example, there is a company that has $97,000 in expenses. When this figure is subtracted from gross profit of $140,000, the result is net operating income of $43,000. Interest income and interest expenses must also be calculated to arrive at a net profit before tax. If a company is operating in the red, it shows up here as negative profit or a loss.

Cash Flow Analysis

Based on the information in the balance sheet and income statement, an accountant can prepare a statement of cash flow. It shows the sources and uses of cash (e.g., net borrowing under credit agreements, cash used in investing, proceeds from long-term debts, and dividends paid). Tracking the flow of money in and out of the business is fundamental to the big picture, but unless safety and health professionals are particularly adept with accounting terms and practices, arranging a statement of cash flow is an area best left to an accountant.

That doesn't mean safety and health professionals shouldn't stay as close to the numbers as possible. Safety and health managers should review balance sheets once a month, and every three months or so put together a statement of changes in financial condition. That will illustrate where funds are coming from and how they are being applied. It also lets safety and health professionals compare working capital from one period to the next.

These simple principles can assist in tracking costs. Perhaps it will give safety and health professionals the tools to turn the safety and health department from a cost center into profit center. All of this results from knowing more than the discipline of safety and health and showing fiduciary responsibility. That can be a formula for success.

Using these financial principles can assist safety and health professionals in understanding and demonstrating the financial value of proposals. Furthermore, this understanding can also help manage budgets and ensure costs are not running out of control. The next section addresses budgets specifically as they relate to safety and health.

BUDGETING FOR SAFETY

A safety and health professional, like managers of other important corporate functions, must have and be accountable for a budget. Budgeting is an important part of the planning function and must be completed as early as possible in the planning process. Before implementation, the budget must be developed and agreed to by all concerned parties.

Budgeting requires the safety and health professionals to identify the costs of performing day-to-day activities. Before safety and health professionals can determine what items should be purchased, the current equipment should be inventoried and budget analyzed. Included in this chapter are discussions of the current costs for safety and health equipment in use, such as safety and health glasses, prescription safety and health glasses, goggles, rubber and other boots, rubber and other gloves, acid suits, flame-resistant clothing, hard hats, hearing protection, or emergency medical equipment. Current budget items also include the cost for capital equipment, such as sprinkler systems and firefighting equipment, as well as other needs, such as training courses and upgrades identified from insurance audits, and others. Once safety and health expenditures are identified, efforts are made to determine efficient ways to expend allocations. For example, some employers use contractors to maintain first-aid cabinets at a monthly cost. By ordering the medical supplies by mail and using staff to stock first-aid cabinets (which will increase visibility), this cost can be minimized.

Although most safety and health professionals may have little or no formal training in budgeting, it is important they participate in this effort. They are the safety and health experts who know what is required for an effective safety and health effort and what is needs to be included. Budgeting without the safety and health professional's input may result in expenditures that do not adequately reflect the needs or best interests of the safety and health function or the company.

Personal levels of experience are also a consideration when developing a budget. Experienced safety and health professionals may be able to work more quickly and correctly than inexperienced ones; however, the higher salary of the experienced person may offset any savings. If the task is performed by a more highly compensated professional, the hours and the overall cost of time expended may offset any savings realized.

Time Factors in Budgeting

Budgeting for various tasks is typically assigned to the most qualified person in the safety and health function. This person might not be the safety and health manager but, rather, the individual responsible for a specific project or task. It is vital to plan effectively for a task completion, factoring in the financial budgetary factors required at each stage. The budget must be linked to certain milestones in order to make it trackable and to span it over the entire duration of the project. Setting the minimum completion time should be tempered by the knowledge that some things are beyond the control of the program. For this reason, it is usually wiser to budget for average time to completion rather than minimum time.

Practical Reasons for Budgeting

Budgeting is one of the most important tasks performed by safety and health professionals. Without enough funding, any job is almost impossible to accomplish. With over-budgeting, unnecessary tasks are performed and the costs of the safety and health program are higher than necessary. Cost overruns can also adversely affect the company's image and that of the people involved.

DEVELOPING THE OPTIMUM BUDGET

The best budget is one that allows all necessary tasks to be accomplished with the appropriate persons at the correct time. The budget matches the skills of the individual who is performing the task with the time and materials needed to complete it. The optimum budget must allow for revision and necessary research to develop documentation. The approved budget reflects the perspective of the company's management.

Methods of Determining a Budget

There are two major types of budgets: fixed and variable. The more common one is the fixed budget. It plans for a specific period and is updated and compared with actual results at the end of that period. The time period may be lengthened or shortened, depending upon the needs of the company.

The variable budget provides for the possibility that actual costs may not be the same as planned costs. This provision allows for restructuring of the budget to align it with actual costs. The variable budget obviates the need for contingency budgets but requires a very good accounting system.

For the purposes of this discussion, the fixed budget is considered, but the concepts hold true for the variable budget as well. There are several methods to determine the budget to be allocated for safety. Three methods are briefly discussed.

The first step in determining a budget is usually to define program elements or tasks. If a task can be defined, it can be properly budgeted; if it cannot be defined, it cannot be adequately addressed in the budget. This method requires significant up-front detail in describing the total program in manageable steps.

An alternative approach is to allocate a specific percentage of the total company budget. A variation of that method is to apportion a percentage of the engineering, operations, or administration budget to the safety and health program budget. This is an easy way to budget, but it is not as good as other methods, because it does not determine how to perform any one task or define the cost of accomplishing it. This method requires little effort to develop the budget but is useless in identifying where resources will be used.

Level of effort on an as-needed basis is a third way that safety and health may be budgeted. This is the, "I'll call you when I need you," method, but it rarely allows the company to perform the tasks correctly and on time. This means that no matter what the tasks are, the allocated budget does not change to accommodate them. There is no measurable effort in terms of significant accomplishment.

Of these three methods, the task definition one is the best one to identify and allocate where to spend and control company resources. It also provides an easy way to identify the problem areas when an overrun or under-budget situation occurs. For this reason, the task definition method of budgeting is discussed at length.

Task Definition Budgeting

The first step required when budgeting is to decide the tasks to be included. An effective comprehensive safety and health program can be broken into four major categories:

1. Workplace safety
2. Environmental safety
3. Product safety
4. Administration

These major categories can be further broken down into subcategories, which can be called "safety and health programs." The various safety and health programs necessary for any company may include the following:

1. Safety and health management
2. Industrial hygiene
3. Hazardous materials/waste
4. Radiation safety
5. Hearing conservation
6. Safety and health training
7. System safety
8. Electrical safety
9. Construction safety
10. Personal protective equipment

This is not a comprehensive list, nor does it include the subcategories of each of these programs. Each of the programs requires that certain tasks be performed in order to reach the goals. Priority is dependent on the product or services of the company and must be considered when developing a budget. The tasks then serve as the basis for the budget. The safety and health professional must determine priorities of tasks and what resources will be required for each task.

DEVELOPING THE BUDGET FOR SAFETY

Several considerations must be kept in mind when budgeting for safety. Among the most important are the personnel assigned and the equipment, materials, and time required for completing a task. Personnel budgeting commonly takes two forms: one portion of the budget is the time, in worker-hours, needed to complete a task; the other is the dollars required to pay for the level of skill required to perform the task. Therefore, identification of the skill level required to do each task is very important to the budgeting process.

Outside services, such as periodic physical examinations, laboratory services, and hazardous waste disposal costs must be included in the budget. The calibration of instruments must also be specified. In addition, some type of contingency budget should be included to ensure that unexpected or

TASK BUDGET							
TASK							

EXPECTED DATE OF COMPLETION
DATE OF FIRST BUDGET REVIEW
DATE OF SECOND BUDGET REVIEW _____

ITEM	MINIMUM	MAXIMUM	AVERAGE	QUOTE	1ST REVIEW	2ND REVIEW
Manpower (number)						
Skill Level (years experience)						
Equipment List ($)						
Materials List ($)						
Other List ($)						
Time to Complete (man-days)						
Contingency ($)						

Figure 13.6 Budget Planning Form

unplanned tasks are completed. Contingency funds should be used only in an emergency.

It is very important to budget the total task. A common mistake is to budget only the personnel necessary to perform the task and to budget the equipment and materials for the total program or department separately. Figure 13.6 is a form that outlines the items for each task.

Overall budget reduction in a company is always a possibility. Sometimes budgets are cut by a certain percentage, with no consideration given to the reduction in the number or scope of the tasks. With a budget that clearly

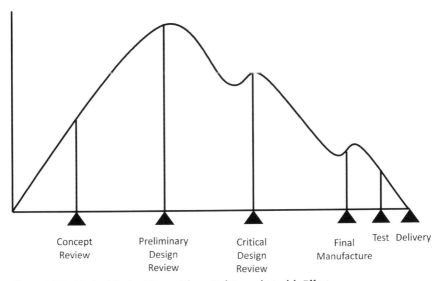

Figure 13.7 Typical Budget Spread for a Safety and Health Effort

defines each task, the safety and health professional can more readily identify areas to cut and obtain concurrence from company management.

Figure 13.7 shows a typical budget spread for a system-safety and health program linked to product design reviews, manufacturing, and testing. Note that most of the effort is expended in the early stages of the program. This is known as *front loading* of the system's safety and health effort and is considered to be the most effective method.

Other types of ongoing safety and health programs may have different budget spreads. The cost peaks relate to activities relevant to performing various tasks within the program. Figure 13.8 shows an industrial hygiene program whose milestones refer to specific air and noise monitoring activities. Note that it is also front-loaded. This is common with increased costs to acquire equipment and the training necessary to use the equipment properly.

There must be a task description and a basis of estimate for each task. The task should be described in detail so that a non-safety and health person can understand it. The description should be as specific as possible and should list the requirements and state where they may be filled. It should show the task to be performed, why it is needed, and the scope of the task.

Budgeting the Total Effort

Although task determination can be done in a top-down manner, all budgeting must be done on a bottom-up basis. To determine where a company will

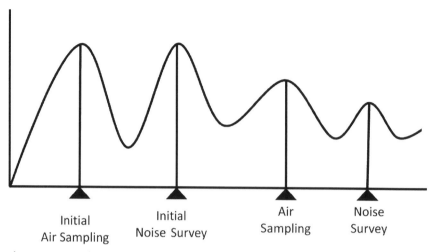

Figure 13.8 Typical Budget Spread for Industrial Hygiene

spend its resources, it must be clear what the company will receive for that investment. Many tasks are required in order to have an effective and complete safety and health department. Each of these tasks needs a budget to ensure that it can be accomplished effectively. In the previous section on task definition budgeting, the safety and health tasks were broken down into four categories. These categories are also very broad and should be broken down further into subcategories to be effective in the budgeting process.

WORKPLACE SAFETY

An effective workplace safety and health program needs several subprograms to be effective. These subprograms should include those required by law and those that make good sense in doing business on a day-to-day basis.

Eye Protection

For an eye protection program to be effective, a survey of eye hazards must be made. Once the eye hazards are determined, controls must be established. When the controls are in place, personnel must be trained to use them. The procedure is typical of most safety and health-related programs. The cost of replaceable materials, such as safety and health glasses and face shields, must be considered. The time required to monitor, train personnel, and document the program must be established. Outside services needed to provide prescription safety and health glasses, fit non-prescription glasses, and should be specified in the budget.

Industrial Hygiene

Industrial hygiene programs provide the quantitative evidence needed to prove that hazardous conditions are properly controlled. Accurate, independent evaluation of the work site can be performed effectively and efficiently by an in-house industrial hygienist or by an outside consultant. Either is appropriate, and depending on the desire of company management, a highly paid, well-qualified, full-time industrial hygienist may not be justifiable.

The budget for an industrial hygiene program must include the equipment necessary to perform the required tasks. Such equipment typically includes air sampling pumps, noise survey equipment (such as sound level meters and noise dosimeter), gas detectors, radiation survey meters, ventilation survey equipment, and heat stress equipment.

The cost of sample analysis also must be included in the budget. Costs vary according to the complexity and the time necessary to run a particular sample.

Radiation Safety

The basis for all radiation safety and health programs is the monitoring of personnel exposure. The cost varies with the number of badges used or how often they are processed. These surveys usually require portable monitoring equipment and wipes. The equipment requires regular calibration and the wipes must be analyzed, usually by an outside service on a fee basis. The monitoring equipment may be leased from the firm doing the calibration or purchased and sent out for calibration on a scheduled basis. The time and materials necessary for these surveys can be substantial.

Hearing Conservation

With the knowledge that noise-induced hearing loss can result in a disability claim, it becomes imperative to reduce noise to acceptable levels and provide protection to employees when noise reduction is not possible. A hearing conservation program identifies sources, gives a pre-placement audiometric examination for all employees, provides periodic audiometric testing for employees working in high-noise areas, fits and issues hearing protection devices, and provides ongoing noise surveys and evaluation of all work sites. Audiometric examinations are usually conducted by an outside firm but can be conducted internally with the proper equipment. Wherever these examinations are conducted, they require properly calibrated equipment used by a trained and certified audiometric technician.

SAFETY AND HEALTH TRAINING

An important element in any comprehensive safety and health program is training. Safety and health training should be designed to prevent accidents and teach the employee the hazards associated with his or her job. When budgeting for training, several factors must be considered. The instructor's time in teaching the class must be taken into account. If an outside instructor is desired, the procedure is fairly simple: a fee is paid and a service is rendered. For inside service, one must account not only for the time needed to present the material but also for the instructor's preparation time.

Refresher training is necessary for employees to remain up-to-date on new developments and for recertification. For example, if the company has a requirement that all electricians be cardiopulmonary resuscitation (CPR) certified, there is an annual requirement. The materials necessary for a training class must also be considered. These include handouts, slides, quizzes, and samples. The widespread use of audiovisual aids has opened a new perspective on training. Videos can be purchased and shown repeatedly, with little instruction required.

Some organizations have developed Internet or web-based safety and health courses to train employees in specific hazards and operations. These can be very effective and can be tailored to a specific company's needs.

Construction Safety

New construction projects for facility renovation or for new buildings and facilities are usually well planned. The safety and health effort required for this activity must also be considered. The problems encountered on construction sites are sometimes unique and different from those in a manufacturing effort. Construction safety and health orders have different requirements from general industry safety and health orders and may require some research by the safety and health department to understand them fully. The time necessary to do this research should be included in the budget.

Subcontractors may cause specific safety and health problems due to the increased use of poorly trained personnel or inadequate equipment. The safety and health department must budget extra time for each subcontractor to ensure the safety and health requirements on a particular job are understood.

A review of drawings is also very important to the safety and health of a construction project. The safety and health engineer can identify particularly

Crane Safety

Cranes and hoists need special handling. They require routine inspections by operators and also by another individual who is qualified to inspect them. Nondestructive testing may be needed periodically to ensure this equipment is safe to operate, which may require the use of a consultant or an outside service.

Respiratory Protection

Many manufacturing companies have some personnel who must wear respirators periodically. The costs of a respiratory protection program must be listed in the budget. All personnel who wear a respirator must pass a physical examination. These costs must be included in the budget.

Training in the limitations of the respirator and proper respirator selection must be provided. An employee who uses a respirator must be provided with one that fits his or her face. Testing for qualitative or quantitative fit must be provided whereby the individual wears the respirator in a test atmosphere. Depending on the type of testing, the equipment can be expensive. Different sizes and types of respirators must be available for fitting and testing of personnel. Consumable materials for the testing and the cleaning of equipment after testing must be included in the budget.

ENVIRONMENTAL SAFETY

The impact of industry on the environment comes in many forms: noise, hazardous materials, oil spills, ionizing and non-ionizing radiation, or thermal energy. The sources of environmental impact must be identified and controlled, and the various programs needed to carry out this task must be established and used by the safety and health department.

Hazardous Materials

Materials used by the company must be identified, their hazards noted, and controls placed to ensure that such materials do not harm personnel or the environment while they are in the facility. A Safety and Health Data Sheet must be on file for all hazardous materials received, or one must be developed and provided to the end users.

hazardous operations and can ensure that he or she is present to make certain such operations are conducted in the safest manner possible.

A hazardous materials spill plan must be developed and used. This plan can require a significant amount of equipment, materials, and personnel. Personnel must be trained initially and annually to respond correctly to different types of spills. Another approach is to hire an outside contractor to respond to spills.

Hazardous Waste

Many companies, regardless of size, must be concerned with the proper disposal of hazardous wastes. The mandatory use of the "manifest system" requires that all waste be identified and a chain of custody established for it. This manifest system and the cradle-to-grave responsibility for waste require a significant amount of time and effort, which must be budgeted.

Annual reporting of waste usage to federal or state regulatory agencies can take a significant amount of time and documentation. If hazardous waste is stored at a company site for any time period, it may require a significant amount of contingency planning. Planning is needed for the proper response to a spill of hazardous waste in order to prevent it from becoming a problem for personnel or the environment. These plans may require a consultant who is an expert in their preparation or much work by in-house personnel. Such plans should be updated annually or as significant changes occur. Their development, implementation, and documentation must be budgeted.

Oil Spill Control

If a company stores a significant amount of oil at its facility, it may be required to have a spill prevention, containment, and countermeasure (SPCC) plan. This plan, mandated by federal regulations, requires extensive research and documentation. It involves inspections of tanks and other storage vessels and documentation of these inspections. If a spill occurs, the SPCC plan directs how the company will respond to the cleanup and how it will provide funds for cleanup costs. Documentation and implementation of SPCC plans can be extensive and expensive.

PRODUCT SAFETY

An effective product safety and health program must identify all the potential hazards in the product or service, warn the user they exist, and follow the product from cradle to grave. An effective safety and health program covers the product from concept through disposal. The safety and health professional should review the drawings, prepare a preliminary design and operations hazard analysis, ensure the labeling and operations manuals describe the hazards, and provide appropriate and detailed documentation. Analytical testing may be required to verify information in the hazard analysis. Stress analyses may have to be conducted and various types and levels of inspection performed to ensure that the product is made as designed. The effective program requires the safety and health professional to review the data and ensure all safety and health requirements are met. Time for consultation with legal personnel, design engineers, operations personnel, and repair and maintenance personnel must be budgeted to ensure all aspects of the product are evaluated.

ADMINISTRATION

One common mistake made in the budgeting process is failing to budget for the administrative portions of the program. The following subsections discuss several items to be considered when budgeting for administrative tasks in a safety and health department.

SAFETY AND HEALTH MANAGEMENT PROGRAM

The safety and health management program must include the tasks necessary to supervise, evaluate, and counsel personnel assigned to the safety and health department. These individuals can include non-technical personnel, such as secretaries, clerks, and technical personnel, including industrial hygienists, health physicians, and safety and health professionals. The program should include the costs of acquiring any new employees, either for replacement or for growth of the department. The tasks associated with acquiring new employees include advertising, recruiting, interviewing, hiring, orientation, and training. The costs associated with planning, organizing, and using new products must also be included in the budget. A hazardous process has a direct and significant effect on the personnel of the safety and health department. The time it takes the safety and health professional to review the process, evaluate personnel hazards, evaluate the environmental impact, evaluate and recommend controls, and plan the monitoring strategy can be substantial.

Materials required for the successful implementation of a safety and health program can be a serious cost factor. When considering materials, the budget for safety and health management should include the administrative supplies and services necessary for the smooth operation of the safety and health department. Materials should be budgeted for on the basis of annual use. This might seem to be a minor cost, unworthy of much effort, but in the area of documentation, the safety and health department must be thorough.

Equipment acquisition or repair must also be included in the budget. For example, the need for an additional computer in the safety and health department or repair of air monitoring equipment are not unusual needs, but they must be anticipated to prepare an accurate budget.

SUMMARY

In terms of budgeting, safety and health professionals must understand not only what activities need to be accomplished but also how much it costs to complete these tasks. This allows for planning for all the expected occurrences and sets the platform for anticipating non-recurrent expenditures. Knowing basic budgeting concepts allows professionals to manage the overall program and its related costs.

REFERENCES AND SUGGESTED FURTHER READING

Risk Mapping

Alexander, D. (1994). "The Economics of Ergonomics." Proceedings of the Human Factors and Ergonomics Society Annual Meeting-1994. *Human Factors Society*, pp. 696–700.

Canada, J.R. and J. A. White. (1980). *Capital Investment Decision Analysis for Man Agement and Engineering.* NJ: Prentice-Hall.

Crandall, R. (1983). *Controlling Industrial Pollution,* Washington, DC: Brookings Institute.

Davis, J.R. (1998). "Effective Decision-Making for Ergonomic Problem Solving." Chapter II in: *Ergonomic Process Management.*

Davis, J.R. (1995). "Automation and Other Strategies for Compliance with OSHA Ergonomics." Proceedings of 1995 International Industrial Engineering Conference, *International Institute of Industrial Engineers*, pp. 592–99.

General Accounting Office. (1997). *Worker Protection: Private Sector Ergonomics Programs Yield Positive Results*, General Accounting Office, GAO/HSAFETY-97-163.

Grant, E.L., Ireson, W.G., and Leavenworth, R.S. (1990). "Aspects of Economy for Regulated Business." Chapter 17 in: *Principles of Engineering Economy.* New York: John Wiley & Sons.

Hansen, M.D., H.W. Grotewold, and R.M. Harley. (1997). Dollars and Sense: Using Financial Principles in the Safety and Health Profession. *Professional Safety and Health*, 42(6): 36–40.

Harford, J.D. (1978). "Firm Behavior Under Imperfectly Enforceable Pollution Standards." *Journal of Environmental Economics and Management*, 5: 26–43.

Harrington, K.H., and Rose, S.E. (February, 1998). *Using Risk Mapping for Investment Decisions*, Mary Kay O'Connor Process Safety and Health Center, First Annual Process Safety and Health Symposium, College Station, TX.

Jones, C.A. (1989). Standard Setting with Incomplete Enforcement Revisited. *Journal of Policy Analysis and Management*, 8(1): 72–87.

National Safe Workplace Institute (NSWI). (1992). *Basic Information on Workplace Safety and health & Health*, p. 16.

Owen, J.V. (1991). Ergonomics in Action. *Manufacturing Engineering*, 106(6): 30–34.

Schelling, T.C. (1983). *Incentives for Environmental Protection*. Cambridge, MA: MIT Press.

Slote, L. (1987). "How to Reduce Work Injuries in a Cost-Effective Way." Chapter 22 in: *Handbook of Occupational Safety and Health*. New York: John Wiley & Sons.

Viscusi, W.K. (1986). "The Impact of Occupational Safety and Health Regulation, 1973–1983." *Rand Journal of Economics*, 17(4): 567–80.

Zeleny, M. (1992). *Multiple Criteria Decisionmaking*. New York: McGraw-Hill.

Time Value of Money

Allison, William W. (1986). *Profitable Risk Control*, American Society of Safety and Health Engineers, Des Plaines.

Blank, Leland, and Tarquin, Anthony. (2012). *Engineering Economy*, 7th Edition, McGraw-Hill, New York.

Brigham, Eugene F. (1989). *Fundamentals of Financial Management*, Dryden Press, Chicago, IL.

English, William. (1988). *Strategies for Effective Workers Compensation Cost Control*, American Society of Safety and health Engineers, Des Plaines.

Grant, Eugene L. and Bell Lawrence F. (1964). *Basic Accounting and Cost Accounting*, McGraw-Hill, New York.

Nikolai, Loren A. and Bazley, John D. (1988). *Intermediate Accounting*, PWS-Kent Publishing, Boston, MA.

Riggs, James L. (1976). *Production System: Planning, Analysis and Control*, Wiley/Hamilton Series in Management and Administration, New York.

Spellman, Frank R., Whiting, Nancy E. (2009). The Handbook of Safety Engineering: Principles and Applications, Government Institutes, *Engineering Economy*, Chapter 10, pp. 119–26.

Tarquin, Anthony J. and Blank, Leland T. (1976). *Engineering Economy*, McGraw-Hill, New York.

Theland, David S. (1988). Management of the Work Environment, Selected Safety and Health & Health Readings, Volume III, Project Minerva, *U.S. Department of Health and Human Services*.

The Cost of Human Life (Office of Management and Budget). http://www.theglobalist.com/the-cost-of-a-human-life-statistically-speaking/.

The Cost of Human Life (Nuclear Regulatory Commission, Occupational Safety and Health Administration and the Environmental Protection Agency). http://www.huffingtonpost.com/elliott-negin/why-is-the-nuclear-regula_b_7415022.html.

Turner, Wayne, C., Mize J.H., and Case, Kenneth E. (1987). *Introduction to Industrial and Systems Engineering*, Prentice-Hall, Englewood Cliff, NJ.

Watts, James P. and Ruder, Mark L. (1992). Las Vegas Safety and Health Workshops, Certified Safety and health Professional, *Home Study Workbook*.

Chapter 14

The Recordkeeping Function— A Legal Perspective

Michael O'Toole

CASE

It is Monday morning and the shift is just getting underway. Bob, the maintenance supervisor, calls and informs Dan, the safety manager, that one of his employees is complaining of back pain and is not sure how to handle the situation. Dan makes his way to the maintenance department to provide assistance to Bob.

Upon arrival, Dan encounters Bob and also Charlie, one of the maintenance mechanics, in the office. When he inquires what is going on, Charlie explains that he has had this back pain for about a week, and it has steadily been getting worse. When asked how he hurt his back, he says that he is really not certain but thinks it may have been when he was cleaning up his work area about ten days ago.

When pressed for further details, all Charlie can report is that he remembers bending over to pick up some trash on the floor when felt a little "tweak" in his lower back. Bob instantly asked why he didn't report this to him at the time, as required by company policy. Charlie said that he really didn't think there was anything significant to report at the time and thought that "the tweak" would resolve itself by the next day. Over the next couple of days, it was more of a "nagging" discomfort and he still didn't think it was important enough to report. Over this past weekend, the pain got progressively more intense, and today it really hurts. Upon returning from the corporate doctor's office, Charlie was given a prescription medication for his back pain and a twenty-five-pound lifting restriction for the next two weeks.

Many supervisors, managers, and safety professionals have faced this or similar situations and are challenged how to handle the case. First, is not to confuse OSHA recordkeeping requirements with that of state Workers

Compensation requirements. Under OSHA's recordkeeping requirements for the logging of injuries and illnesses, it is the employers' decision, based on their investigation, whether or not to record a case.

INTRODUCTION

Critical components of the safety management system (SMS) are recordkeeping and reporting. SMS is dependent on accurate records and reports. Information is the lifeblood of the system. Somewhat vital, often necessary, usually costly, and frequently confusing, the recordkeeping and reporting requirements under OSHA, MSHA, and DOT must be considered. It is not always a clear line between what is required by regulations and what is simply required for efficiency. Literally hundreds of records may be required for a variety of reasons and purposes—far too many to be discussed in detail. Therefore, this chapter will be primarily focused on the recordkeeping and reporting requirements for safety and health from the Occupational Safety and Health Administration (OSHA), the Mine Safety and health Administration (MSHA), and the Department of Transportation (DOT), as it relates to over-the-road trucking. These recordkeeping requirements may be most challenging for small- to medium-sized businesses. But, keep in mind that the same requirements basically apply equally to all organizations, with a few narrow exceptions.

The chapter falls naturally into a couple of logical sections; the first deals with the reporting and recordkeeping requirements of the Occupational Safety and Health Act (OSHA). Although those requirements appear to be fairly straightforward, there have been a number of subtle changes over the past ten–fifteen years that could be troublesome if an organization is ignorant of those changes. The next section will be the major reporting and recordkeeping requirements for those employers covered by the Mine Safety and Health Act (MSHA). (There are more specific and detailed recordkeeping requirements for Coal Operators and for Metal—Non-Metal Operators than would appropriate for this chapter. Although about as clear as the OSHA requirements, violation of MSHA's reporting and recordkeeping requirements typically carry more severe financial penalties with several potential civil penalties. The last section will briefly cover the required records required by fleet operators by the Department of Transportation (DOT).

RECORDKEEPING

Safety and health-related records are vital to different entities for several different reasons. First, employers keep them because they are required. Second, and too often overlooked, is the fact that the survival of the business depends on records. Without critical records, employers simply don't know how the business is performing. Without records on accidents and injuries, employers may be unable to remedy hazards or address risks and build a sound accident prevention program. Moreover, without the legally required records on safety and health, key employees in organizations may be subject to civil penalties. For example, failure to collect and maintain information about prior workplace exposures to some chemicals may result in the company being held liable for a former employee's current or future health problems, even decades after the alleged exposure took place. If companies fail to keep proper records in connection with issues like employee health records, exposure records, or even manufactured or transported products and wastes, a court may consider this as evidence of laxness and make a product liability award far beyond the company's ability to handle.

Unfortunately, wisdom gained in safety and health recordkeeping is too often a matter of hindsight. The burden of recognizing what records or supplemental documentation may be needed, as well as having the awareness of what specific records are required, is really up to the company's management. An excuse of not recognizing the need to create and maintain key records is unacceptable during an agency administrative hearing, in a court of law, or at a stockholders' meeting. It is sometimes impossible to draw a line between "what is good to have" versus "what is absolutely necessary" in the way of records.

Recordkeeping under OSHA

In general, OSHA requires companies to record information on illnesses and injuries that pertain to the workplace. Basically, it is necessary to record relevant information that involves any fatality, lost workday, amputation, job restriction, job transfer, or certain other injuries that do not involve lost workdays. All cases of occupational illnesses are recorded, with or without medical treatment.

This is not to say that information on non-work-related injuries and illnesses need not be reported. These cases may not meet the OSHA, MSHA, or DOT recording or reporting criteria but may clearly require documentation related to states' Workers' Compensation laws. In the past, injuries sustained on the plant parking lot while the worker was out to lunch or participated in company-sponsored sports activities were considered "work-related" by OSHA. Although excluded from OSHA recordkeeping requirements today, the above examples would be considered "work-related" under most states' Workers Compensation laws. If there is any question at all as to whether an accident is or is not work-related, a report should be duly made and thorough documentation should be retained until there is little likelihood of action.

Handling the OSHA recordkeeping function is hardly trivial, since the Act states that making a false statement can be punished by a fine of as much as $10,000 and/or six months in prison. MSHA, under similar circumstances, can subject the company to fines, as well as civil and/or criminal penalties.

OSHA REPORTING REQUIREMENTS

All three agencies have specific requirements and specific events that trigger an employer's requirement to contact the agency.

The Occupational Safety and Health Administration requires a covered establishment to contact the agency within eight hours if any one of the following events occurs:

- The death of an employee that results from a workplace exposure.
- Hospitalization (IE admittance) of any employee as the result of a workplace exposure.
- Amputation that results from a work exposure.
- Loss of an eye within twenty-four hours of exposure.

Both federal and state OSHAs have telephone contact numbers for employers to report one of the above events. Should one of the "reportable" events occur on the weekend or after business hours, the employer is still obligated to contact the local OSHA office, and as a minimum, leave a message within the eight-hour time requirement.

TYPES AND FORMS OF RECORDS

Under OSHA, employers are required to keep accurate records on at least three forms in addition to supplemental accident records. These required forms are a log of occupational illnesses and injuries (OSHA 300), an annual summary of occupational illnesses and injuries (OSHA 300a), and a supplemental record of each occupational illness or injury (OSHA 301). The employer may elect to use and complete an internal accident/illness record form as long as it contains at least the same information as the OSHA 301. Forms are available on request from the nearest OSHA office, Bureau of Labor Statistics office, or from state OSHA agencies. In addition, employers can download an electronic version from several sites, including www.OSHA.gov.

Where state law requires no specific format, the record or log can be of a company's own design as long as it includes the required data. Figures 14.1, 14.2, and 14.3 show Forms 300, 300a, and 301, respectively, indicating how they should be filled out.

In addition to the written program, OSHA standards also require accurate and current records to be maintained regarding compliance with several regulations. Most of these records will be related to either training or evaluation of the effectiveness of the process. For example, employers need to have written programs and records on who and when training occurred for the Hazard Communication standard, the Permit Required Confined Space standard, and the Lock-Out/Tag-Out standard. For these and other OSHA standards, the employer must periodically (such as annually) evaluate the effectiveness of the employer's compliance with the standard.

Each of these has its own type of backup records, all of which can be examined in more detail in the OSHA standards themselves—also included on the OSHA website. Employers whose establishments operate under federally approved state OSHA plans must meet or exceed the same recordkeeping requirements as those states covered by federal OSHA, so that meeting the state requirements will simultaneously meet federal standards. If an employer has fewer than eleven employees, the employer is exempted from the federal recordkeeping (paperwork) requirement, but compliance with the standards is still a must! (NOTE: there are several states that require records are kept for a smaller number of employees.) Additionally, there are certain types of businesses or industries exempted from maintaining most of OSHA's required paperwork. Current examples are gasoline stations, florists, and legal establishments. The complete list of partially exempt industries can be found at: https://www.osha.gov/recordkeeping/ppt1/RK1exempttable.html.

The Log and Summary of Occupational Injuries and Illnesses (OSHA 300a) is a summary form that must be posted from the 1st of February and remain in place until the last day of April.

The Supplementary Record (OSHA 301 or the employer's equivalent) provides additional information. This form must be completed and present in the establishment within seven workdays after the injury or illness is reported. Entries are later transferred to the log (OSHA 300) from this form within the same seven day requirement. The material from the OSHA Form 300 is transferred to the (OSHA 300a) Annual Summary of Occupational Injuries and Illnesses (see figure 14.2). This form is then posted in the workplace from February 1 to April 30 for results of the previous year.

There are several specification standards that either require documentation by standard or by fiat. These are requirements for employers to ensure that certain tools or equipment be inspected regularly and/or prior to use. Some of these requirements imply that the inspection be documented. For example, fire extinguishers are to be inspected monthly to ensure they are physically present where they are supposed to be, as well as to confirm the extinguisher has adequate pressure indicated on the gauge. Other routine inspections requiring documentation to verify the inspection include items such as mobile equipment, wire ropes and slings, fall arrest systems, electrical extension cords, and welding and cutting equipment.

Recordkeeping is an important part of a radiation protection program if an employer has qualifying equipment. Ionizing and nonionizing radiation are covered by separate OSHA standards. A summary of the radiation-related records include the following:

1. A daily record of the contamination levels at the entrance of each radiation restricted area.
2. A record of daily-accumulated exposures on pocket devices.
3. A record of weekly or semi-monthly accumulated exposures on film badges.
4. A record of non-routine exposures monitored by special devices.
5. A record of exposures for employees working under special work permits.
6. A record of visitors' exposure records.

Discretion is allowed in the design of training records, the formats of accident/incident reports, and the records and minutes of plant safety committee meetings. In the latter case, for example, the various safety committees may include in their minutes such matters as safety projects, safety promotional activities, and any items requiring management action. While specific details may be required for the records, they also allow unusual freedom for the company safety consultant or manager to provide direction.

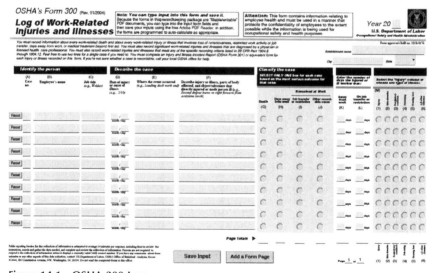

Figure 14.1 OSHA 300 Log

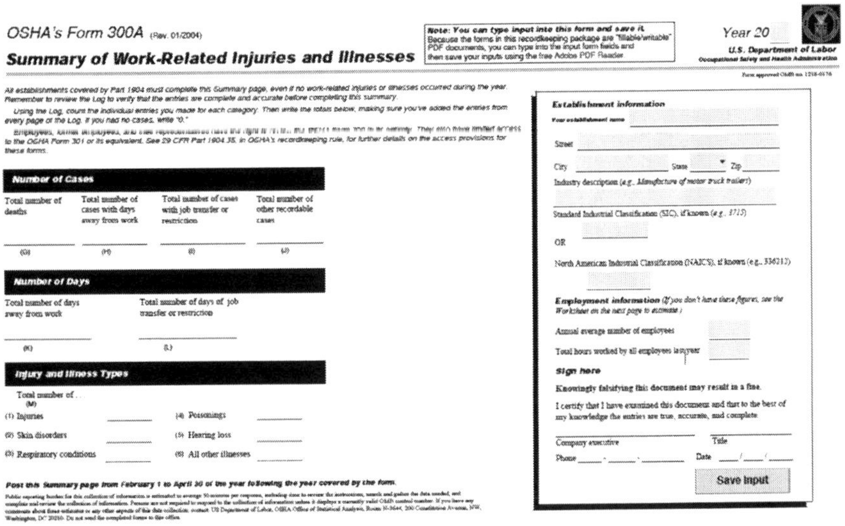

OSHA's Form 300A (Rev. 01/2004)

Summary of Work-Related Injuries and Illnesses

Note: You can type input into this form and save it. Because the forms in this recordkeeping package are "fillable/writable" PDF documents, you can type into the input form fields and then save your inputs using the free Adobe PDF Reader

Year 20

U.S. Department of Labor
Occupational Safety and Health Administration

Form approved OMB no. 1218-0176

All establishments covered by Part 1904 must complete this Summary page, even if no work-related injuries or illnesses occurred during the year. Remember to review the Log to verify that the entries are complete and accurate before completing this summary.

Using the Log, count the individual entries you made for each category. Then write the totals below, making sure you've added the entries from every page of the Log. If you had no cases, write "0."

Employees, former employees, and their representatives have the right to review the OSHA Form 300 in its entirety. They also have limited access to the OSHA Form 301 or its equivalent. See 29 CFR Part 1904.35, in OSHA's recordkeeping rule, for further details on the access provisions for these forms.

Number of Cases

Total number of deaths	Total number of cases with days away from work	Total number of cases with job transfer or restriction	Total number of other recordable cases
(G)	(H)	(I)	(J)

Number of Days

Total number of days away from work	Total number of days of job transfer or restriction
(K)	(L)

Injury and Illness Types

Total number of . . .
(M)

(1) Injuries		(4) Poisonings	
(2) Skin disorders		(5) Hearing loss	
(3) Respiratory conditions		(6) All other illnesses	

Post this Summary page from February 1 to April 30 of the year following the year covered by the form.

Establishment information

Your establishment name

Street

City State Zip

Industry description (*e.g., Manufacture of motor truck trailers*)

Standard Industrial Classification (SIC), if known (*e.g., 3715*)

OR

North American Industrial Classification (NAICS), if known (e.g., 336212)

Employment information (*If you don't have these figures, see the Worksheet on the next page to estimate.*)

Annual average number of employees

Total hours worked by all employees last year

Sign here

Knowingly falsifying this document may result in a fine.

I certify that I have examined this document and that to the best of my knowledge the entries are true, accurate, and complete.

Company executive Title

Phone Date

Save Input

Figure 14.2 OSHA 300A Summary

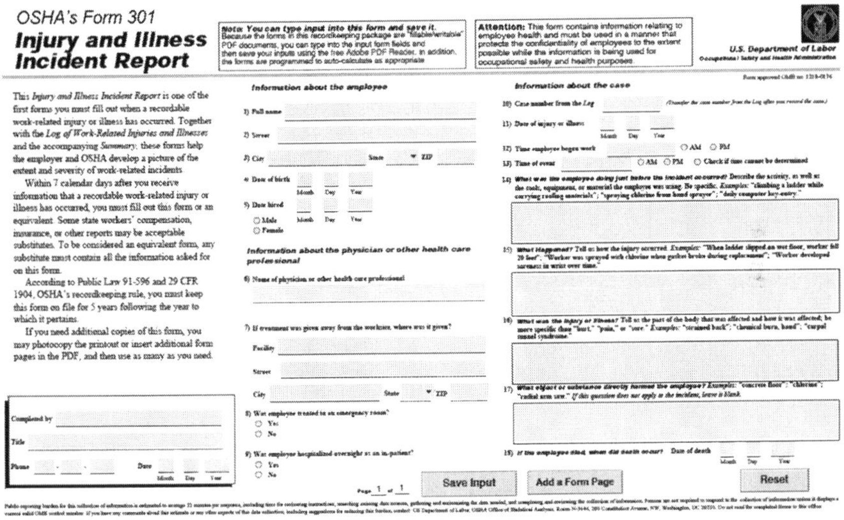

OSHA's Form 301

Injury and Illness Incident Report

Note: You can type input into this form and save it. Because the forms in this recordkeeping package are "fillable/writable" PDF documents, you can type into the input form fields and then save your inputs using the free Adobe PDF Reader. In addition, the forms are programmed to auto-calculate as appropriate.

Attention: This form contains information relating to employee health and must be used in a manner that protects the confidentiality of employees to the extent possible while the information is being used for occupational safety and health purposes.

U.S. Department of Labor
Occupational Safety and Health Administration

Form approved OMB no. 1218-0176

This *Injury and Illness Incident Report* is one of the first forms you must fill out when a recordable work-related injury or illness has occurred. Together with the *Log of Work-Related Injuries and Illnesses* and the accompanying *Summary*, these forms help the employer and OSHA develop a picture of the extent and severity of work-related incidents.

Within 7 calendar days after you receive information that a recordable work-related injury or illness has occurred, you must fill out this form or an equivalent. Some state workers' compensation, insurance, or other reports may be acceptable substitutes. To be considered an equivalent form, any substitute must contain all the information asked for on this form.

According to Public Law 91-596 and 29 CFR 1904, OSHA's recordkeeping rule, you must keep this form on file for 5 years following the year to which it pertains.

If you need additional copies of this form, you may photocopy the printout or insert additional form pages in the PDF, and then use as many as you need.

Completed by

Title

Phone Date Month Day Year

Information about the employee

1) Full name

2) Street

3) City State ZIP

4) Date of birth Month Day Year

5) Date hired Month Day Year

Male
Female

Information about the physician or other health care professional

6) Name of physician or other health care professional

7) If treatment was given away from the worksite, where was it given?

Facility

Street

City State ZIP

8) Was employee treated in an emergency room?
Yes
No

9) Was employee hospitalized overnight as an in-patient?
Yes
No

Information about the case

10) Case number from the *Log* (*Transfer the case number from the Log after you record the case.*)

11) Date of injury or illness Month Day Year

12) Time employee began work AM PM

13) Time of event AM PM Check if time cannot be determined

14) ***What was the employee doing just before the incident occurred?*** Describe the activity, as well as the tools, equipment, or material the employee was using. Be specific. *Examples:* "climbing a ladder while carrying roofing materials"; "spraying chlorine from hand sprayer"; "daily computer key-entry."

15) ***What Happened?*** Tell us how the injury occurred. *Examples:* "When ladder slipped on wet floor, worker fell 20 feet"; "Worker was sprayed with chlorine when gasket broke during replacement"; "Worker developed soreness in wrist over time."

16) ***What was the injury or illness?*** Tell us the part of the body that was affected and how it was affected; be more specific than "hurt," "pain," or "sore." *Examples:* "strained back"; "chemical burn, hand"; "carpal tunnel syndrome."

17) ***What object or substance directly harmed the employee?*** *Examples:* "concrete floor"; "chlorine"; "radial arm saw." *If this question does not apply to the incident, leave it blank.*

18) ***If the employee died, when did death occur?*** Date of death Month Day Year

Page 1 of 1 Save Input Add a Form Page Reset

Figure 14.3 OSHA 301 Incident Report

Definitions

Certain terms recur frequently in OSHA standards and directions, particularly those that pertain to recordkeeping. For clarity's sake, here are some brief definitions of these terms:

Occupational injury is any injury such as a cut, fracture, sprain, or amputation which results from a work accident or from an exposure involving a single incident in the work environment. (Note: Conditions resulting from animal bites, including insect or snakebites, or from a one-time exposure to chemicals are considered to be injuries.)

Occupational illness of an employee is any abnormal condition or disorder (other than one resulting from an occupational injury) caused by exposure to environmental factors associated with employment. It includes acute and chronic illness or disease that may be caused by inhalation, ingestion, or direct contact.

Medical treatment includes treatment administered by a physician or by registered professional medical personnel under the standing orders of a physician. Medical treatment does not include first aid treatment (one-time treatment and subsequent observation of minor scratches, cuts, burns, and so forth, which do not ordinarily require medical care), even though provided by a physician or registered professional personnel. Medical treatment includes actions such as the setting of bones, application of sutures, and ordering prescription medications.

Establishment is a single physical location where business is conducted or where service or industrial operations are performed (e.g., a factory, mill, store, hotel, movie theater, restaurant, farm, ranch, bank, warehouse, or office). Where distinctly separate activities are performed at a single physical location (such as construction activities operated from the same physical location as a lumber yard), each activity shall be treated as a separate establishment.

Work environment comprises the physical location, equipment, materials processed or used, and the kinds of operations performed in the course of an employee's work, whether on or off the employer's premises.

Injury and Accident Statistics

Although OSHA-associated statistics on illness and injury have been kept since the passage of the OSHAct, the available information greatly understates the size of safety and health problems. Prior to OSHA, there was no consistent or systematic attempt to collect data on the incidence of occupational illness and disease; even the data on accidents was grossly inadequate. Although the OSHAct brought new recordkeeping requirements, there is every reason to believe that sufficient information to properly identify problems or take effective corrective actions is still not collected.

One of the main reasons for OSHA recordkeeping is to size up the magnitude and nature of safety and health problems. However, official statistics are just the tip of the iceberg as far as the magnitude of occupational safety and health problems is concerned. While this lack of information is a national problem, its greatest impact is on smaller scale businesses, which lack guidance and resources for safety and health efforts.

The business world is understandably and typically reluctant to accept more reporting requirements and paperwork, since they question the cost-effectiveness of such measures. Cost versus benefits analysis on new reporting requirements has already resulted in a lowering of those requirements deemed to be a "burden" to the smaller business. A ready, practical solution is not likely for the near future. Some individual organizations have found that proper recordkeeping enables them to pinpoint safety and health hazards or problems, thus allowing them to improve future performance or reduce costs. For these businesses, the cost versus benefits question has already been answered. But for many organizations, it has yet to be seen or demonstrated.

More and more organizations are attempting to identify and focus on what are termed "leading indicators" of safety performance. Although there are books written about data use and safety, suffice it to say this is a potentially positive step in accident prevention efforts. With that being said, the use of data generated from the OSHA 300 and 300a forms appears to be of less value to those organizations in their accident prevention programs.

Quick Reference

The variety of records, forms, logs, and reports may seem almost too complex and extensive to grasp. How is the manager to know if all the proper forms and records are being kept? As a convenient way of answering this question, the Occupational Safety and Health Administration has a publication OSHA 2254-09R 2015 *Training Requirements in OSHA Standards*. This document is available on the OSHA website and should be downloaded for reference.

For example, the document covers the requirements for Exit Routes and Emergency Planning (SubPart E), as well as those requirements related to the use of personal protective equipment (SubPart I). The document covers not only requirements for General Industry, but also Maritime, Construction, and Agriculture.

Location, Retention, and Access to Records

All employers must maintain records for each establishment or workplace. Where different activities are performed at the same physical location, each activity must be addressed under the requirements of the OSHAct. For example, where one would have construction activities (CFR 1926) carried out in a lumberyard (CFR 1910), the action or activity would determine the applicable standard or regulation.

Transportation, construction, maritime, as well as manufacturing and agriculture firms may operate at widely dispersed locations. The place to which workers report or are assigned each day is considered a separate establishment. Separate records of illnesses or injuries must be maintained and available at each establishment. For example, an organization manufactures widgets and has a distribution warehouse ten miles away. Each would be considered a separate establishment and will be required to have required postings (IE OSHA Poster) and required records (IE OSHA 300, 300a 301, and SDS's). However, the organization can consolidate and even use digital records as long as they are readily available during normal business hours. If the employees do not regularly report to the same place, such as sales workers, technicians, or maintenance employees, a central location from which they are supervised will suffice as the establishment.

Forms required to be posted, such as OSHA 300a, must be posted even if there have been no injuries or illnesses. In that case, the forms will simply show zeroes under each heading. In the case of Annual Summary totals, the person responsible for the summary will verify it by signing it at the bottom, thus indicating that the totals are true and complete.

All required records must remain in the establishment (or be available electronically) for five years after the year to which they relate. If a business changes hands, the new owner must preserve the records for that required period. Exposure records, such as industrial hygiene sampling reports for noise, dust, gases, fibers, etc., and data analysis, must be maintained for thirty years post last exposure. Employee medical records required by specific standards, such as audiograms, chest x-rays, pulmonary function tests, and blood lead levels, must be maintained for the duration of employment plus thirty years. It should be noted that certain states or individual agencies (such as a state's Workers Compensation Board) might have even more stringent requirements.

Employees, or their authorized representative, have a right to view or receive a copy of their medical and/or exposure records, even after employment has been terminated. Affected employees have a right and should be told about the results of industrial hygiene monitoring conducted in their work area, such as dose or measured exposure levels to noise, carbon monoxide, or silica lead, solvents, and so on.

Authorized federal or state agencies have the right to inspect or copy OSHA records at any reasonable time. Exposure and medical records may be seen and copied by responsible representatives of OSHA or NIOSH (National Institute for Occupational Safety and Health).

Medical Records

The question of medical records not only pertains to OSHA, MSHA, and DOT, but also has general applications. The various federal agency requirements for accurate recordkeeping placed upon employers to keep reports of work-related deaths, injuries, and illnesses for varying lengths of time may appear confusing and overly burdensome, especially if an organization is covered by more than one of the agencies. However, some of these same records can aid management by providing data for use in job placement, establishing health and physical standards for certain jobs, documenting workers compensation cases, providing data for epidemiological study, and assisting management in the task of improving company safety and health practices.

When dealing with employee medical records, employers must be conscious of the Health Insurance Portability and Accountability Act of 1996 (HIPAA). This set of regulations was developed to protect the privacy and security of certain health information. The U.S. Department of Health and Human Services (HHS) published what are commonly known as the HIPAA Privacy Rule and the HIPAA Security Rule.

The Privacy Rule, or *Standards for Privacy of Individually Identifiable Health Information*, establishes national standards for the protection of certain health information. The *Security Standards for the Protection of Electronic Protected Health Information* (the Security Rule) were established for protecting certain health information that is held or transferred in electronic form. The Security Rule operationalizes the protections contained in the Privacy Rule by addressing the technical and non-technical safeguards that organizations called "covered entities" must put in place to secure individuals' "electronic protected health information" (e-PHI).

The Security Rule defines "confidentiality" to mean that e-PHI is not available or disclosed to unauthorized persons. The Security Rule's confidentiality requirements support the Privacy Rule's prohibitions against improper uses and disclosures of PHI. The Security rule also promotes the two additional goals of maintaining the integrity and availability of e-PHI. Under the Security Rule, "integrity" means that e-PHI is not altered or destroyed in an unauthorized manner. "Availability" means that e-PHI is accessible and usable on demand by an authorized person.

Employers may generate medical and exposure records as they attempt to comply with various regulations under OSHA, MSHA, and/or DOT. Ensuring compliance with HIPAA regulations and the requirements of the three agencies is often a very thin line. Since this guideline is not intended to comprise legal advice, if in doubt when handing employee medical or exposure records, an attorney should be consulted.

Exposure records generated in a workplace will depend largely on the type of industry and potential to generate exposures to physical or chemical hazards.

The employer must maintain records generated by exposure monitoring for thirty years. Some typical exposure records might include the following:

- Noise monitoring
- Air monitoring for gases, mists, vapors, dust, or fumes that may be present in the workplace
- Passive or active monitoring for ionizing radiation

Medical records that might be generated as the result of regulatory agency requirements will also vary depending on the industry and potential exposure to harmful agents in the workplace. The employer must maintain these medical records for a period of the employee's employment plus thirty years. Examples of medical records that must be maintained by the employer include but are not limited to the following:

- Audiograms
- Chest x-rays, pulmonary function tests, and EKGs required as part of the employer's respiratory protection program
- Blood lead levels
- Vision screening
- Drug test results required by DOT

This requirement does not extend to health insurance claims records, first-aid treatment, and medical records of employees with less than one year of employment, as long as those records were offered to the employee at the time of separation.

RECORDKEEPING UNDER THE MINE SAFETY AND HEALTH ADMINISTRATION

Many companies in the United States have operations covered by the the Mine Safety and Health Administration (MSHA). Their regulations require the mine operator to contact the agency within fifteen minutes of becoming aware of any of the following events:

- The death of an individual at a mine
- An injury to an individual at a mine which has a reasonable potential to have caused death
- An entrapment of an individual for more than thirty minutes or which as a reasonable potential to cause death
- An unplanned inundation of a mine by a liquid or gas

- An unplanned inundation of a mine of gas or dust
- In underground mines, an unplanned fire not extinguished within ten minutes of discovery; in surface mines and surface areas of underground mines, an unplanned fire not extinguished within thirty minutes of discovery
- An unplanned ignition or explosion of a blasting agent or an explosive
- An unplanned roof fall at or above the anchorage zone in active workings where roof bolts are in use or in an unplanned roof or rib fall in active workings that impairs ventilation or impedes passage
- A coal or rock outburst that causes withdrawal of miners or which disrupts regular mining activity for more than one hour
- An unstable condition at an impoundment, refuse pile, or culm bank which requires emergency action in order to prevent failure or which causes individuals to evacuate an area or failure of an impoundment, refuse pile, or culm bank
- Damage to hoisting equipment in a shaft or slope which endangers an individual or which interferes with use of the equipment for more than thirty minutes
- An event at a mine which causes death or bodily injury to an individual not at the mine at the time the event occurs.

If an accident occurs, an operator shall immediately contact the MSHA District Office having jurisdiction over its mine. If an operator cannot contact the appropriate MSHA District Office, it shall immediately contact the MSHA Headquarters Office in Arlington, Virginia, by telephone at (800) 746-1553.

If an accident occurs, an operator shall immediately contact the MSHA District Office having jurisdiction over its mine. If an operator cannot contact the appropriate MSHA District Office, he or she shall immediately contact the MSHA Headquarters Office in Arlington, Virginia, by telephone at (800) 746-1553. The operator shall contact MSHA as described at once without delay and within fifteen minutes. If communications are lost because of an emergency or other unexpected event, the operator shall notify MSHA at once without delay and within fifteen minutes of having access to a telephone or other means of communication.

MSHA appears to recognize the difficulties inherent in determining what constitutes "reasonable potential to cause death," especially in cases where a decision must be made in a matter of minutes or where an all-hands rescue or recovery may be in progress. The agency suggests that experience and common medical knowledge point to concussions, cases requiring cardiopulmonary resuscitation (CPR), limb amputations, major upper body blunt force trauma, and cases of intermittent or extended unconsciousness, as among the type of injuries which have a reasonable potential to cause death.

Even with MSHA's guidance, fifteen minutes is often not enough time to analyze whether an event is a "defined" accident. The "fifteen-minute rule" improperly prioritizes accident reporting over emergency response. Despite its recognition that small mines may have difficulty complying with the fifteen-minute requirement because all personnel could be engaged in emergency response, MSHA refused to include in its final rule any exceptions for small mines. Nonetheless, MSHA will continue to vigorously enforce this provision at large and mid-size mines.

It is, therefore, more critical than ever to have a plan in place establishing procedures for mine personnel—primary and alternates—to make an immediate report to MSHA in the event of an accident. Mine operators should also ensure their site personnel are trained and re-trained on the new accident reporting requirements and mandate a company rule that the instant a member of management thinks an event might meet MSHA's revised fifteen-minute reporting requirement, notify MSHA. On the other hand, if one believes MSHA has improperly cited the mine for allegedly violating Section 50.10, engage the agency in some thoughtful discussion, and seek counsel. Although MSHA is serious about aggressively enforcing the fifteen-minute notification requirement, it remains a possibility that having poorly written citations vacated is likely.

The 30 CFR 50.1 implements Sections 103(e) and 111 of the Federal Coal Mine Health and Safety Act of 1969, 30 U.S.C. 801 et seq., and Sections 4 and 13 of the Federal Metal and Nonmetallic Mine Safety Act, 30 U.S.C. 721 et seq., and applies to operators of coal, metal, and nonmetallic mines. It requires operators to immediately notify the Mine Safety and Health Administration (MSHA) of accidents, requires operators to investigate accidents, and restricts disturbance of accident related areas.

This part also requires operators to file reports pertaining to accidents, occupational injuries, and occupational illnesses, as well as employment and coal production data with MSHA and requires operators to maintain copies of reports at relevant mine offices. The purpose of this part is to implement MSHA's authority to investigate and to obtain and utilize information pertaining to accidents, injuries, and illnesses occurring or originating in mines.

Each operator shall maintain at the mine office a supply of MSHA Mine Accident, Injury, and Illness Report Form 7000-1. These may be obtained from the MSHA District Office. Each operator shall report each accident, occupational injury, or occupational illness at the mine. The principal officer in charge of health and safety at the mine, or the supervisor of the mine area in which an accident or occupational injury occurs or may have originated, shall complete or review the form in accordance with the instructions and criteria in §§50.20-1 through 50.20-7.

Each operator shall report each occupational injury or occupational illness on one set of forms. If more than one miner is injured in the same accident or is affected simultaneously with the same occupational illness, an operator shall complete a separate set of forms for each miner affected. To the extent that the form is not self-explanatory, an operator shall complete the form in accordance with the instructions in §50.20-1 and criteria contained in §§50.20-2 through 50.20-7.

Each operator of a mine shall maintain a copy of each investigation report required to be prepared under §50.11 at the mine office closest to the mine for five years after the concurrence.

Each operator shall maintain a copy of each report submitted under §50.20 or §50.30 at the mine office closest to the mine for five years after submission. Upon request by the Mine Safety and Health Administration, an operator shall make a copy of any report submitted under §50.20 or §50.30 available to MSHA for inspection or copying.

DEPARTMENT OF TRANSPORTATION

Corporations must also consider regulations established by the Department of Transportation (DOT). The secretary of Transportation is responsible for regulating interstate commerce. As part of its responsibility, the DOT has issued advisory standards which, while not mandatory, are useful to companies engaged in transport. One part in particular, "Cargo Security Advisory Standards," deals with seal accountability, high value commodity storage, and internal accounting procedures. These areas suggest that certain records are directly connected with safety, as well as security.

For the fleet safety director, there are three major types of records to maintain:

1. Driver records
2. Accident records
3. Maintenance and compliance records

Driver records, generally speaking, should include the following items:

1. Application for employment
2. Medical examiner's certificate
3. Responses from agencies and past employers on driver's record
4. Certificate of driver's road test and a copy of the classified license
5. Certificate of written test
6. Certification of traffic violations, submitted annually
7. Annual record review copy
8. Safety and service awards
9. Accident record
10. Commendation letters
11. Violation of rules and regulations
12. Disciplinary actions
13. Training and retraining records
14. Road patrol reports

Accident information records are developed to help determine the causes of accidents, as well as their solutions. They frequently serve as evidence in defending suits or supporting damage claims. Accident reports must be filed with federal, state, and local agencies (depending on the situation) and to the workers compensation agency when injuries are involved. In the latter case, an accident report must also be filed according to OSHA requirements. Maintenance records are also a vital part of fleet safety records. The American Trucking Association has developed a system suitable for large or small fleets for either manual or computerized application. The manager is urged to familiarize himself with these guidelines.

Recordkeeping under the DOT

The DOT requires a trucking or bus company must report a traffic crash if it involves any of the following:

- Any truck that has a gross vehicle weight rating (GVWR) of more than 10,000 pounds or a gross combined weight rating (GCWR) of more than 10,000 pounds used on public highways, OR
- A fatality: any person(s) killed in or outside of any vehicle (truck, bus, car, etc.) involved in the crash or who dies within thirty days of the crash as the result of an injury sustained in the crash; OR
- An injury: any person(s) injured as the result of the crash who immediately receives medical treatment away from the crash scene; OR
- A tow-away: any motor vehicle (truck, bus, car, etc.) disabled as a result of the crash and transported away from the scene by a tow truck or other vehicle:

Required Reporting under DOT Federal Regulation

The Hazardous Materials Regulations (HMR; 49 CFR Parts 171-180) require certain types of incidents be reported to the Research and Special Programs Administration (RSPA). Section 171.15 of the HMR requires an immediate telephonic report (within twelve hours) of certain types of hazardous materials incidents and a follow-up written report. Section 171.16 requires a written report for certain types of hazardous materials incidents within thirty days. Each type of report is explained below. (The full text of these sections is at the end of the instructions.)

Who Must Complete the Report?

Any person in possession of a hazardous material during transportation, including loading, unloading, and storage, incidental to transportation, must report to the DOT if certain conditions are met. This means that when the conditions apply for completing the report, the entity having physical control of the shipment is responsible for filling out and filing DOT Form F 5800.1. For example, if a shipper is carrying hazardous material, the consignee is unloading the material and there is an incident involving this material, the consignee is responsible for filling out and filing the form. However, if the consignee is unloading the hazardous material and causes a hazardous materials incident involving a consignment intended for someone else, the shipper is responsible for filling out and filing the form.

Hazardous material—a substance or material that has been determined to be capable of posing an unreasonable risk to health, safety, and property when transported in commerce and that has been so designated. The term includes hazardous substances, hazardous wastes, marine pollutants, elevated temperature materials, materials designated as hazardous under the provisions of § 172.101, the Hazardous Materials Table (HMT), and materials that meet the defining criteria for hazard classes and divisions in Part 173.

Hazardous substance—a material, including its mixtures and solutions, that—

1. Is listed in Appendix A to § 172.101;
2. Is in a quantity, in one package, which equals or exceeds the reportable quantity (RQ) listed in Appendix A to § 172.101; and
3. When in a mixture or solution—
   i. For radionuclides, conforms to paragraph 7 of Appendix A to § 172.101.
   ii. For other than radionuclides, is in a concentration by weight which equals or exceeds the concentration corresponding to the RQ of the material.

The term *hazardous substance* does not include petroleum, including crude oil or any fraction thereof, which is not otherwise specifically listed or designated as a hazardous substance in Appendix.
A to § 172.101, and the term does not include natural gas, natural gas liquids, liquefied natural gas, or synthetic gas usable for fuel (or mixtures of natural gas and such synthetic gas).

Hazardous waste—any material that is subject to the Hazardous Waste Manifest Requirements of the U.S. Environmental Protection Agency specified in 40 CFR Part 262.

Marine pollutant—a material that is listed in Appendix B to § 172.101 (also see § 171.4) and, when in a solution or mixture of one or more marine pollutants, is packaged in a concentration that equals or exceeds:

1. Ten percent by weight of the solution or mixture for materials listed in Appendix B; or

2. One percent by weight of the solution or mixture for materials that are identified as severe marine pollutants in Appendix B.

Undeclared hazardous material —means a hazardous material that is

1. Subject to any of the hazard communication requirements in subparts C (Shipping Papers), D (Marking), E (Labeling), and F (Placarding) of Part 172 of this subchapter, or an alternative marking requirement in Part 173 of this subchapter (such as §§ 173.4(a)(10) and 173.6(c)); and
2. Offered for transportation in commerce without any visible indication to the person accepting the hazardous material for transportation that a hazardous material is present, on either an accompanying shipping document, or the outside of a transport vehicle, freight container, or package.

A Written Report (DOT Form F 5800.1) Must Be Submitted.

Under § 171.16, a written report must be submitted within 30 days after any of the following:

- An incident that was reported by telephonic notice under § 171.15;
- An unintentional release (see definitions) of a hazardous material during transportation including loading, unloading, and temporary storage related to transportation;
- A hazardous waste is released;
- An undeclared shipment with no release is discovered; or
- A specification cargo tank 1,000 gallons or greater containing any hazardous materials that—
  1. received structural damage to the lading retention system or damage that requires repair to a system intended to protect the lading retention system, and
  2. did not have a release.

A Telephone Report Must Be Made.

Under § 171.15, telephone notice must be provided within 12 hours after the incident occurs when one of the following conditions occurs during the course of transportation and is a direct result of the hazardous material:

- A person is killed.
- A person receives an injury requiring admittance to a hospital.
- The general public is evacuated for one hour or more.

- One or more major transportation arteries or facilities are closed for one hour or more.
- The operational flight plan or routine of an aircraft is altered.
- Fire, breakage, spillage, or suspected radioactive contamination occurs involving a radioactive material.
- Fire, breakage, spillage, or suspected contamination occurs involving an infectious substance other than a diagnostic specimen or regulated medical waste.
- There is a release of a marine pollutant in quantities exceeding 450 liters (119) gallons for liquids or 400 kilograms (882 pounds) for solids.
- A situation exists of such a nature that in the judgment of the person in possession of the hazardous material, it should be reported to DOT's National Response Center (NRC), even though it does not meet the above criteria.

Employers may decide that the situation should be reported even though it does not meet any of the criteria. It is important to request the NRC report number when you make a telephonic report.

OSHA Recordkeeping Case

In the case presented at the beginning of the chapter, an employee is reporting back pain from a non-specific activity ten days prior to his reporting to his supervisor. OSHA does provide the employer some guidance related to the decision-making process when deciding to record a case on the OSHA 300 Log of Occupation Injuries and Illnesses.

The first step is to determine if there is an injury or illness that would require the case to be recorded. Cases to be recorded typically will have one of the following conditions:

- The employee received medical treatment beyond first aid. OSHA provides ample examples on their website to guide employers on this issue.
- The employee was unable to return to work for at least one day as the result of their medical condition. The day of injury is not counted here.
- The employee receives restrictions from a licensed medical professional that restricts their work activities or motion. Examples are a lifting restriction below what an employee is normally expected to lift, or the physician restricts the employee's use of ladders or climbing stairs.
- The employee is transferred to another position.

The next step is to determine if the case is work-related. Cases that are work-related are as follows:

- Cases caused by events or exposures in the work environment
- Cases contributed to by events or exposures in the work environment
- Cases significantly aggravated by events or exposures in the work environment

Now to conclude the case at the beginning of the chapter 14 to record or to not record? There are three alternatives for addressing this case, any of which are acceptable; however, these alternatives are not intended to constitute legal advice.

Alternative #1—Do not record the case on the OSHA 300. In this instance, the employer has to make some assumptions about the case. The employee reported an injury that appears to be work-related, received medical treatment beyond first aid, and has had medical restrictions on work. The employer could decide the employee was, in fact, not injured at work, since he could not identify a specific activity that "caused" the injury, and the employee failed to report the alleged incident in a timely manner. In this case the employer would be assuming the employee's presentation of the facts was less than credible. Upon inspection of the OSHA 300, 300a, and 301 forms, an OSHA inspector could conclude the employer was overly conservative in the interpretation of CFR 1904.7—OSHA's Recordkeeping Guidelines. If the compliance officer detects a pattern of this kind of under-reporting, the compliance officer could issue citations for failure to properly comply with this requirement.

Alternative #2—Record the case and move on. This is perhaps the simplest and most straightforward of the alternatives, but could also present a potential management challenge. Many organizations tie at least a portion of supervisors' and managers' performance review to safety outcomes. One of the outcomes often used is the number of OSHA recordable injuries on the OSHA 300 log for the year. OSHA has found cases in the past that as the result of certain compensation systems, there is actually an incentive to not record cases on the OSHA 300 log resulting in under-reporting of occupational injuries. Again, should an OSHA compliance officer detect a pattern of under-reporting, citations would likely follow.

Alternative #3—The employer will record the case on the OSHA 300 log and then take a pencil and draw a line through the case. The employer would not include this case in the totals transferred onto the OSHA 300a at the end of the year. Using this approach with a case that the employer questions the work-relatedness, should eliminate suggestions of under-reporting of cases, since the employer, in fact, listed the case on the OSHA 300 log. If an OSHA Compliance Officer questions the case and suggests this is actually work-related, the employer would simply need to erase the pencil line and include the case on the OSHA 300a.

The case presented at the beginning of this chapter was purely fictional and presented to demonstrate a somewhat challenging situation employers may face from time-to-time. Alternative #2 above is the absolute best choice in this author's opinion and experience. The employer must be aware of pressure to under-report by either local management or employees due to some incentive, be it financial, social, or otherwise. In cases where the employer has real doubts about the work-related nature of the case, Alternative #3 provides a potential path to maintain accurate records and minimize the likelihood of citations and associated fines.

RECORDKEEPING AND SMS

An SMS generally requires records in addition to those mandated by law. Consideration must be given as to whether or not safety goals are being met. This may not be accomplished through OSHA, MSHA, or DOT recordkeeping alone. The questions of what should be measured and how to determine the effectiveness of SMS must be answered. Meaningful data must be fed back into the system so adjustments and system improvements can be made. Records support decision-making, so appropriate data must be available to management. Typical additional records include internal and external audit reports, employee self-reporting results data and follow-up actions, and any indications needed to determine whether goals are being met (Stolzer, 2009).

CONCLUSION

This chapter has presented a brief summary of the recordkeeping requirements connected with safety and health, particularly as they impact the average manager. It is meant to serve as a general guideline to provide some clarity and scope of the field. This brief review serves to underline the importance of thorough documentation and recordkeeping, even beyond the minimum compliance requirements

Consideration of preventive action and decision-making functions in the SMS requires records and reports far beyond the scope of those called for by OSHA, MSHA, or DOT. The conscientious manager will be quick to grasp the fact that the cost of this recordkeeping may be more than justified by the amount it could save the company in the long run, should citations and associated fines follow. Additional records may also need to be maintained in order to effectively evaluate and improve the SMS.

Even this very brief view of recordkeeping requirements makes one thing obvious: recordkeeping is a major task today, involving considerable expertise and detail, not to mention a significant amount of time, paperwork, and money. The Small Business Regulatory Flexibility Act of 1980 had provided some small amount of relief from federal government recordkeeping regulations. Among other things, this requires each federal agency to analyze the impact of its rules and regulation on small, privately owned businesses. One portion of the Act calls for careful consideration by the issuing agency regarding the burdens represented by any new regulation, and a review of existing rules to see if the reporting load may be reduced. It is a long-term process, but it does offer some hope, at least for the small business. For most managers, the recordkeeping task remains one of the most time-consuming and growing areas of administrative work.

REFERENCES

Accident Prevention Manual for Industrial Operations, 7th ed. (1974). Chicago: National Safety Council.

Anton, T.J. (1979). *Occupational Safety and Health Management.* New York: McGraw-Hill.

Ashford, N.A. (1977). *Crisis in the Workplace.* Cambridge, Mass: MIT Press.

Gray, I. (1975). *Product Liability.* New York: American Management Association.

Grimaldi, J.V. and Simonds, R.H. (1975). *Safety Management.* Homewood, IL: Richard D. Irwin.

Lenz, Matthew, Jr. *Risk Management Manual.* Santa Monica, CA: Merritt Corp.

Lollar, Felix, *OSHA Reference Manual.* Santa Monica, CA: Merritt Corp.

Petersen, D. (1978). *Techniques of Safety Management*, 2nd ed. New York: McGraw-Hill.

Pope, W.C. (1979). *Organizing Your Safety Department.* Alexandria, VA: Safety Management Information Systems.

Rosenfield, H.N. (November, 1980). *Wire From Washington.* National Safety News.

Stolzer, A. (2009). *How to Recognize an SMS When You See One.* [Video File]. Retrieved from: https://erau.instructure.com/courses/31947/pages/1-dot-4-video-recognizing-an-sms?module_item_id=1305496.

Training Requirements of the Occupational Safety and Health Standards. (n.d.). Washington: Department of Labor.

Whitman, L.E. (1979). *Fire Prevention.* Chicago: Nelson-Hal.

Wynholds, H. and Bass, L. (1977). *Product Liability.* Princeton, NJ: ECON, Inc.

Chapter 15

Industrial Hygiene Within the SMS

Tracy L. Zontek and Burton R. Ogle

CASE

A 200,000 square foot municipal building in a large metropolitan area was the target of a bioterrorist attack. The building was contaminated with deadly *Bacillus anthracis* (anthrax) spores, and it was believed the spores were weaponized by milling and mixing them with an additive to keep the spores separated and more easily suspended in the air. The Incident Command System (ICS) protocol was deployed and experts from the U.S. Army, Centers for Disease Control and Prevention, U.S. Homeland Security, Federal Bureau of Investigation, et al. gathered in the nearby command center. Over 3,000 building occupants, including workers and visitors, were decontaminated and quarantined. No one or group took responsibility for the attack; therefore, the target was unclear. ICS *technical experts* from the Environmental Protection Agency (EPA) were put in charge of decontaminating the building. Over the next four days, chlorine dioxide (CLO2) was pumped throughout the building utilizing the duct work. U.S. Public Health Service (USPHS), Environmental Health Officers (EHO), fitted with Level-4 personal protective equipment were deployed to decontaminate areas which would not be reached by the CLO2. Afterward, EHO personnel were deployed throughout the building to monitor the ambient air for anthrax spores. Spores were found in high numbers in all of the air and surface samples. Technical experts from EPA were dumfounded the anthrax could survive their decontamination efforts. USPHS EHOs trained in industrial hygiene suggested that since spores are resistant to heat, cold, chemicals, etc., that the humidity should be raised in the building in an effort to vegetate the spores, making them more susceptible to the disinfectants. Over the next three days, facility engineers were able to increase the humidity from 20 percent to 75 percent in the building. Following the humidity adjustment, the identical decontamination protocol was utilized, as before with CLO2, along with spot disinfection. Subsequent air sampling demonstrated the new protocol was highly effective in that no anthrax was present in the samples; the building was declared to be fully decontaminated.

Questions to consider:

1. Why did the technical experts not collaborate with USPHS industrial hygienists prior to the initial decontamination efforts?
2. What resources should the operations management team use to identify qualified professionals?
3. What types of resources can the operations management team provide to prevent, mitigate, and control hazards?

Introduction

Recent events clearly demonstrate the importance of addressing health and safety issues that affect both the workforce and the community at large. Dramatic changes are taking place, almost daily, as a result of new processes and new chemical compounds being introduced into the workplace and throughout the world—nanoscale materials, nonionizing radiation due to increased use of cell phones, and globalization of goods and services. The impact of the media and Internet has led to a more aware citizenry that is becoming highly vocal on environmental factors, industrial pollution, potential exposures during natural and terrorist events, and occupational exposures. As different countries and regions have varied regulations on occupational health practices, the impacts of globalization on products and services create emerging issues for industrial hygienists and occupational health professionals. A systems approach must be taken to protect employees and also to develop a sustainable business and supply chain.

This chapter traces the development of industrial hygiene as a major professional discipline and provides an understanding of the overall function of the industrial hygienist. Ten major functions are identified to help provide the manager with a clearer understanding of the valuable resources of the industrial hygienist and how best to utilize this expertise. This will help to assure the most rewarding returns from the organization's resources—financial and personnel—in productivity, worker morale, and community relations.

INTRODUCTION TO INDUSTRIAL HYGIENE

The field of industrial hygiene is continually changing and has emerged from one of investigation and response to incidents to a discipline dedicated to preventive program management. An industrial hygienist is dedicated to improving workplace safety though the art and science of anticipation, recognition, evaluation, and control of environmental factors and stressors that may cause occupational injuries and illnesses or impaired health and well-being of employees or members of the community.

Industrial hygienists typically have at least a baccalaureate degree in engineering, chemistry, physics, or a closely related biological or physical science and experience in the field. It is not uncommon for industrial hygienists to complete graduate work in a more specialized area of industrial hygiene. Areas that industrial hygienists, depending on level of training, may competently practice include indoor air quality; evaluating and controlling occupational and environmental exposure to physical, biological, and chemical agents; emergency response planning and community right-to-know; occupational diseases; cumulative trauma disorders; ionizing and nonionizing radiation; hazardous materials and waste management.

The culminating certification of an industrial hygienist is administered by the American Board of Industrial Hygiene (ABIH). ABIH offers a certification examination that, if successfully passed, allows the candidate to be known as a *Certified Industrial Hygienist (CIH)*. Candidates must have both academic training and professional experience under a CIH to be qualified to take the examination. The certification examination covers topics such as air sampling and instrumentation; analytical chemistry; basic science; biohazards; engineering controls/ventilation; ergonomics; health risk analysis and hazard communication; program management; biostatistics and epidemiology; community exposure; noise; non-engineering controls; ionizing and nonionizing radiation; thermal stressors; toxicology; and work environments/industrial processes. The pass rate on the certification examination is typically 40 percent.

Other occupational health and safety professionals may have certification through the Board of Certified Safety Professionals (BCSP). The culminating certification under BCSP is the *Certified Safety Professional*. Similar to a CIH, in order to qualify for this examination, the candidate must have academic and professional experience in the discipline. The pass rate on the certification examination is typically 55 percent. BCSP also offers a number of certifications for younger professionals, as well as supervisors and front-line workers including Occupational Health and Safety Technologist, Construction Health and Safety Technician, Safety Trained Supervisor, Safety Trained Supervisor Construction, and Certified Environmental, Safety and Health Trainer. Supporting employees to obtain safety certification enhances their ability to perform effectively and may provide legal support later. Further investments in employees lead to a safer workplace and lower workers compensation costs. There are approximately 300 certification programs and titles in the United States related to environment, safety, and health with various degrees of rigor and credibility. Certifications offered by the ABIH and BCSP are internationally recognized and respected.

While the discipline of industrial hygiene may at first seem singularly scientific, the art of industrial hygiene is expressed in working with people and identifying problems and creative solutions to new problems. One of the major endeavors of any industrial hygienist is education and training for those they are protecting, and their first loyalty and protection always is to the worker. Behavior-based safety, a safety partnership between employees and management, steeped in psychology, organizational culture, stress research, and industry best practices is a critical component of a successful industrial hygienist and safety program within the safety management system (SMS). Furthermore, the occupational health and safety team should be part of the management team and the overall SMS. Industrial hygienists, as part of the occupational health and safety team and overall SMS, must also understand management perspectives, such as finance, product life cycle, public relations, marketing, strategic planning, and production. Industrial hygienists must face the challenge to acquire knowledge in areas outside the industrial hygiene discipline and to develop the skills necessary to becoming effective members of the management team.

RESPONSIBILITIES OF HEALTH AND SAFETY PROFESSIONALS

In reviewing the general approach to industrial hygiene, one must keep in mind that the conservation of human resources is a continuing process that involves all levels of management and every employee. The ultimate goal of the occupational health and safety professional is appropriately stated in the preamble to the Occupational and Safety Health Act of 1970, "to assure safe and healthful working conditions" for "every working man and woman in the Nation."

Management Responsibility

In the context of providing safe and healthful working conditions, it is management's responsibility to establish overall policies and guidelines to meet this goal, most significantly support and resources. It is the responsibility of the safety and health team to see that these goals are met and to provide guidance and insight into the best techniques and practices.

A safety and health team plays an integral part in a business's overall health and safety program and SMS, in that it is responsible for (1) conducting an effective health and safety program by coordinating educational, engineering, and enforcement activities; (2) providing educational materials; (3) assisting supervisors in teaching safety and health rules and procedures;

(4) conducting surveys of potentially hazardous areas to ensure that proper practices and procedures are followed; and (5) recommending changes to keep pace with technological advancements.

Supervisor Responsibilities

Similarly, it is each supervisor's responsibility to maintain safe working conditions within his or her department or group and to directly implement the health and safety initiatives. Program objectives should include (1) maintaining a work environment that assures the maximum safety for employees; (2) instructing each new employee in the safe performance of his or her job; (3) ensuring meticulous housekeeping practices are developed and employed at all times; (4) furnishing all employees with proper personal protective devices and enforcing the use of such equipment; and (5) informing the health and safety professionals of any operation or condition which appears to present a potential hazard to employees.

Employee Responsibilities

Everyone is individually responsible for contributing toward the success of a health and safety program. Each individual's responsibility includes (1) observing all safety rules and making maximum use of prescribed personal protective equipment and clothing; (2) following practices and procedures established to conserve health and safety; (3) notifying his or her supervisor immediately when certain conditions or practices are discovered which may cause personal injury or property damage; (4) developing and practicing good habits of personal hygiene and housekeeping; and (5) reporting near misses and accidents in a timely fashion.

One of the first considerations in a properly oriented occupational health and safety program is creating an "awareness" in each employee through an effective educational and training effort. Thus, everyone feels that a safe environment is their responsibility, and they are empowered to make changes for the better. The industrial hygienist, as part of the health and safety team, is the lynchpin to coordinate and execute these goals while minimizing risk to environmental factors and work stressors.

ENVIRONMENTAL FACTORS AND WORK STRESSORS

In order to recognize environmental factors and stressors influencing health, the industrial hygienist must be familiar with work operations and processes. The categories of stressors of interest are as follows:

1. *Chemical stressors*—liquid, dust, fume, mist, vapor, or gas;
2. *Physical stressors*—nonionizing and ionizing radiation, noise, vibration, and extremes of temperature and pressure;
3. *Biological stressors*—insects, mites, molds, yeasts, fungi, bacteria, blood borne pathogens, and viruses; and
4. *Ergonomic stressors*—body position in relation to task, monotony, boredom, repetitive motion, worry, work pressure, and fatigue.

The industrial hygienist recognizes such stress may immediately endanger life and health or cause significant discomfort and inefficiency. Industrial hygiene is recognized as a science (and an art) devoted to the recognition, evaluation, and control of those environmental stressors—chemical, physical, biological, and ergonomic—that can cause sickness, impaired health, or significant discomfort to employees or residents of the community.

One of the objectives of the industrial hygienist is to determine the level of environmental demands in the workplace and to determine workers' limitations, as well as tolerance for the stress. In the general case, long continued work under extremely low-stress conditions produces lack of alertness, and work under high-stress conditions produces fatigue and increased risk. Both factors play a leading role in causing accidents. The goal then is to determine, for each worker, an optimal stress level resulting in increased productivity and performance and decreased accidents.

The variety of substances and processes that present occupational health hazards steadily increases. Recently developed raw materials and methods of manufacture, and new combinations of their older counterparts, create new environmental stressors. Improved techniques for the prevention and control of existing hazards and stressors are continually being developed.

THE ART AND SCIENCE OF INDUSTRIAL HYGIENE

Evaluation of the magnitude of environmental factors and stressors arising in or from the workplace is done by the industrial hygienist, aided by quantitative measurement of the chemical, physical, biological, or ergonomic stressors and experience and perspective into the various political, regulatory, non-regulatory and best practices that surround the management and control of these stressors. The basic tenets of industrial hygiene are anticipation, recognition, evaluation, and control. In the United States, the Occupational Health and Safety Administration sets regulations that are the baseline for employee health and wellness. As companies seek a more systems approach to productivity, the industrial hygienist may need to consider the source of raw materials and chemicals, cradle-to-grave environmental impacts, hazardous waste management, financial impact of different types of controls; ultimately, serve as part of the procurement, engineering, marketing, safety, medical, maintenance, and management teams to understand how each decision may affect another area of business.

Functions of the Industrial Hygienist

The following functions examine the types of work an industrial hygienist may perform depending on the size and scope of the organization.

1. Direct the industrial hygiene program.
2. Examine the work environment and environs by
    a. Studying work operations and processes to get full details of the nature of the work, materials, and equipment used, products and by-products, number of employees, and hours of work;
    b. Measuring the magnitude of exposure or nuisance to workers and the public. In doing so, he or she must:
        i. Select or devise methods and instruments suitable for such measurements;
        ii. Conduct such measurements in accordance with approved/standard methods; and
        iii. Study and test material associated with the work operation.
    c. Studying and testing biological materials, by chemical and physical means, when such examinations will aid in determining the extent of exposure.
3. Interpret results of the examination of the work environment and environs in terms of ability to impair health, nature of the impairment, workers' efficiency, and community nuisance and/or damage, and present specific conclusions to appropriate parties such as management and health officials.
4. Determine the need for, or effectiveness of, control measures, and when necessary, recommend procedures which will be suitable and effective for both the environment and environs.
5. Prepare rules, policies, standards, and procedures for the healthful conduct of work and the prevention of nuisance in the community.
6. Present expert testimony before courts of law, hearing boards, workers compensation commissions, regulatory agencies, and legally appointed investigative bodies covering all matters pertaining to industrial hygiene.

7. Prepare appropriate text for labels and precautionary information for materials and products used by workers and the public in compliance with the United Nations' Globally Harmonized System of Classification and Labeling of Chemicals (OSHA Hazard Communication Program).
8. Conduct programs for the training and education of workers and the public in the prevention of occupational disease and community nuisance.
9. Direct epidemiologic studies among workers and industries to discover the presence of occupational disease and to establish or improve threshold limit values or standards as guides for the maintenance of health and efficiency.
10. Conduct research to advance knowledge of the effects of occupation upon health and to advance the means of preventing occupational health impairment, community air pollution, noise, nuisance, and related problems.

CONCLUSION

The art and science of industrial hygiene professionals will ensure that each facility has effectively anticipated, recognized, evaluated, and controlled the environmental stressors. It is the role of management to ensure that the industrial hygiene program is professionally staffed, adequately resourced, and supports health and safety over operational objectives.

REFERENCES

American Board of Industrial Hygiene. (2014). *About ABIH*. Retrieved from: http://abih.org/.

American Conference of Governmental Industrial Hygienists. (2016). *Defining the Science of Occupational and Environmental Health*. Retrieved from: http://www.acgih.org/home.

American Industrial Hygiene Association. (2016). *AIHA Protecting Worker Health*. Retrieved from: https://www.aiha.org/about-aiha/Press/Pages/default.aspx.

American National Standards Institute. (2015). *ANSI*. Retrieved from: http://www.ansi.org/.

American Society of Safety Engineers. (2014). *A Global Professional Membership Organization*. Retrieved from: http://www.asse.org/assets/1/7/ASSEFactSheet_1014.pdf.

DeReamer, R. (1980). *Modern Safety and Health Technology*, John Wiley and Sons: New York.

National Institute for Occupational Safety and Health (NIOSH). (2015). *Providing National and World Leadership to Prevent Workplace Illnesses and Injuries*. Retrieved from: http://www.cdc.gov/niosh/.

National Safety Council. (1983). *Accident Prevention Manual for Industrial Operations*, National Safety Council: Itasca, IL.

Occupational Safety and Health Administration. (n.d.). *About OSHA.* Retrieved from: https;//www.osha.gov/about.html.

Occupational Safety and Health Review Commission. (n.d.). *About the commission.* Retrieved from: http://www.oshrc.gov/.

Pierce, J.O. (1985). *Chapter 17—Industrial Hygiene and Radiation Safety* in Safety Management Planning, Ferry, T. (editor). Government Institutes: Rockville, MD.

Chapter 16

Radiation Safety

Tracy L. Zontek and Burton R. Ogle

CASE

Dietta Meryl is exposed to relatively low levels of radiation every day. She knows she is under the limit generally permitted for employees. In recent months Dietta and her husband have been discussing starting a family. She is unsure if the limits to which she can be exposed now are going to possibly cause problems with a newly conceived fetus. Dietta approaches her manager and asks for his guidance.

INTRODUCTION

In addition to problems directly related to industrial hygiene, many industries are finding an increasing and expanding use of radioactive materials and processes involving ionizing and nonionizing radiation. Within the SMS, it is important for management to have at least a basic understanding of the operations, materials, and safety precautions necessary to safely utilize them. Radioactive materials, whether raw, refined, or waste, must be protected from those who might use them for terrorist activities. Management has the principle responsibility to ensure radioactive material/waste inventories are tracked as carefully as possible. This section provides an overview of radiation professionals, radiation safety standards, and regulations and information regarding radiation management of hazards and controls. In a global society, awareness that radiation regulations are also addressed by international standards is critical.

The use of radioactive materials has become widespread throughout industry, healthcare, commercial applications, and consumer-based goods.

Essentially, every home has one or more smoke detectors powered by the isotope *Americium* (Am-241); *Cobalt* (Co-60) is widely used in industrial radiology applications; and brachytherapy, a common treatment for many types of cancer, often utilizes *Iridium* (Ir-192) to shrink or eliminate malignancies. Cell phones and microwaves are daily essentials. Ultraviolet and infrared radiation from the sun or anthropogenic sources are continuous impacts. The misunderstanding and fear associated with *radiation* has grown, and it is more important than ever for management to have a basic understanding of the problems that must be addressed when working with radioactive materials and other sources of ionizing and nonionizing radiation. This discussion of *radiation safety management* briefly covers the structure of a basic radiation safety program necessary to handle either radioactive sources or radiation-producing devices.

The safety procedures and legal requirements for using radioactive material have largely evolved along with the growing use of such agents. The radiation safety field in the United States dates back to the days of the Manhattan Project during World War II, and the first major compendium of federal law on the subject was developed in 1954. The concurrent development of regulations and safety procedures with the evolution of the *radiation industry* has produced an excellent safety record throughout the more than sixty years of work with ionizing radiation. Ionizing radiation has enough energy that during an interaction with an atom, it can remove tightly bound electrons from the outer orbit of an atom, causing the atom to become charged or ionized. Non-ionizing radiation does not possess the energy to remove electrons from the outer orbit of an atom. While the past sixty or so years have mainly focused on ionizing radiation in the field of radiation safety, non-ionizing sources are becoming more abundant in both commercial and consumer goods and also pose a threat to human health and welfare. Natural particles of ionizing radiation consist of alpha, beta, and gamma radiation. Anthropogenic ionizing radiation sources, such as x-rays, pose hazards essentially identical to gamma radiation. Examples of non-ionizing radiation are microwaves, radio waves, infrared and ultraviolet radiation. It is a mistake to believe that ionizing radiation is somehow more dangerous than non-ionizing radiation since, if improperly utilized, both can be lethal.

RADIATION SAFETY PROFESSIONALS

For decades, radiation has been beneficial to humans, from treating cancer to generating electrical power. When utilized recklessly, radiation can cause injury to human beings and other animals and destroy welfare (plants, aquatic ecosystems, soil, etc.). Safe management of nuclear reactors, nuclear weapons, high-energy particle accelerators, x-ray machines, and radionuclides used in biomedical research and therapy is essential. Accidents with radioactive substances may lead to irreparable environmental contamination and significant mortality and morbidity. Many industries, medical facilities, defense plants, and research laboratories demand professionals who understand radiation hazards and their prevention and control. Below are some of the more prominent professionals in the field of radiation safety.

- *Health Physicists*—Health physicists work in a variety of disciplines including research, industry, education, environmental protection, and enforcement of government regulations. Although the health physicist usually concentrates in one of these disciplines, a professional health physicist typically performs duties in several areas. In research, health physicists investigate principles by which radiation interacts with matter and living systems. Health physicists also study environmental levels of radioactivity and the effects of radiation on biological systems on earth and in space. This information is used in many ways, ranging from designing radiation-detection instrumentation to establishing radiation protection standards. Industrial or applied health physicists draw upon their technical knowledge and varied experience to advise and make recommendations to management regarding methods and equipment for use in radiation work. The health physicist also assists engineers and scientists in designing facilities and new radiation control programs. As the primary consultant during any radiation emergency, a health physicist commonly has total control of the involved area. Health physicists working in education develop and instruct training programs for future health physicists. They also provide any necessary training for radiation workers and the general public. These individuals instruct workers and other health physicists on the level of risk associated with particular radiation sources and methods used to reduce risk. One goal is to help individuals understand the relative degree of risk of radiation exposure. In most cases, the risk is no greater than that found in other industries. Using an array of radiation detectors, a technician assesses the level of internalized radioactivity in an individual. Health physicists who work in regulatory enforcement must have knowledge and experience concerning all types of radiation hazards in order to establish guidelines for adequate radiation control. These guidelines help society receive the greatest benefits from radiation sources at the lowest possible exposure (Health Physics Society, 2016). Managers should be aware of the certification designation for health physics—the Certified Health Physicist (CHP), which designates that the individual has demonstrated field content and maintains continuing education in the field. The CHP designation is overseen by the American Board of Health Physics.

- *Medical Physicist*—Individuals who practice health physics in the healthcare arena are often referred to as *medical physicists*. Certification is designated through related experience and examination by the American Board of Medical Physics. The ABMP certifies physicists and related scientists to practice clinical medical physics. The certification is known as Medical Health Physicist (MHP).
- *Radiation Protection Technologist*—A radiation protection technologist (RPT) is a person engaged in providing radiation protection to the radiation worker, the general public, and the environment from the effects of ionizing radiation. The RPT has a basic understanding of the natural laws of ionizing radiation, the mechanism of radiation damage, methods of detection, and hazards assessment. The RPTs' tasks are accomplished by providing supervisory, administrative, and/or physical control and utilizing sound health physics principles in compliance with local and statutory requirements and accepted industry practices. The RPT mitigates hazards associated with radioactive material and ionizing radiation-producing devices, always adhering to the "as low as reasonably achievable (ALARA)" philosophy. RPT are eligible to sit for the examination for certification after five years of experience offered by the National Registry of Radiation Protection Technologists (NRRPT, 2016).

STANDARDS

The procedures (and often the regulations) have been continuously updated as more knowledge has been acquired. Radiation standards are established by individuals states (typically to regulate x-ray devices); federally (as with the Nuclear Regulatory Agency—NRC), and internationally. Below is a brief examination of the sources of these standards.

International Standards

Three international organizations recommend radiation protection levels: the International Commission on Radiological Protection (ICRP), the International Atomic Energy Agency (IAEA), and the International Commission on Radiation Units and Measurements (ICRU).

The Second International Congress of Radiology established the ICRP in 1928. Although initially concerned with the safety of medical radiology, it now covers safety for all sources of radiation. Its mission is "to deal with the basic principles of radiation protection and to leave to various national protection committees the responsibility of introducing the detailed technical regulations, recommendations, or codes of practice best suited to the needs of their individual countries." The ICRP is the principal source of recommendations on safe radiation levels. Members come from many countries and include scientists, physicians, and engineers. Organized in 1956 to promote the peaceful uses of nuclear energy, the IAEA is a specialized agency of the United Nations.

The IAEA publishes both standards and recommendations in addition to books on nuclear science and technology written by consultants or groups of experts invited from member states. Created in 1925, the ICRU develops international recommendations regarding quantities and units of radiation and radioactivity, procedures for their measurement and application in clinical radiology and radiobiology, and physical data needed to ensure uniformity in reporting on their applications.

U.S. Standards

U.S. groups involved with recommending radiation standards include the National Council on Radiation Protection and Measurements (NCRP) and federal and state agencies. The NCRP began its work in 1929 as the Advisory Committee on X-Ray and Radium Protection. Congress chartered the organization in 1964 as the NCRP to address the scientific and technical aspects of radiation protection. The nonprofit corporation is not a federal agency; although, its recommendations are part of the basis of federal, state, and local regulations dealing with radiation hazards. The organization draws its members from public and private universities, medical centers, national and private laboratories, the government, and industry solely on the basis of their scientific expertise.

The U.S. Environmental Protection Agency is responsible for recommending federal guidance on radiation protection for use by federal agencies in their regulatory processes and for establishing standards to protect the general environment from radioactive material under a variety of authorities, including the Clean Air Act, Safe Drinking Water Act, and Superfund and Atomic Energy Act.

The independent NRC is the federal agency responsible for regulating commercial nuclear technologies. The NRC prescribes and enforces separate limits on the amount of radiation that workers and members of the public can receive from all pathways, such as air and water. These regulations apply to operators of nuclear power plants, as well as industrial and medical facilities licensed to use man-made radioactive materials. The NRC bases its regulations on recommendations made by the NCRP, the ICRP, and on the EPA's federal guidance and standards.

ELEMENTS OF A RADIATION SAFETY PROGRAM

Management should not attempt to develop, implement, and manage a radiation safety program; however, managers need to be aware of when they are required, who best in their organization will oversee the program, personnel who might be subject (included in) to the program, and when they need to hire a consultant to bring in the expertise for a radiation safety program. A generic radiation safety program can be broken down into four basic elements or components as outlined below.

1. An *administrative program* to assure the use of radioactive material and/or radiation is in compliance with all applicable state and federal laws and regulations.
2. An *operational program* to assure the safety and health of employees and the public who may potentially be exposed to the radioactive material and/or radiation used in the facility.
3. A *recordkeeping program* to show compliance with all laws and regulations; to document employee exposures; to document releases to the environment; and to generate a file of evidence for potential defense of liability suits.
4. An *education and training program* to teach employees how to safely handle radioactive material and/or radiation and in some cases to inform the public of both what is being done and the fact that it is being done safely.

Obviously, the detailed design of a comprehensive radiation safety program is beyond the scope of company management and in many cases will require the services of a professional consultant such as a health physicist.

Administrative Program

Within the SMS the internal organization responsible for a company's radiation safety program will, in most cases, be directed by the radiation safety officer (RSO) whose name will appear on the license to obtain and use radioactive material and who will be technically and often administratively responsible for the program. This person may range from a part-time technical consultant (for a small program) to a full-time, technically trained professional. It is important that the RSO not be a part of or report directly to the user group within the company. That is, the RSO must be independent of the people who actually use the radioactive material and/or ionizing radiation. This point cannot be made too strongly. There are many examples where the RSO was located in the organization in a position reporting to one of the users, and these arrangements have created internal conflict.

In most cases, in addition to the RSO, a Radiation Safety Committee (RSC) must also be established, but very simple uses of one or two sources will usually not require such a committee. The RSC should have a minimum of three members, not including the RSO. The members should be chosen from fairly high levels of management and may include some of the direct users of the sources of ionizing radiation. The RSC reviews plans and procedures proposed by the RSO, approves uses of radiation sources within the organization, and provides management support to the RSO in his or her dealings with users.

The basic law governing the use of radioactive material is found in 10 CFR 20, administered by the Nuclear Regulatory Commission (NRC). Federal law allows the NRC to enter into agreements with the individual states to enable a state to set up and administer its own radiation control program. There are currently some 26 "agreement states" that have entered into such an arrangement. Note the NRC is authorized to control only man-made radioactive material. Any regulations governing the use of radiation-producing machines or natural radioactive materials (e.g. thorium and radium) will always be under state regulations, regardless of whether or not the particular state is an agreement state.

In order to obtain (purchase) and use radioactive material, an organization must have a license issued either by the NRC or the agreement state. If this license is from the NRC, it will specify compliance with 10 CFR 20; if it is issued by the state, it will specify compliance with that state's appropriate laws. In all cases state laws will be equal to or more restrictive than federal regulations. The license will also include specific requirements dealing with the unique conditions for each particular applicant. The RSO will be named on the license, and he or she will be responsible for making sure the organization meets all of the requirements established by that license.

The quantity, variety, and complexity of use of radioactive material will determine whether a consulting, part-time, or full-time RSO is required, and by the same token, the size and qualifications of the support staff will be determined by the requirements of a particular license. The license never specifies staffing requirements. The conditions of the license must be met, and the licensing agency leaves it to the discretion of management as to how to meet those conditions.

The NRC and/or agreement states perform periodic inspections and audits (usually annually or bi-annually) of the licensee's premises and programs. Any violations of the general or specific license conditions are citable. Such citations may include a monetary fine, if the violation is considered flagrant or likely to cause a real safety hazard. If a large number of very serious violations (e.g., endangering the public near the facility) are found, the license may be suspended or revoked.

The acquisition of radiation-producing machines (e.g., x-ray machines) does not necessarily require a license. However, particular states may require registration of such machines, the licensing of machine operators, and annual or bi-annual inspection programs; and generally, if an organization uses only radiation-producing machines, it will not be required to have a formally designated RSO or a highly organized and complex radiation safety program. Radiation exposure monitoring of workers and periodic calibration and/or surveys of machines will, however, be necessary.

Operational Program

When radiation is an issue within the SMS, a properly designed radiation safety program assures appropriate controls and procedures are in place, so that radiation workers will not be exposed to harmful amounts of ionizing or nonionizing radiation from external or internal sources. In addition, the program must contain elements to assure radioactive material is not routinely released to the environment (except for certain "allowed" situations) and that members of the general public are not exposed to sources of ionizing radiation from either machines or radioactive material. In 1956 the National Commission on Radiological Protection (NCRP) recommended permissible doses of radiation for both radiation workers and members of the general public. These doses are legally permissible and, based on the best scientific evidence, will produce no observable effects over a worker's fifty year working lifetime. Nonetheless, the basic philosophy of any radiation safety program is to keep all doses "as low as reasonably achievable." In recent years, the NRC has required all radiation safety programs licensed by either the NRC or an agreement state must contain an approved ALARA program. However, ALARA is not a number and what may be reasonably achievable for one licensee may not be reasonable for another.

External Exposure Control

Whether the source of ionizing radiation is a machine, a large sealed source (e.g., an industrial radiography source), or an unsealed one of radioactive material (e.g., unsealed sources used in research), the initial consideration in designing a radiation safety program is to minimize the probability an employee can be exposed to harmful doses of radiation through routine work or by accident. Planned exposures, for either repetitive, routine, or occasional maintenance type operations may be controlled by using one or more of the following three procedures:

Time—Limit the time a worker is exposed to the source of radioactivity. This technique is probably most useful for non-routine operations where work steps can be carefully preplanned and timed and/or where several workers do various parts of the job in order to keep all personnel doses in the permissible range. In addition, when working with radioactive sources (particularly unsealed sources), it is good practice to run through the procedure at least once without the presence of radioactive material to reduce workers' exposure time, while increasing their knowledge of how to perform the job.

Distance—Increasing the distance from a radioactive source reduces the exposure rate by the square of the ratio of the two distances (i.e., doubling the distance reduces the exposure rate by a factor of 4; tripling distance reduces exposure rate by nine times). This rapid decrease of the dose rate with distance is the reason why remote handling tools (even relatively short tongs) are so effective in reducing dose to workers.

Shielding—Any material interspersed between the radiation source and a person will absorb some fraction of the radiation and reduce the dose to that person. For most ionizing radiation used in industry (except neutrons), the more dense the material, the better the shield. Radiation shields may take the form of fixed lead or concrete walls, lead-filled steel casks for holding a source, or portable shields that can be positioned as needed.

Accidental exposures to radioactive sources are generally prevented by access controls of one type or another. Simple locks with controlled access to the key(s) or electrical and/or mechanical interlocks are often used to restrict access to the high radiation area. For example, radiation detectors interlocked to access doors are frequently used to prevent accidental entry into x-ray machine rooms, accelerator vaults, or radioisotope irradiation facilities. High radiation alarms are also used to warn personnel of a dangerous condition.

Internal Exposure Control

One of the more serious radiation safety problems is presented by work using unsealed (i.e., open) sources of radioactive material that could find its way into the worker's body. This is serious, since once radioactive material is deposited in the human body it cannot easily be removed or "flushed out" by medical treatment. The material stays in the body and continues to irradiate the tissues until it either decays or is removed by normal metabolic processes. Foreign materials can get into the body by inhalation, ingestion, absorption (through the skin), or through a wound (e.g., a puncture). Therefore, when working with an unsealed source of radioactive material, it is necessary to evaluate the potential for each of these possible pathways into the body and then to take appropriate steps to minimize that potential. Detailed planning for such eventualities may require the services of a professional health physicist. It is useful to consider the concept of "contamination control" at this point. "Radioactive contamination" is simply radioactive material that is in some undesired or unexpected place. Such "contamination" can generally be easily cleaned up, if it is known to exist. The problem arises when radioactive material is unknowingly transferred from where it is supposed to be, to some other location where it is not supposed to be. In this case, the potential for ingestion or inhalation of the material, either by a worker or a member of the general public, rises sharply. Hence, the need for very careful contamination control measures to assure that the unsealed radioactive material does not get transferred to an unexpected location.

Environmental Discharges

In general, a radiation safety program governing the use of unsealed radioactive material should try to prevent the discharge of any radioactive material to the environment (sewer, air, ground, etc.). Even though the regulations do allow discharge of radioactive material into the atmosphere and/or sewer lines, under carefully controlled and *monitored* conditions, it is only prudent to plan not to do so. This is particularly true in today's political climate; wherein, any pollutant discharge to the atmosphere, even while legal, may not be worth the adverse public reaction.

Monitoring

The program areas discussed so far require extensive and detailed monitoring to document that exposures have been within permissible limits and have also met whatever ALARA guidelines the company (licensee) has adopted. External dose monitoring is done by using some type of integrating dosimeter, such as a film badge or thermoluminescent dosimeter (TLD). These dosimeters are changed, usually monthly or quarterly, depending on the perceived potential for actually receiving doses that approach the permissible limit. Personnel radiation dosimeters are available from certified commercial vendors. The dosimeters are similar, but the type of record reporting varies greatly (e.g., from quarterly reports to in-house display terminals connected to the vendor's cloud). It is important for management to ascertain that all employees assigned a personnel dosimeter wear it at all times while on the job, and that they take reasonable care of the device. The records generated by the personnel dosimetry system are primary in the defense of any liability claim alleging radiation injury. If such records are not available or are incomplete, defense of a liability claim becomes much more difficult, partly because lack of a record reflects poorly on management's commitment to safety. Internal dosimetry involves much more difficult and complex measurements than for external dosimetry. Various bioassay techniques (urinalysis, thyroid count, whole body count, etc.), as appropriate, are used for obtaining data on the amount of radioactive material deposited in a person's body. The description of such a program is beyond the scope of this chapter, except to say that complete, accurate records are again of paramount importance.

Finally, if there is any potential for the release of radioactive material to the environment, it is essential to keep records documenting measurements that substantiate what *was or* was not released. In all monitoring programs, results showing a reading below detection limits are just as important as positive results and should be recorded.

PREGNANCY AND RADIATION

A special problem occurs if there are potentially pregnant women in the population of employees exposed to ionizing radiation doses. The problem is not with all fertile women, but only with those who have decided to bear a child. The concern arises because the human fetus is most sensitive to radiation during the first trimester of growth, and in particular, during the first four to six weeks of life—precisely when the woman may not know she is pregnant. The generally accepted safe dose to the fetus is a total of 0.5 rem (similar to the annual permissible dose to a member of the general population). Therefore, if a pregnant or potentially pregnant woman limits her monthly dose to 0.05 rem, then her fetus can receive no more than 0.450 rem (0.05 rem × 9 months).

To achieve this dose limitation, a woman may want to restrict her exposure to ionizing radiation or transfer to another job while she is pregnant or may become pregnant. It is the responsibility of the employer to assure all women working with ionizing radiation sources are given this information.

RECORDKEEPING

The list below is not intended to be all-inclusive, but does give the reader a sense for the types of records required. The following records are examples of data which should be kept to facilitate any defense against potential liability suits:

1. Records detailing the type of training each employee has received, on what dates, who the instructor was, etc.
2. Work area survey records, detailing measurements of area radiation fields, use of protective equipment (gloves, anti-contamination clothing, fume hoods, etc.), and results of contamination surveys.
3. All disposal of radioactive waste must be carefully documented.

4. Each receipt, transfer, and disposal of a source of radiation (including machines).
5. Calibration records of all instruments used to measure or survey radiation fields or amounts of radioactive material (e.g., in effluent discharges).
6. A running inventory of radioactive sources on hand, along with leak test records for any sealed sources of radioactive material.

In view of the increasingly litigious nature of society, recordkeeping has become an essential part of any safety program. The various regulatory agencies, for example, NRC, EPA, and OSHA, are assessing large monetary fines at a rapidly increasing rate for lack of proper records. It is almost a truism that "any action or measurement not properly recorded has not been done."

EDUCATION AND TRAINING PROGRAMS

In the preceding sections of this chapter, the basic components of a radiation safety program have been briefly described. However, unless the employees and often the neighboring public are adequately informed of what is being done to assure their safety, much of what is done will be relatively ineffective. Liability lawsuits often start on the basis of someone's perception of what has or has not been done to assure that people are protected from the harmful effects of some material, agent, or product. In order to reduce the potential for such suits and to put the risks of ionizing radiation exposure into proper perspective relative to other everyday risks, all employees (both actual users as well as non-radiation workers) should be given some education about the properties of ionizing radiation. The licensing agencies specifically require that all users must undergo training at a level appropriate to their use of, or exposure to, ionizing radiation. Such training may be as simple as a one or two-hour orientation lecture or as extensive as a two to three-day course. It is also a good idea to require a brief quiz at the end of any training session, and it is always necessary to keep, as a permanent part of each employee's personnel record, information pertaining to the date, length, and subject matter of each training session attended.

Minimum information to be given to those employees working near sources of ionizing radiation (but not actually using such sources) should include the following:

1. Properties of radiation
2. Methods of controlling exposure, that is, time, distance, and shielding
3. Permissible doses
4. Dose rate conditions likely to be found at the particular facility

5. Biological effects of radiation
6. Relative risks of radiation exposure and relation to natural background radiation

The people who actually work with radiation sources must receive all of the information listed above, as well as the following:

1. Principles of operation and use of survey equipment
2. Use of personnel dosimeters
3. Control of contamination from unsealed sources (if appropriate to the specific facility)
4. Field use of radiation sources (if business involves field radiographic operations)
5. Review of all applicable laws and regulations pertaining to the facility license to use ionizing radiation

Users should also receive refresher training periodically. Annual retraining is usually considered adequate. It may be necessary to prepare an environmental impact statement (EIS) if the facility in question will discharge any radioactivity to the environment (air or water), and this could involve public hearings. The process will present management an opportunity to educate the public about the program procedures used to assure the safety of the public in the vicinity of the facility. Even if public hearings are not required, a public information program may be desirable to dispel fears which generally arise from lack of information and/or misunderstandings about the hazards of radiation sources. In summarizing this section, it is appropriate to reiterate the essential need for education and training of radiation workers and fellow employees. While such training costs money in terms of time away from productive work, there is no question that it is necessary even if the law did not require it. Training provides workers with the information they need to intelligently assess the real risks of their job and gives them a basis for making informed and safe decisions.

CONCLUSION

This chapter presented a brief overview of the basic components required in the part of an SMS designed for handling radiation sources. More detailed information can be obtained from the Nuclear Regulatory Commission website or from individual state health departments. In practice, every radiation safety program is specifically designed to meet the particular needs of the organization. Except for the simplest cases, this requires the services of radiation safety experts. Often such people can be located through the local chapter of the Health Physics Society or sometimes through the offices of county or state health departments (radiological safety branches).

REFERENCES

Bureau of National Affairs, The. *Occupational Safety and Health Reporter.* Vol. 10, Number 25, November 20, 1980.

DeReamer, Russell. (1980). *Modern Safety and Health Technology.* New York: John Wiley and Sons.

Grimaldi, John V. and Rollin H. Simonds. (1975). *Safety Management*, 3rd ed. Homewood: Richard D. Irwin, Inc.

Health Physics Society. (2016). Retrieved from: http://hps.org/publicinformation/hpcareers.html.

Health Physics Society. (2015). McCall, Brenda. *Safety First—At Last.* New York: Vantage Press, 1975.

National Registry of Radiation Technologists. (2016). Retrieved from: http://www.nrrpt.org/index.cfm/m/2/lt/About%20the%20NRRPT/.

National Safety Council. (1974). *Accident Prevention Manual for Industrial Operations*, 7th ed. National Safety Council.

Occupational Safety and Health Reporter. Vol. 11, Number 4, June 25, 1981.

Occupational Safety and Health Reporter. PS Vol. 11, Number 5, July 2, 1981.

Patty, Frank A., ed. (1963). *Industrial Hygiene and Toxicology*, 2nd rev. ed. New York: Interscience Publishers, Inc.

United States Department of Health, Education and Welfare. (1973). *The Industrial Environment—Its Evaluation and Control.* Washington: United States Government Printing Office.

Chapter 17

Employee Health and Wellness

Tracy L. Zontek and Burton R. Ogle

CASE

A moderate-size manufacturer of heavy equipment operates around the clock with a typical three-shift worker rotation. First- and second-shift workers focus on the primary assembly of the equipment including frame, chassis, and drivetrain while the third shift primarily focuses on maintenance and repair of new and used heavy equipment. All shifts work with the same basic assembly tools, including robotic assembly machines, mills, lathes, and hydraulic presses. End-of-shift duties include clean up of work areas and assembly tools. Each assembly area is to be left clean and tidy so the following shift will begin their workday in an organized environment. Management and most supervisors all work the first shift with some overlap with the second. A skeleton supervisor crew and on-call managers work the third shift. The third shift is primarily composed of newer, less-experienced employees who hope to someday get moved to one of the earlier shifts. Over time, the cleanliness and organization of the assembly areas deteriorate as workers throughout the day pay less and less attention to keeping the work area tidy. The second shift gripes that the first-shift workers are "too good to clean-up after themselves." The third-shift workers, fearing they might be fired if they complain, often inherit equipment covered in cutting oil, grease, metal dust, and shavings, and general debris. The third-shift workers must spend an hour or more each day cleaning up the work area before they can begin their scheduled projects. These employees are continuously reminded that their production is falling behind; still, no one feels comfortable enough to inform management of the state of the work area when they take over their shift. Over time, third-shift workers become stressed, have the highest incidence of lost time work injuries, and have the lowest production. Minor injuries and complaints (skin irritation and difficulty breathing) go unreported since the occupational health nurse only works the first shift. Third-shift workers feel they are in a dead-end job and will never be recognized for their abilities to contribute to the company.

Questions for consideration:

- What could the occupational health medical team do to address this situation?
- What health concerns should management address?
- Is the company guilty of violating OSHA regulations?
- What "stressors" could be reduced that would improve morale and productivity?
- What additional safety issues might arise from "shift work"?

OCCUPATIONAL HEALTH MEDICINE

A critical area of injury and illness prevention and treatment is the occupational health medical program. In the past, occupational health medicine had a predominantly compliance-based approach; within the safety management system (SMS), it is a more comprehensive methodology encompassing everything that can contribute to worker satisfaction and productivity. It is a process, not a check in the box, "In compliance or out of compliance." In most developed countries, the large, dirty, and dangerous industries with devastating injuries and illnesses are a thing of the past. Today, more specialized materials, rapid research and development, and stress related to the constantly changing technology, create a new set of challenges. Occupational medicine has paralleled these changes by incorporating an individual's physical, mental, and social well-being into its framework. The occupational medical team can provide a significant contribution to the sustainable development of an organization by ensuring a healthy, well-trained staff that is stable, has low turnover, and are contributing members of society, enhancing the social capital of the community.

The occupational medical program, providing screening and surveillance, is an essential part of every occupational health and safety team (safety professionals, industrial hygienists, technicians) within the SMS, in order to ensure a comprehensive and global view of employee wellness and, in turn, lower insurance costs, improve employee health and increase productivity. Occupational health and safety medicine has gone beyond the requirements for workers compensation related to occupational disability and medical care services to a more holistic, systems approach for overall employee wellness and health. Furthermore, it has followed the changes in traditional medicine, enacting an evidence-based approach to provide the best practices in employee wellness and health care. Workplace occupational health programs have demonstrated decreased absenteeism, reduced cardiovascular risk, reduced health care claims, decreased turnover, and decreased musculoskeletal injuries while increasing productivity and organizational effectiveness (Mossinik and Licher, 1998; Oxenburgh, 1991). In order to reap these benefits, the occupational medical program must be carefully constructed.

Elements of an Occupational Medical Program

The elements of an occupational medical program within the SMS needed to maintain or improve employee health require consideration of a number of factors. The size, scope, and staffing needed will largely depend on the company's requirements; frequency; and level of hazardous biological, physical, and chemical agent use and the use/availability of outside health-care resources. An occupational medical program can serve a multitude of functions, such as: determining if employees are medically and physically able to complete job duties without injuring themselves, others, or property; determining if an employee injury or illness could be the result of workplace factors; identifying employees who may be at risk due to physical or psychological issues; and providing general screening tools for employee health promotion (U.S. Department of Defense, 2008). A strong occupational medical program will not only assist with regulatory compliance but also integrate with the occupational health and safety program and overall fitness of the company. Not all occupational medical program intervention is predicated on regulatory compliance, but this may have a significant impact during program consideration. For example, employees who experience pain due to ergonomic issues are not necessarily covered by a specific legal regulation, but these conditions can cause employee injuries/illnesses, require lost time, and lower overall productivity and morale. When developing an occupational medical program, there is a balance between recognizing needed services, community resources, and the bottom line. The cost and benefit must be determined while identifying the target population (all employees or only certain job titles), type and frequency of testing, and how the data will be used both individually and as whole (Weissman, 2014).

The elements of an occupational medical program typically consist of screening and surveillance. Screening is a tool used to provide early identification, diagnosis, and if needed, treatment. Screening is a prevention tool that can minimize costs and demonstrate commitment to employee health and wellness; in order to promote a healthy workforce, it is generally administered to employees who are not currently exhibiting symptoms. Screening is often encompassed in employee wellness programs. In contrast, medical surveillance is typically more clinically focused in response to specific regulations and/or materials an employee is using in the workplace and in response to workplace injuries/illnesses.

EMPLOYEE WELLNESS PROGRAMS

As SMS workplaces strive to cultivate a strong and robust workforce, manage health insurance costs, increase productivity, and decrease absenteeism and illness, many have developed or improved employee wellness programs. The U.S. Department of Health and Human Services has provided millions of dollars in grants to assist companies with health promotion programs.

Employee health promotion or wellness programs, such as those that include nutrition and exercise counseling, smoking cessation, alcohol and substance misuse, and flu shots have found health care costs reduced by $176 per employee and a return on investment of $1.65 for every dollar spent (Naydeck, Pearson, Ozminkowski, Day and Goetzel, 2008). Furthermore, a similar study determined that for every dollar spent on workplace disease prevention and wellness programs, $3.27 was saved on medical costs and $2.73 was saved on absenteeism costs (Baicker, Cutler, and Song, 2010). Common health issues can strongly impact fitness for work. For example, 34.9 percent of U.S adults are obese; this condition can lead to issues such as heart disease, stroke, Type 2 diabetes, and cancer (Ogden, Carroll, Kit, and Flegal, 2014). Medical costs for obese people are $1,429 higher than those for normal weight people (Finkelstein, Trogdon, Cohen, and Dietz, 2009). Medical screening through employee health promotion programs can provide early intervention, diagnosis, and treatment, thus improving the overall health of the workforce. In addition, it can aid companies in determining employee fitness for duty.

Besides just physical wellness, employers must be cognizant of individuals' psychological wellness and the overall organizational culture. Psychological wellness is strongly influenced by perceived levels of autonomy, control, and, most importantly, stress. An employee health wellness program can provide assistance for issues that originate in and outside the workplace to identify sources of stress and develop coping mechanisms to ensure workers are psychologically prepared.

The Impact of Stress

Stress is a normal physiological reaction of the body to a perceived threat. For example, when one encounters a perceived threat—a bear chases one away from a favorite fishing hole—the hypothalamus sets off an alarm system in the body. Through a combination of nerve and hormonal signals, this system prompts the adrenal glands to release a surge of hormones, including *adrenaline* and *cortisol*. Likewise, occupational hazards such as noise, vibration, thermal exposures, and high everyday stress have long been linked to the release of adrenaline. Adrenaline, the "fight-or-flight" hormone, was useful when humans needed to avoid being eaten by wild animals; however, a daily, regular dose of adrenaline is injurious to cardiovascular health. More recently, stress has been directly linked to the hormone cortisol, which also contributes to poor cardiovascular health. Cortisol levels vary throughout the day. As stress levels go up, cortisol levels go up. Over time, this increase in cortisol weakens the heart muscle and is believed to be a major contributor to our number one killer—heart disease. Employee wellness programs that focus on a healthy diet and regular exercise are believed to help curb the effects of cortisol. Also released by the hypothalamus, the hormone *oxytocin* is believed to combat the effects of elevated cortisol levels. Social support services such as counseling, workplace friendships, and other social connections are the best way to maintain levels of oxytocin. Employees, who face on or off-the-job stress they are unable to cope with or control, are at increased risk from the effects of these hormones and may experience decreased work performance.

Karasek's (1979) demand-control-support (DCS) model is the theoretical basis of the stress-injury relationship. The premise of this model is workers in high demand jobs, who have little control or social support, are unable to channel their stress response into an effective coping mechanism and some type of strain (e.g., cardiovascular disease, decreased performance, depression, and injury) results (Kristensen, 1995). Job stress has been associated with depression and other psychological symptoms (Motowidlo et al., 1986; Norbeck, 1985; Revicki and May, 1989). Depression, in turn, has been associated with lower work performance (Motowidlo et al., 1986). Dewe (1989) reported work overload was the main cause of tension and tiredness in nurses. Low job control (e.g., participation in planning of work tasks) was found to double the risk of cardiovascular disease (Bosma et al., 1997; Johnson and Hall, 1988). Psychosocial work demands were associated with back pain (Bru, Mykletun, and Svebak, 1996) and reported muscle tension (Theorell, Harms-Ringdahl, Ahlberg-Hulten, and Westin, 1991). Occupational stress has been linked to unsafe behaviors, near misses, and adverse health outcomes.

In contrast to employee wellness and health promotion programs, occupational medical surveillance is the more traditional form of occupational health in the workplace.

OCCUPATIONAL MEDICAL SURVEILLANCE

In an effective SMS, an occupational medical surveillance system can provide a continuous feedback loop to the occupational health and safety team regarding exposures to chemical, biological and physical agents, effectiveness of controls, and breaks in program implementation and value. Surveillance may be administered to individuals or cohorts of employees in a single instance or more likely over time (e.g., annually). Detailed record keeping and analysis of data collected can serve as an early alert system for potential employee health issues. These tools can be used to inform individual employees of potential health issues, identify sentinel cases of disease, or be analyzed at a population level to potential high-risk health effects. Workers who responded to the 9/11 World Trade Center disaster provide a keen example of the importance of baseline data and medical monitoring over time, as disease progression may occur over years.

When developing or improving an occupational medical program, a company can use its own employee records or data collected from others (Bureau of Labor Statistics, NIOSH, and peer-reviewed literature) to determine common occupational fatal and nonfatal injuries/illnesses. For example, those who worked in the wholesale and retail trade in 2006 represent 15.5 percent of the work population yet accounted for 20.1 percent of nonfatal injuries and illnesses (Anderson, Schulte, Sestito, Linn, and Nguyen, 2010). This study identified the top two job tasks that lead to injuries: overexertion and contact with objects/equipment. This information could be used for consideration of materials handling, development of an ergonomics program, handling of musculoskeletal training, and a medical surveillance program for those in similar industries. The U.S. Bureau of Labor Statistics administers an Injuries, Illnesses, and Fatalities program that annually compiles data on the rate and number of work-related injuries, illnesses, and fatal injuries. It also addresses how these statistics vary by incident, industry, geography, occupation, and other characteristics that can be used as an additional source of information for the occupational medical team to identify emerging issues and common denominators to workplace injuries and illnesses.

Medical surveillance is required by the Occupational Health and Safety Administration (OSHA) for a number of regulations. Companies whose occupational medical program is limited in scope should first consider compliance with the following regulations. Table 17.1 provides a list of the OSHA regulations with required medical surveillance, the regulatory standard, and an example of testing. Users must still refer to and understand the entire standard to confirm compliance.

Whether the intent is to provide basic medical services to comply with regulatory mandates or to develop a larger program that uses the systems approach to identify and control many factors that affect employee health, a qualified team of medical professionals must be assembled to meet company goals and objectives.

Table 17.1 OSHA Regulations with Required Medical Surveillance

Hazard	29 CFR 1910 (General Industry), 1926 (Construction) and 1915 (Shipyards)	Example of Required Medical Surveillance
Acrylonitrile	1910.1045(n); 1926.1145; 1915.1045	Fecal occult blood
Arsenic (Inorganic)	1910.1018(n); 1926.1118; 1915.1018	Nasal and skin exam
Asbestos	1910.1001(l); 1926.1101(m); 1915.1001	Pulmonary function tests
Benzene	1910.1028(i); 1926.1128; 1915.1028	Complete blood count and differential
Bloodborne Pathogens	1910.1030(f)	Must offer Hepatitis B vaccine
1,3-Butadiene	1910.1051(k); 1926.1151	Lymph node examination
Cadmium	1910.1027(l); 1926.1127; 1915.1027; 1928.1027	Males over 40, prostate palpation
Carcinogens (Suspect)	1910.1003-1016(g); 1926.1103; 1915.1003-1016	Determination for increased risk (e.g., treatment with steroids or cytotoxic agents)
Chromium(VI), Hexavalent Chromium	1910.1026(k); 1926.1126(i); 1915.1026(i)	History of respiratory system dysfunction
Coke Oven Emissions	1910.1029(j)	Urine cytology
Compressed Air Environments	1926.803(b)	Preplacement exam
Cotton Dust	1910.1043(h)	Pulmonary function tests
1,2-dibromo-3-chloropropane	1910.1044(m); 1926.1144; 1915.1044	Reproductive and genitourinary exam
Ethylene oxide	1910.1047(i); 1926.1147	Pregnancy and fertility testing (female/male) if requested
Formaldehyde	1910.1048(l); 1926.1148; 1915.1048	Evidence of irritation or sensitization of the skin
HAZWOPER	1910.120(f); 1926.65	Fitness for duty and ability to wear PPE
Hazardous Chemicals in Laboratories	1910.1450(g)	When required by other standards (e.g., formaldehyde)
Lead	1910.1025(j); 1926.62	Zinc protoporphyrin (blood)
Methylene chloride	1910.1052(j); 1926.1152	Cardiovascular system exam (blood pressure and pulse)
Methylenedianiline	1910.1050(m)	Urinalysis
Noise	1910.95(g); 1926.52	Audiometric testing
Respiratory Protection	1910.134(e); 1926.103	Specific protocol to evaluate respiratory system
Vinyl chloride	1910.1017(k); 1926.1117	Special attention to detecting enlargement of the liver, spleen, or kidneys

Constructing an Occupational Medical Team

The most important part of the occupational medical team is the culture that has been developed and is manifest in all levels of management within a company. A company can recruit the most qualified occupational medical specialists and top-notch facilities, but if the commitment to employee health is not demonstrated at every level of an organization, it will fall short. In an effective SMS, cooperation is critical among employees, supervisors, management, and the occupational health team. Company policies, beliefs, and actions should support and empower those in the occupational medical team and occupational health and safety team with the tools to anticipate and prevent potential occupational injuries and illness, identify employee wellness issues, and empower employees to take an active role in their own health and safety. This approach requires a team of experts in many specialties—no one group has all the skills to achieve these goals.

An occupational medical team consists of a variety of professionals: occupational health physicians; nurses; psychologists; industrial hygienists; toxicologists; ergonomics specialists; epidemiologists; safety professionals; and human resources, including insurance, worker compensation, and budgeting specialists. Similar to a safety professional earning the Certified Safety Professional (CSP) credential, the American Board of Preventive Medicine offers certifications for physicians in occupational medicine and the American Board of Occupational Health Nurses offers professional certification in the field for nurses. These specialties have evolved in the face of changing needs in the workplace and the increased focus on prevention and overall public health. There are many continuing education courses in the medical community focused on preventive and occupational health care. It is important to consider the legal scope of practice by state for each medical professional (physician, nurse, nurse practitioner, and physician's assistant) that is part of the occupational medical team. In the past, the physician may have led the occupational health medical team—today, those with the best team management skills that have frequent exposure to employees may serve in this function. Often, an occupational health nurse serves as a team leader and in many other roles: clinician, emergency responder, manager, care coordinator, health educator, counselor, and researcher (WHO, 2001). The professionals in the occupational medical team must understand the interactions of preventive and clinical practice with safety, industrial hygiene, toxicology, industrial psychology, and other surveillance to effectively fulfill respective roles on the occupational medical team.

A large company may choose to create its own occupational medical team on-site or have a centralized team at headquarters, smaller units at individual facilities, or a mobile examination service (e.g., mobile audiometry units).

The on-site occupational medical team may minimally have a nurse based on the size, need, and funding; nurse practitioners, physician assistants or physicians may also be part of the daily team. Advantages of an on-site occupational medical team are the ability to provide screenings, first aid, and basic sick visits at the facility, resulting in lower commute time, reduced wait time, and reduced time that employees may need off from work. For Kansas-city based Cerner Corporation, the convenience of an on-site medical clinic translated to about $1.2 million savings among 5,700 workers (Alsever, 2011).

One recent development is that of local occupational medicine providers, such as urgent care facilities, sometimes aligning with larger health providers to offer occupational health services (physical examinations, hearing tests, pulmonary function tests, vision screening, x-rays, drug testing, and management of occupational injuries and illnesses). These types of partnerships can perform services on an individual, case-by-case basis or enter into a formal contract with companies to provide comprehensive occupational medical services, including the collection and analysis of data. This arrangement can provide companies, even those that are small to mid-size, a wide array of services without the cost of employing many different medical professionals in-house. The availability and quality of these medical resources in the community should be considered when constructing an occupational medical team. Another emerging trend is that of telemedicine, where patients can access medical professionals virtually at their convenience and without having to leave their home or workplace. This can be particularly effective for employer worksites that are remote (oil rigs or service workers who travel extensively, such as those in the telecommunications or trucking industry); timeliness and access to medical professionals would greatly increase in these circumstances.

One key issue when using outside occupational medical services is the privacy of employee confidential medical information. For example, if screening is performed for high blood pressure, how is this data presented to and/or used by the employer? Will it be presented in aggregate with no identifying factors or with individual specific data? The data cannot legally be used to discriminate against those with disabilities or chronic illness, so a careful consideration of data management must be completed. Furthermore, with the rapid expansion of data collection via electronic devices (Fitbit, mobile phone) and online sharing, the protection of health information has become a tangled topic that will likely see additional research and legislation.

Whether the occupational medical team is on- or off-site, the confidentiality and management of employee data is an important consideration. The Health Insurance Portability and Accountability Act (HIPAA) restricts sharing of medical information to only those with a legitimate need (physician review, tracking disease, workers compensation). The Privacy Rule of

HIPAA protects and limits the health information that an employer (including managers and supervisors) may request; however, it does not protect employment records. For example, a manager or human resources professional can ask for documentation to administer sick leave or disability from an employee, but an employee's health care provider cannot provide this information without employee authorization (U.S. Dept. of Health and Human Services, n.d.). Workplace medical screening and surveillance also must be completed in compliance with the Americans with Disabilities Act (ADA) of 1990. Employees have the right to seek the care of medical professionals outside the employer's requirements or recommendations; however, workers compensation and health insurance coverage may impact the financial responsibility for treatment.

For occupational medical team members who are not on-site, plant visits should be made available so the medical providers have first-hand knowledge of the conditions and stressors in the workplace. This may stimulate them to provide suggestions on preventive measures and offer more focused care to employees.

The first allegiance of physicians, nurses, industrial hygienists, and safety professionals is always to the worker (patient), and company policy and procedure must value and understand this tenet of professional ethics. Medical staff recommendations, such as additional testing or more specialized care, should be dictated only by professional medical expertise and not company pressures to return workers to work before they are ready to ensure increased productivity and lower lost time and associated costs.

Functions of the Occupational Medical Team

The basic functions of the occupational medical team may include the following: pre-placement health examination; periodic medical evaluations or OSHA-regulated medical surveillance; diagnosis, treatment, and follow-up of work-related injuries and illnesses; and medical examinations when employment is terminated. In addition, those medical teams that engage in health promotion of employee wellness may also provide additional services, often in conjunction with the company's health insurance provider.

Prior to completing a health examination, the occupational health team should evaluate employee job descriptions, tasks, and subsequent chemical, physical, and biological hazards in the workplace. This analysis should include likely routes of exposure, frequency and duration of exposure, and intensity or toxicity of exposure. With this information the occupational medical team can evaluate target organs, potential interactions, contraindications for individuals with preexisting health conditions, and latency of health effects. This will provide a much more detailed and focused initial assessment of the employee with regard to expected job duties. In addition, it allows the occupational medical team to consider chemicals with multiple routes of exposure (chemicals that can be absorbed through skin and inhaled leading to impaired liver function) or hazards that have additive or synergistic effects in the body (noise and some solvent exposure increases hearing loss).

The employee's health must be considered from a systems view of the entire workplace. More widely used in the European community than in the U.S., biological monitoring (urine, blood, and exhaled breath) provides a more holistic determination of workplace contaminants or their metabolites. Many of these tests are extremely time-sensitive, for example, determining workplace exposure to carbon monoxide requires either a blood sample or exhaled air sample to be taken at the end of the shift and analyzed for carboxyhemoglobin. These tests (identified in the American Conference of Governmental Industrial Hygienist's *TLV Booklet*, under Biological Exposure Indices) need preplanning from both a clinical perspective and from the employee and management. It is critical to explain these tests to employees in an easy-to-understand language, obtain informed consent, and assure the employee that no other substances or diseases will be identified from this testing (except those permitted by the employment contract—illegal drugs but not HIV status). Employee medical information should be handled in a dignified and empathetic manner. The development of policies and procedures in advance aids in the smooth administration of medical screening and surveillance for employee health.

When the occupational medical team identifies a potential employee overexposure, these policies and procedures guide proper protocol. The employee is always notified in writing of the results, the significance of the results, potential health effects if exposure continues, any changes in work practices that are necessary, and if any medical treatment is needed. The occupational health and safety team may then reevaluate the workplace to identify if there have been process or employee work practices that have changed. In the meantime, the occupational medical team may retest or identify a more accurate and specific test to confirm the initial result. If the results lead to a work restriction, accommodation, or removal, the occupational medical team must be specific in work restrictions or accommodations about the employee's ability to complete tasks, without jeopardizing personal health or that of others. This information is also confidential and should only be released to personnel who have a true need—those who handle workers compensation claims, supervisors, or managers who need to understand changes in job duties and accommodations, or first-aid and emergency responders. Medical restrictions will include ways to eliminate or limit the duration, frequency, or intensity of employee exposure. This may be accomplished through engineering, administrative, or personal protective equipment controls. In some instances an employee may be medically removed from a specific job, temporarily or permanently. This must be completed with compliance with the ADA while maintaining employee status, wages, and benefits.

MEDICAL EVALUATIONS

All evaluations will begin with a detailed medical history provided by the employee. Pre-placement health examinations can assess parameters related to a specific job (ability to lift forty pounds) and should only be administered after the individual has been given a conditional offer of employment. The purpose of the pre-placement medical examination is to ensure that an individual is fit to perform the job tasks without injury to himself, others, or property; it is not to diagnose or treat previously undiscovered medical problems. The examination should be specific to the job functions of the employee and, if possible, should follow regulatory or consensus-based recommendations to avoid potential bias and discrimination in the process. In order to do this fairly, the initial job description must provide precise and detailed depiction of the skill requirements to perform the essential functions of the job. This may aid in providing accommodation under the ADA. A pre-placement health examination could include any of the following services: medical history, physical examination, vision screening, audiometric testing, drug screening, blood and urine analysis, respiratory system testing, tuberculosis testing, review of immunization record, or chest x-rays. This information may be later used if an employee transfers to a new position within the organization. Industries that must comply with specific regulations may have more focused and directive testing. Screening for hypertension is required by the Federal Aviation Administration (FAA) for pilots and by the Department of Transportation for those who operate a commercial motor vehicle. Pre-placement screening completed by the occupational medical team can also aid in educating employees about potential hazards and controls of their assigned tasks, in addition to informing employees of adverse health outcomes if control measures are not consistently and correctly employed.

Periodic medical examinations are the basis for many occupational medical teams to identify, diagnose, treat, and manage work-related injuries and illness. According to the Bureau of Labor Statistics (BLS) (2014), in 2013 the most common nonfatal occupational injuries and illnesses requiring days away from work in the private sector were (in descending order): overexertion and bodily reaction; falls, slips, or trips; contact with object or equipment; transportation incidents; and violence and other injuries by persons or animal. Often the first line of diagnosis and treatment is the occupational health team who will triage, exam, and determine the need for additional and/or more specialized care.

Periodic medical examinations should begin with a discussion of any changes in the employee's personal medical history as well as changes in job duties, hazards, or controls in the workplace. For example, employees who work with isocyanates may become sensitized and develop respiratory symptoms; while working with these compounds did not affect them initially, the occupation medical team can be aware of potential issues that may develop. Employee updates on health status and work tasks can then become part of the feedback loop to determine if hazards have been adequately recognized and controlled. For example, employees who are required to wear chemical protective clothing to control exposure may suffer from heat stress and dehydration as a result of this control. It is not uncommon for unintended effects to occur, and employees can best describe outcomes that may be affecting their health or inhibiting their ability to complete their tasks. Periodic medical examinations may be routine to identify emerging issues or triggered by regulatory compliance, such as an annual hearing test.

Upon voluntary or involuntary separation from the company, employees should complete a final medical evaluation to document health status and system function at this time. Many industrial hazards, specifically chemicals, may have adverse health effects that manifest many years later. It is important to have detailed records of employee job duties, hazards, controls, and periodic medical surveillance to protect the employee from currently unknown health outcomes that may be an issue later (e.g., asbestos). This documentation can also potentially limit the company's liability from future claims. Employee medical records must be kept for the duration of employment plus thirty years, with a few exceptions (OSHA, 2001). The medical record keeping system is critical to both regulatory compliance and managing occupational injury and illness risk and occurrences. There are many specific examples of regulatory-driven medical surveillance. At a minimum the OSHA Form 300, Log of Work-Related Injuries and Illnesses OSHA; Form 300A, Summary of Work-Related Injuries and Illnesses; and OSHA Form 301, Injury and Illness Incident Report, must be maintained and filed as appropriate. As previously discussed, medical records are confidential and subject to HIPAA regulations in the United States.

Examples of Screening and Surveillance

An occupational health medical team is the front line to conduct needs assessments and screening of common health issues, identify initial clusters of illness and injury, prioritize new and existing health and safety issues, and track health and safety intervention effectiveness related to regulatory issues and surveillance.

Screening

Preventive health programs or employee screenings are vital components of many health promotion programs. The most common screening and employee assistance activities are related to smoking cessation, healthy nutrition, physical fitness, drug and alcohol abuse, stress management, and work/life issues (counseling, child care). These items are often found in employee wellness programs.

Besides the traditional support system an employee wellness program may encompass, new programs are being developed to tackle common issues of diet, exercise, and weight loss. Personal health tracker devices or wearable technology (pedometers, Fitbits, cell phones) can be used to track movement, sleep, and levels of exercise. Companies have set up individual or group challenges to encourage employees to get more exercise and build comradery. On-site fitness classes or workout equipment can make increasing physical fitness more convenient and less expensive than a gym membership. More flexible work stations (sit/stand desks and adjustable furniture) can encourage employees to move more, potentially reducing cumulative trauma disorders, while still being productive. Safe walking trails can be identified and mapped near the facility to encourage activity breaks or set up routine walking meetings.

An on-site or contracted dietician can provide information on nutrition and serving sizes. The dietician can also prepare samples of nutritionally balanced meals and recipes along with ideas for healthy snacks and lunches. If there is a cafeteria on-site, calorie counts can be posted; healthier options can be more inexpensive. Many vending machines also list calorie counts, and companies can choose to eliminate snacks and beverages that are high in sugar, sodium, and/or fat, thus decreasing temptation for employees. Water fountains providing filtered cold water for easy refill of water bottles can increase healthy drink options.

Stress is one of the biggest issues in the workplace, whether is it is personal or work-related. Offering instruction or classes in meditation, biofeedback, yoga, or Tai Chi can provide an outlet for employees to more effectively manage stress and develop coping mechanisms. In addition, a workplace garden or quiet space (no work allowed) can provide a place to re-focus. Other modalities such as a massage therapist or chiropractor can be offered or contracted. Personal stress issues related to commuting, child care, elder care or flexible hours can be alleviated by on-site programs or partnering with companies or agencies in the local community.

Ultimately the goal of screening and developing an employee wellness program is to identify and help manage as many issues as possible to ensure employees are healthful, productive, and satisfied in their employment. The extent and duration of the chosen programs will vary and evolve with emerging health issues, technology, and culture.

Surveillance

Periodic medical surveillance can provide documentation for regulatory requirements, identify changes within individuals over time, and also serve as an early alert of trends and patterns related to industrial exposure. Prevention and early identification of medical issues can save money over treatment, lost time, and retraining in the event of long-term illness or the inability to return to work.

For those working with suspect or confirmed carcinogens, prevention and early detection is imperative. The OSHA carcinogen standards (e.g., methyl choromethyl ether, benzidine, and ethyleneimine) require medical surveillance of employees who work with carcinogens by providing a detailed medical history, occupational history, and family history (as related to inheritable disorders). This history may focus on specific body systems (liver, kidney, respiratory) in relation to potential health effects caused (e.g., enlargement or abnormalities of the liver when working with vinyl chloride). A longitudinal approach to collecting medical data can prove useful for early detection and investigation, especially since it is estimated that less than 2 percent of chemicals that are used in commerce have been tested for carcinogenicity (CDC, n.d.). In addition, it is estimated that approximately 20,000 cancer deaths and 40,000 new cases of cancer each year in the United States are attributable to occupation (CDC, n.d.). Occupational medical programs can assist in the early detection of cancer by collecting data and providing front-line information on the type and duration of exposure to suspect carcinogens, as well as confounding factors that may contribute to illness. Increasing advances in molecular biology, analytical chemistry, and genetics will provide additional tools for the occupational medical team, particularly for evolving issues such as work with nanoscale materials.

Medical surveillance for those exposed to nanoscale materials is being conducted in order to document baseline health data for employees (e.g., urinalysis, blood chemistry, spirometry, chest x-ray, or symptom questionnaires) and provide an opportunity to detect changes in advance of debilitating health effects. The primary challenge of creating a medical surveillance program for nanoscale materials is the lack of toxicological data. The properties of materials in the nanoscale can be significantly different in terms of physical properties than in normal scale, and while the research, development, and application is exploding (sunscreen, stain-free clothing, lighter, stronger automotive parts, designer drug delivery system), potential health effects lag behind. In addition, new nanoscale materials are being developed at a rapid pace, encompassing many different structures (nanohorns, nanotubes) and with various dopants (e.g., nickel, arsenic). While the health effects are currently unknown, the occupational medical team can collect data to identify changes early in potential disease progress and better protect employee health.

An emerging issue identified through medical surveillance was the cluster of severe lung disease caused by the butter flavoring (diacetyl) in a microwave popcorn production plant (Kreiss, 2012). Once identified, medical surveillance, including respiratory questionnaires and spirometry, was used to document disease (constrictive bronchiolitis) and its prevalence and progression among affected employees. An exposure assessment study was also performed to identify the specific chemicals and employee exposure levels by job tasks. In this example, the entire medical and occupational health and safety team was required to identify, investigate, and determine ways to control and prevent additional cases. From this cluster of cases, other agencies (NIOSH, California Division of Occupational Safety and Health) conducted further studies at other plants identifying additional cases. Medical surveillance in this example, constrictive bronchiolitis, is important as the disease can develop and progress quickly and timely intervention is critical (Kreiss, 2012).

A final example of the utility of an occupation medical program is for employees who conduct frequent manual lifting or other tasks that may lead to musculoskeletal disorders in the course of their job duties. Research has shown that in individuals who have experienced low back pain, the causes are multifactorial; factors such as the employee's medical, administrative, and socioeconomic situation, as well as psychological and behavioral factors, can all contribute (Petit, Fassier, Rousseau, Mairiaux, and Roquelaure, 2015). These types of injuries and illnesses can lead to short and long term lost work days and disability. Early intervention and coordination of health care providers (physician, nurse, physical therapist) with occupational health and safety professionals, including ergonomics specialists, can serve to identify job factors that may exacerbate or cause this type of injury/illness in other workers. Cumulative trauma disorders can result in long-term treatment and lost time; prevention is the key to the bottom line.

ROLE OF THE OCCUPATIONAL MEDICAL TEAM IN EMERGENCY RESPONSE

Whether a facility has a limited number of first-aid responders or an entire occupational medical team, the integration of these resources into the emergency response plan is significant. The occupational medical team, depending on its size and scope, may plan and implement the following functions during an emergency: triage and immediate medical care; policies and procedure setting for the types of injuries/incidents that the on-site medical team may offer (applying a bandage or administering stiches); determination of location, number, and training needed for Automated External Defibrillators (AED); assurance that antidotes and other medical interventions are on-site for acute hazards (calcium gluconate for hydrofluoric acid exposure); management of first aid supplies and personal protective equipment; administration of drills and mock emergency medical events; and assurance that those with physical disabilities will be able to follow the evacuation procedure. This list is a starting point for integration of the occupational medical team during emergency planning and response.

Off-the-Job Safety

The occupational medical team may also be faced with non-occupational injuries and illnesses, from at-home or sports injuries to contagious diseases. The National Safety Council (NSC) estimated there are 15x more fatalities off-the-job than on-the-job (NSC, 2013). The commitment to plan, develop, and maintain a healthy workforce must address the potential for injuries/illnesses that occur outside the work environment. The screening measures previously discussed can provide some ability for early intervention of common illnesses (hypertension, heart disease, decreased lung function); however, employees may still engage in hazardous activities outside the workplace. These hazardous activities are often in the guise of normal homeowner daily chores or health/recreation. For example, yard maintenance may expose individuals to high levels of noise, dust, and flying objects (lawnmower, leaf blower, power washer, or chain saw). Employees' attempts to decrease stress and increase overall health may lead them to partake in walking/running/hiking, intramural sports, or new adventure-based activities (zip lines). Companies who choose to develop an off-the-job safety program see the competitive advantage to preserve their human resources, as well as maintain a civic responsibility to their community and actively exhibit the company's core values.

While there are 15x more fatalities off-the-job than on, there have been some positive strides in the past decade. The occupational (33 percent) and highway (36 percent) death rates have decreased from 1992 to 2012. In contrast, the home and community death rate has increased 68 percent over the same timeframe (National Safety Council, 2014). The increase in the home and community death rate is predominantly due to poisoning and choking (NSC, 2014). Poisonings were the leading cause of death in eighteen states;

the number of deaths due to unintentional opioid prescription painkiller overdoses is rising. This increasing use of opioid prescription painkillers also is an alarming issue for worker safety on-the-job. Electronic devices (e.g., cell phones) present new challenges to on- and off-the-job safety.

In off-the-job safety, NSC estimated that 26 percent of all motor vehicle crashes are due to cell phone use (NSC, 2014). New state laws have made it illegal for drivers to text and, in some cases, use their phones while driving; company policies should follow these initiatives. The impact of hands-free systems in vehicles and voice activation of GPS, phone, and media devices are areas ripe for future injury and illness research. It is estimated that motor vehicle crashes cost employers $60 billion annually in medical care, legal expenses, property damage, and lost productivity (OSHA, n.d.). Furthermore, motor vehicle crashes can increase the cost of workers compensation, private health and disability insurance, as well as the overhead needed to administer these internal and external programs (OSHA, n.d.). The costs of off-the-job accidents can be estimated using an example of a motor vehicle accident (see table 17.2). This method was developed as an easy to use system for companies to estimate the costs through collaboration between OSHA, the Network of Employers for Transportation Safety, and the National Highway Transportation Administration.

The preservation of human resources may be the strongest motivator for companies to maintain off-the-job safety programs. Employee injuries affect the employees and also may create a void in the company requiring overtime and associated pay, the loss of invaluable resources, scientific knowledge, and/or institutional memory. The increasing integration of wellness programs, health insurance costs, and workers compensation provide a strong justification for management commitment to a quality off-the-job safety program.

One way for companies to address off-the-job safety is to factor it into their evaluation of the overall health and safety audit. Employees who work safely at work are more likely to work safely at home. Ultimately, a company wants to develop a stellar environmental health and safety culture that empowers each employee to vigilantly consider the risks and hazards and mitigate as working conditions evolve. Developing employees with this type of mind set allows them to easily transition the same level of analysis to household, recreation, and everyday life situations. Unfortunately, this is not always the case. One challenge is the acceptance by employees of the off-the-job safety initiatives. Employees must realize the value of an off-the-job safety program to embrace it when no one is watching. These key components should already be part of the company health and safety program and can assist to developing an off-the-job safety culture:

Table 17.2 Method to Estimate Transportation Accident Costs

Direct Costs to the Organization	Estimated Cost
Workers' compensation benefits	
Health-care costs	
Increases in medical insurance premiums	
Auto insurance and liability claims and settlements	
Physical and vocational rehabilitation costs	
Life insurance and survivor benefits	
Group health insurance dependent coverage	
Property damage (equipment, products, etc.)	
Motor vehicle repair and replacement	
EMS costs (ambulance or medivac helicopter)	
Vehicle towing, impoundment and inspection fees	
Municipality or utility fees for damage to roads, signs or poles	
Direct Total	

Indirect Costs	
Supervisor's time (rescheduling, making special arrangements)	
Fleet manager's time to coordinate vehicle repair, replacement, etc.	
Reassignment of personnel to cover for missing employees (less efficient)	
Overtime pay (to cover work of missing employees)	
Employee replacement	
Re-entry and retraining of injured employees	
Administrative costs (documentation of injuries, treatment, absences, crash investigation)	
Inspection costs	
Failure to meet customer requirements resulting in loss of business	
Bad publicity, loss of business	
Indirect Total	

TOTAL = Direct total + Indirect total	

- Leadership and commitment of the program extending through all levels of management; supervisors; and front-line workers by setting policy, defining activities, and setting standards.
- Resources to ensure safety is the priority over production in every instance.
- Managers and supervisors trained in safety fundamentals, setting the example of safe behavior, directing program activities, and evolving to meet changing conditions.
- Engineers designing work areas and processes to comply with safety codes, industry standards, and best practices built on years of accident research and international consensus standards.
- Purchasing staff and materials control groups researching safety and health factors to buy the safest possible equipment.
- Preventive maintenance properly performed to keep equipment and facilities in safe operating condition.

- Regular workplace inspections to detect unsafe conditions and practices.
- People, oriented to hazards in the workplace and trained in safe work methods.
- Reference libraries maintained, periodicals screened, and seminars attended to keep safety knowledge up-to-date.

Employers can monitor off-the-job safety though analysis of public records (traffic infractions, arrest records) and through the use of social media sites. A more effective and less invasive option is to provide a mechanism of reporting that is supportive, aiding in future prevention, rather than punitive. In order to identify common off-the-job accident types, companies can use information from the National Safety Council, Centers for Disease Control and Prevention, or state/community level data. A well-designed survey could also be used to identify off-the-job injuries, near misses, and interests/hobbies/family activities/concerns from employees. Supervisor and employee focus groups and interviews can also aid in identifying common off-the-job injuries and near misses. If using a survey, interview, or focus group, it is important to search for peer-reviewed, validated instruments to avoid common errors when gathering data. As with any SMS employee reports, the opportunity should exist to return surveys anonymously, so employees complete truthfully and not feel targeted or judged regarding their outside work activities. Furthermore, when engaging in face-to-face data collection, there are a number of steps to consider. First, the purpose should be explained- the company is interested in doing more to help them keep safe off-the-job and would like to know more about the person's activities and interests. This should be presented in an objective and empathetic manner; the employee should feel free to conclude the interview at any time with no risk to job or benefits. Next, general questions should start the process regarding home and recreation activities. Then, specific items that are common (residence, appliances, hobbies, etc.) are asked about and then information about what equipment, materials, and safeguards are used is requested. Next, a determination of how well the person recognizes hazards and the potential dangers in the activity is sought. It is imperative to be objective, not judgmental. This is not necessarily an opportunity for education unless asked; neutrality in the questions and how respondents' answers are addressed is imperative. It is helpful to use some type of form to ease data entry and analysis. Notes to demonstrate the person's responses are important, but the flow of the interview should not be interrupted with excessive note taking. Summarizing key comments assures the interviewer understands them. The door is left open by providing a way for a person to provide additional information about exposures or near-accidents. The end of the interview can serve as a summary and a point where education about potential hazards and controls can take place.

The development of a form to create easy data collection and analysis may use some of the following common off-the-jobs injuries/illnesses grouped by home, transportation, and public. Home safety slips and falls are prevalent. Contributing factors to home slips and falls may include: footwear and clothing; floors, stairs, and other walking surfaces; bathtubs and wet enclosures; ice, mud, and other weather hazards; running and rushing; stools, ladders, and climbing devices; illumination, open and/or weak floor surface covering; protective railings and edge warnings; and using cell phone or other electronic device when walking. Another area of home injury is related to lifting and carrying leading to back strain and musculoskeletal injuries and illnesses. These may be prevented or alleviated with proper lifting techniques as well as mechanical lifting devices. Finally, home chemical safety issues may include household cleaning products; insecticides and pesticides; corrosives; poisonous plants (people or pets); prescription drugs; e-cigarettes and vapes; and recreational drug use (a number of states have legalized marijuana). Electrical safety home items of interest may include: color coding for electrical connections; grounding and bonding; defective electrical equipment; high voltage lines; and electrical issues during flooding. Fire safety in the home may be impacted by: flammable and combustible materials; heat sources; electric appliances (blankets); fire detection; and emergency evacuation. Also, basic home maintenance can require use of lawn mowers, leaf blowers, trimmers, hand tools (saw, hammer), and power tools.

Transportation safety may include private motor vehicle operation; driving during emergencies or severe weather; vehicle maintenance; influence of drugs and alcohol; recreational aircraft or boating.

Off-the-job public safety factors may include recreation; children's play; competitive sports; individual sports; outdoor recreation; hobbies and crafts; playing in a band; recreational vehicles; travel; and civil disturbances.

Once companies have developed a baseline of the off-the-job safety issues relevant for their employees (this may vary from facility to facility), they can also consider their geographic location and cater safety messages throughout the year based on popular activities (skiing, hunting, hiking, recreational water activities). The company safety committee can serve as a launching point of an off-the-job safety program. Early initiatives can tie into National Fire Safety Month or Breast Cancer Awareness initiatives that are taking place in the community. Another step is to involve worker's families and friends through coloring (or other) contests, programs at schools, churches or civic organizations in the community, and participation in fund raising or other community events. The key is to integrate off-the-job safety as part of bigger issues that are of concern and interest to employees. Better yet, the initiatives should tie into employee hobbies and interests (hunter safety related to tree stands, ergonomic issues related to iPods and video games).

Employees can then receive the information, self-identify their own interests, and learn about issues while perceiving that their employer is concerned with their total health and safety. For example, DuPont Sustainable Solutions has developed a training program, *Take Safety Home*, which adapts standard workplace occupational health and safety to a home setting. This program covers topics such as the use of personal protective equipment, chemical handling and storage, safe operation of power tools, fire prevention and management, back safety, and slips, trips, and falls. For companies just beginning an off-the-job safety program, this may provide an example of topics and training modules to consider implementing.

Safety tool box talks and meetings can include a variety of topics applicable to off-the-job safety: vehicle safety, choking, poisoning, fall prevention, first aid, hearing conservation, eye protection, safe work with chemicals, fire safety, electrical safety, ladder safety, power equipment and tools, and proper lifting. Seasonal topics also work well: sunburn prevention, heat stress, water safety and drowning, boat safety, lawn and garden safety, and holiday safety. Many companies also allow employees personal protective equipment (safety glasses, hearing protection) at a free or reduced cost to encourage use off-the-job.

The focus and scope of the off-the-job safety program may vary and evolve depending on the availability of company resources and depth of control intervention that will be undertaken. Some risks are adequately controlled with knowledge of a safe practice, requiring education. Other hazards need employees to develop additional skills; the off-the-job safety program may provide ways to get supervised practice, such as through a training session, recreation club, or sports team. Often the hazard can be controlled with personal protective equipment, and the off-the-job safety program may have a source of equipment (low or no cost), proper fitting, and instruction on personal protective equipment use. Other safeguards may need to be employed, such as electrical insulation or machine guards, and the program must provide guidelines for purchasing these items.

One key challenge of off-the-job safety programs is that each employee must serve as the executive, middle manager, supervisor, safety coordinator, engineer, purchasing manager, human relations director, trainer, and researcher—all at the same time. Employees must have adequate knowledge to recognize, anticipate, and control hazards on their own. Many hazards faced outside of work have crossover with occupational hazards.

The development of a comprehensive employee health and wellness program is multifaceted involving stakeholder commitment at every level. The National Institute of Occupational Safety and Health (NIOSH) has developed *Total Worker Health*—policies, programs, and practices that integrate protection from work-related safety and health hazards with promotion of injury and illness prevention efforts to advance worker well-being off-the-job.

CONCLUSION

The integration of an occupational medical team into the overall SMS can serve to create a healthy and productive workforce through employee wellness programs, medical surveillance, and off-the-job safety programs. The size and scope of each company's occupational medical program will vary according to hazards, need, locations, and culture. Ultimately, prevention will save human and fiscal resources and contribute to the overall wellness of employees.

REFERENCES

Alsever, J. (2011). *Do-it-yourself Health Care in Corporate America*. Fortune August 2, 2011. Retrieved from: http://fortune.com/2011/08/02/do-it-yourself-health-care-in-corporate-america/.

Anderson, V.P., Schulte, P.A., Sestito, J., Linn, H. and Nguyen, L. S. (2010). Occupational Fatalities, Injuries, Illnesses, and Related Economic Loss in the Wholesale and Retail Trade Sector. *Am. J. Ind. Med.,* 53, 673–85. doi: 10.1002/ajim.20813.

Baicker, K., Cutler, D. and Song, Z. (2010). Workplace Wellness Programs can Generate Savings. *Health Aff,* 29(2), 304–11.

Bosma, H., Marmot, M., Hemingway, H., Nicholson, A., Brunner, E., and Stansfeld, S.A. (1997). Low Job Control and Risk of Coronary Heart Disease in Whitehall II (Prospective Cohort) Study. *British Medical Journal,* 314, 558–65.

Bru, E., Mykletun, R.J., and Svebak, S. (1996). Work-related Stress and Musculoskeletal Pain among Female Hospital Staff. *Work & Stress,* 10(4), 309–21.

Centers for Disease Control and Prevention. (n.d.). NIOSH Data & Statistics by Industry Sector. Retrieved from: http://wwwn.cdc.gov/niosh-survapps/Gateway/Default.aspx?c=CAN&s=***.

Dewe, P.J. (1989). Stressor Frequency, Tension, Tiredness and Coping: Some Measurement Issues and a Comparison Across Nursing Groups. *Journal of Advanced Nursing,* 14(4), 308–20.

Finkelstein, E.A., Trogdon, J.G., Cohen, J.W. and Dietz, W. (2009). Annual Medical Spending Attributable to Obesity: Payer and Service Specific Estimates. *Health Aff,* 28(5), 822–31.

Goetzel R.Z. and Ozminkowski R.J. (2008). The Health and Cost Benefits of Work Site Health-promotion Programs. *Annu Rev Public Health,* 29, 303–23.

Johnson, J.V. and Hall, E.M. (1988). Job Strain, Work Place Social Support, and Cardiovascular Disease: A Cross-sectional Study of a Random Sample of the Swedish Working Population. *American Journal of Public Health,* 78(10), 1336–42.

Karasek, J. and Robert A. (1979). Job Demands, Job Decision Latitude, and Mental strain: Implications for Job Redesign. *Administrative Science Quarterly*, 24, 285–308.

Kreiss K. (2012). Respiratory Disease among Flavoring-exposed Workers in Food and Flavoring Manufacture. *Clin Pulm Med*, 19, 165–73.

Kristensen, T.S. (1995). The Demand-control-support Model: Methodological Challenges for Future Research. *Stress Medicine*, 14, 155–68.

Mossinik, J. and Licher, F. (1998). Costs and Benefits of Occupational Safety and Health. *Proceedings of the European Conference on Cost and Benefits of Occupational Safety and Health*. Amsterdam, NIA TNO B.V.

Motowidlo, S.J., Packard, J.S., and Manning, M.R. (1986). Occupational Stress: Its Causes and Consequences for Job Performance. *Journal of Applied Psychology*, 71(4), 618–29.

National Safety Council. (2013). *Injury Facts 2013*. Itasca, IL: National Safety Council.

National Safety Council. (2014). *Injury Facts 2014*. Itasca, IL: National Safety Council.

Naydeck, B.L., Pearson, J.A., Ozminkowski, R.J., Day, B.T. and Goetzel, R.Z. (2008). The Impact of the Highmark Employee Wellness Programs on 4-Year Healthcare Costs. *J Occup Environ Med*, 50(2), 146–56.

Norbeck, J.S. (1985). Perceived Job Stress, Job Satisfaction, and Psychological Symptoms in Critical Care Nursing. *Research in Nursing and Health*, 8, 253–59.

Ogden, C.L., Carroll, M.D., Kit, B.K. and Flegal, K.M. (2014). Prevalence of Childhood and Adult Obesity in the United States, 2011–2012. *JAMA*, 311(8), 806–14.

Oxenburgh, M. (1991). *Increasing Productivity and Profit through Health and Safety, Australia*, CCH Australia Ltd.

Petit, A., Fassier J., Rousseau, S., Mairiaux, P. and Roquelaure, Y. (2015). French Good Practice Guidelines for Medical and Occupational Surveillance of the Low Back Pain Risk among Workers Exposed to Manual Handling of Loads. *Annals of Occupational and Environmental Medicine*, 27(18).

Revicki, D.A. and May, H.J. (1989). Organizational Characteristics, Occupational Stress, and Mental Health in Nurses. *Behavioral Medicine*, 15, 150–63.

Theorell, T., Harms-Ringdahl, K., Ahlberg-Hulten, G., and Westin, B. (1991). Psychosocial Job Factors and Symptoms from the Locomotor System—a Multicausal Analysis. *Scandinavian Journal of Rehabilitation Medicine*, 23(3), 165–73.

U.S. Bureau of Labor Statistics (BLS). (2014). *News Release: Nonfatal Occupational Injuries and Illnesses Requiring Days Away from Work, 2013*. Retrieved from: http://stats.bls.gov/news.release/archives/osh2_12162014.pdf.

U.S. Department of Defense (DoD). (2008). *Occupational Medical Examinations and Surveillance Manual*, DoD 6055.05-M. Retrieved from: http://www.dtic.mil/whs/directives/corres/pdf/605505mp.pdf.

U.S. Dept. of Health and Human Services (DHHS). (n.d.). *Health Information Privacy*. Retrieved from: http://www.hhs.gov/ocr/privacy/index.html.

U.S. Department of Labor, Occupational Health & Safety Administration (OSHA). (2001). *Access to Medical and Exposure Records.* Retrieved from: https://www.osha.gov/Publications/pub3110text.html.

U.S. Department of Labor, Occupational Health & Safety Administration (OSHA). (n.d.). *Guidelines for Employers to Reduce Motor Vehicle Crashes.* Retrieved from: https://www.osha.gov/Publications/motor_vehicle_guide.html.

U.S. Department of Labor, Occupational Health & Safety Administration (OSHA). (n.d.). *Medical Screening and Surveillance.* Retrieved from: https://www.osha.gov/SLTC/medicalsurveillance/.

Weissman, D.N. (2014). Medical Surveillance for the Emerging Occupational and Environmental Respiratory Diseases. *Current Opinion in Allergy and Clinical Immunology,* 14(2), 119–25.

World Health Organization. (2001). *The Role of the Occupational Health Nurse in Workplace Health Management. EUR/01/5025463.* Retrieved from: http://www.who.int/occupational_health/regions/en/oeheurnursing.pdf.

Chapter 18

Emergency Response Planning

Kimberlee K. Hall

CASE

A critical part of the safety management system is emergency response planning. It is, in fact, required under the Federal Aviation Administration (FAA) Safety Management Systems Final Rule. Of course, included within most emergency response planning is consideration of potential disasters. The objectives of emergency and disaster planning are to (1) prevent injuries and fatalities, (2) reduce damage to buildings and materials, (3) protect the surrounding community and environment, and (4) facilitate the continuation of normal operations. At first glance, these objectives seem relatively simple and straightforward, with common sense dictating that prevention be the primary objective. However, emergency and disaster planning must also address preparedness and response measures in the event of an actual emergency, as they often require coordination of a range of information, services, and materials from within the organization and with the local community. Preparing for and responding to every foreseeable emergency is unavoidably complex, but it can mean the difference between a quick return to business and an emergency that has disastrous consequences for the company, its employees, and the community.

Effective emergency and disaster planning first and foremost begins with facility owners, operators, and managers. Numerous resources are available to aid in the development and implementation of emergency response plans, and the content should address all identified hazards present at a facility. Plans that address legally mandated categorical requirements, including evacuation procedures, means of accounting for employees, and reporting requirements, are vital in the prevention of, preparation for, and response to an emergency in a way that protects the company and its employees. Another significant characteristic of effective emergency response planning is the communication of identified hazards and developed plans to external agencies providing information, services, and materials during an emergency. Effective communication helps to ensure a proper response that protects the health and safety of first responders and the community.

West Fertilizer Plant Fire and Explosion: A Study in Emergency Response Planning

The anticipation of on-site hazards and preparation for emergencies when no legal mandate exists also deserves consideration in emergency response planning. Thorough assessment and communication of all on-site hazards, whether or not they are legally reportable, can go a long way in preventing injuries and fatalities during an actual emergency response and in protecting the public from those hazards. Nowhere was this precaution made clearer than in West, Texas, on April 17, 2013, when an intense fire and explosion occurred at the West Fertilizer Plant. Around 7:30 pm, a fire was reported at a wooden warehouse building at the storage and distribution center at West Fertilizer Plant. First responders arrived and immediately began to fight the fire but were unaware that the building contained approximately thirty tons of ammonium nitrate. Within twenty minutes of first responders arriving at the scene, a massive explosion resulting from the detonation of the ammonium nitrate killed fifteen people and injured more than 200 while destroying nearby schools, homes, and businesses.

The preliminary findings of the U.S. Chemical Safety Board (CSB) from its investigation of the West Fertilizer Plant fire and explosion identified several factors as contributing to the disaster, including the storage of the explosive ammonium nitrate in wooden bins and in a combustible setting, the lack of sprinkler systems or automatic fire suppression systems, and confusing or contradictory code provisions for the safety of ammonium nitrate. Fire emergencies are one of the more frequently observed types of emergencies across a variety of settings, and emergency response plans often address this hazard by adhering to fire codes and OSHA standards. McLennan County, where the facility was located, had not adopted a fire code and therefore the facility was not required to follow National Fire Protection Association (NFPA) recommendations for the storage of ammonium nitrate. The OSHA standard for "Explosive and Blasting Agents" is similar to NFPA codes and would have applied to the storage of ammonium nitrate and the presence of a fire suppression system but only if greater than 2,500 tons were present (CSB, 2014). The failure of the West Fertilizer Plant to recognize the existence of an ammonium nitrate hazard and plan for a fire and explosion emergency, regardless of regulatory standards or guidance, was identified as contributing to the intensity of the fire and explosion.

Preliminary findings of the fire and explosion investigation also identified failures in communication between Local Emergency Planning Committees (LEPCs) and emergency responders in McLennan County. Under the Emergency Planning and Community Right-to-Know Act (EPCRA), facilities that store or use hazardous substances are required to provide information for reporting to State Emergency Response Commissions (SERCs) and LEPCs. LEPCs then develop emergency response plans and communicate with local first responders to coordinate a response plan. West Fertilizer Plant filed an EPCRA Tier II report with the McLennan County LEPC in 2012 identifying the presence of ammonium nitrate. The reporting of ammonium nitrate at the facility should have triggered the development of an emergency response plan by the LEPC, yet the CSB preliminary report concluded that there "is no indication that West's filing with local authorities resulted in an effort to plan for an ammonium nitrate emergency" (CSB, 2014). The absence of a developed ammonium nitrate emergency response plan by the LEPC and failure to communicate the reported hazard to local first responders left firefighters in harm's way and unable to respond to the disaster accordingly.

The West Fertilizer Plant disaster emphasizes that effective emergency response plans are those that fully recognize hazards, implement preventive and preparedness measures, and communicate and coordinate response efforts. This disaster has brought to light many holes in current regulatory standards and guidance for the safety of ammonium nitrate, such as lack of reporting coverage under either OSHA's Process Safety Management (PSM) standard or EPA's Risk Management Program (RMP) rule. OSHA and EPA are leading efforts to address hazards and safety at facilities like the West Fertilizer Plant, but regulations and standards in general are often slow to develop, if they are developed at all. Emergency response planners should err on the side of caution and be proactive in emergency prevention and preparedness, rather than wait for legal mandates to require that specific hazards be addressed and reported. The development of a correct, complete, and comprehensive emergency response plan that addresses both regulatory requirements and nonregulatory concerns ultimately allows the facility to fully meet the objectives of emergency response planning and protect the company, its employees, and the community.

INTRODUCTION

Emergencies, by their very nature, are the unexpected, unpredictable events that disrupt the best of plans and operations. The senior manager, however well organized and far-seeing, can do nothing to prevent a freak event, a natural disaster, or any of the dozens of emergency situations that arise beyond human control. No one likes to think that they are helpless to prevent something, especially when that something can be completely catastrophic. A single disaster can destroy an entire organization, cost millions of dollars in damage and human misery, and take years from which to recover.

However, there are effective countermeasures and preventive plans that can keep such events from doing unlimited damage or keep a bad situation from becoming worse. Here the manager does have some degree of control. Many emergency preplanning measures in themselves can keep an event from becoming a disaster in the first place. For example, immediate evacuation of an area filling with toxic vapors may keep anyone from being killed or injured. The quick, efficient action of notification and suppression systems may keep a fire from getting out of hand and igniting highly explosive or toxic chemicals. Such is the nature and intent of emergency and disaster planning. This chapter will concentrate on such control measures and the preparedness and planning that can be brought into immediate action in an emergency to lessen the impact on employees, the community, and the organization.

Preventing disasters is a prime concern, but it is only half of the picture. Of nearly equal importance to the manager is making certain that business can be conducted, resumed, or directed into new channels *after* an emergency or disaster. The precautions that assure survival must be brought together in an emergency/disaster plan, yet if such a plan really attempts to deal with all possible emergencies and all their ramifications, it is going to be unavoidably complex. Such complexity can be addressed by focusing on the following four areas that must be considered in order to successfully address an emergency or disaster through planning (table 18.1):

1. Types of emergencies or disasters
2. Necessary steps in developing a plan
3. Elements contained in a complete plan
4. Functionaries, agencies, and individuals involved

Each of these four areas must be thoroughly examined and coordinated with each other. While this poses a large task for management to undertake, it may be possible that not every organization needs to plan for every conceivable emergency. For example, a facility located on high ground or in a desert area may not have to consider the possibility of a flood, or a facility may not need to plan for a hurricane if it is located in an area where the possibility of a hurricane is extremely remote. Model plans and procedures may also already exist for some emergencies, such as fires, and those plans may only need to be adapted to a particular situation or incorporated into a larger plan.

Although a facility may not need to plan for those emergencies with a remote possibility of occurrence and/or may be able to take advantage of

Table 18.1 Considerations in Developing an Emergency/Disaster Response Plan

Type of Emergency	Planning Steps	Plan Elements	Responsible Parties
• Fire • Explosion • Flood • Hurricane • Tornado • Earthquake • Terrorism • Accident • Chemical spills/ vapor release • Radiation • Extreme weather • Energy emergencies	• Assess probability • Assess hazards • Designate a plan coordinator • Develop the emergency response plan • Train personnel • Test the plan • Revise the plan • Coordinate with external emergency responders	• Mission statement • Purpose of the plan • Authority • Control center • Emergency equipment and supplies • Facility maps and location of data • Hazard assessment • Auditing of general operations and procedures • Communication during an emergency • Evacuation procedures • Shelter procedures • Transportation • Employee welfare • Special procedures • Equipment shutdown • Restoration, repair, and salvage • Personnel assignments • Training • Documentation • Appendices	• Chief operating officer • Emergency response plan • Coordinator • Functional committee • Fire agencies • Security • Medical services • City, county, and state • Governments • Federal government • Mutual aid

existing model plans, we don't want to paint an overly optimistic picture. A small emergency or small omission in a disaster plan can rapidly lead to other more catastrophic events. One of the major aims of a disaster plan is to hold down or contain the effects of an event. This is generally termed the *ameliorative function* of a disaster plan.

Types of Emergencies and Disasters

Ordinary routine activities may actually present the potential for great losses. However, it is the unexpected events or the ones that are beyond the normal capacity to handle that are of interest. A *disaster* is defined as "a great, sudden misfortune resulting in a loss of life, serious injury, or property damage." Strictly speaking, such a misfortune can occur even on an individual level. Having one's house burn down to the ground is a personal emergency or disaster. In current usage, however, the term generally refers to a sudden occurrence that kills or injures a relatively large number of people. Whatever the scope of the disaster, planning is needed to assure survival and to provide for amelioration of the consequences. The following discussion focuses on various emergencies and the steps involved in preparing for and responding to them.

Fire—Fire is probably the most common emergency or disaster situation, but that does not make it the most harmless. Having automatic fire protection systems, such as sprinklers or fire doors, does not remove the need for immediate response. The automatic devices may fail, they may not protect the particular area involved, or a fire may simply be beyond their capacity to control. If a single person evaluates and manages all risks, that person should be considered the plant fire chief. At larger plants, fire protection may be such a complex question that it is regarded as a separate function or department. This is particularly true for operations that involve an unusual risk of fire or explosion (petrochemical industries, those that make use of highly flammable or explosive mixtures).

Time is of the essence when a fire starts, and small fires should be kept small and handled as soon as possible. Plans should provide for quick action by trained personnel with immediate and automatic notification of larger resource units. The first few minutes are especially critical. Warning and evacuation should begin immediately no matter how small the fire. Waiting until the fire is out of hand could be disastrous. Familiarity with the notification and response action is also necessary to ensure quick action is taken should a fire emergency occur. Routine testing of the fire plan is crucial in keeping small fires from becoming catastrophes.

Explosion—Explosions very rarely occur unaccompanied by other emergencies. In some cases, the explosion happens as the result of a fire or other unexpected event. In other cases, the explosion itself causes a fire and other attendant disasters. The onset is very sudden and there is seldom any advance warning. Therefore, the need of fire and other emergency action is urgent, requiring a faster reaction than normal. The suddenness of an explosion should never be lost sight of in any aspect of disaster planning.

Flood—For planning purposes, floods should be divided into two categories: flash and normal floods. Like explosions, flash floods are very sudden, seldom give advance warning, and require faster-than-usual action. Normal floods generally build more slowly, threatening days before they actually occur. There is usually enough time to take protective measures. Where floods are frequent, permanent protective measures are often possible. For example, dikes might protect an operation located in a probable flood area. Floods may also be accompanied by other disasters, including windstorm and fire. A complete flood plan should include measures for dealing with these attendant disasters, as well.

Hurricane—Hurricanes are generally confined to certain geographical regions, and those regions are usually alerted to a hurricane's possible occurrence well in advance. Well-developed meteorological tracking and warning systems go into action as much as several days in advance, and so should the organization's emergency plan. When the danger reaches a certain point, the operation can close down in an orderly manner and precautions can be taken to minimize the expected wind and water damage. Some measures can't be taken on short notice, such as battening down openings, putting shutters in place, or anchoring tall chimneys and signs. Because these activities may take a considerable amount of time, they should be undertaken as soon as possible. Hurricane emergency plans should also take into account the fact that the state and local emergency services may well be overtaxed and unable to respond to a call. The facility should be prepared to handle attendant damage or emergencies with its own resources.

Tornado—Tornado tracking and warning systems also allow for some time to prepare for a tornado. While the specific path of a tornado is highly erratic and unpredictable, tornado warnings are generally broadcast to the entire area of possible threat. This means that there is usually time to make preparations before a tornado strikes. Even if a tornado does not touch down in a given area, there may be strong wind conditions that can cause a broad path of damage, interruption of utilities, and unavailability of local emergency services. Tornado plans should include procedures for getting personnel to a safe shelter location immediately if advance warning is given, protection procedures for personnel to take if only a few seconds warning is given, and trained emergency crews who can address hazards created by the tornado without injury to themselves.

Earthquake—Earthquakes happen so quickly and so unpredictable that there is seldom time for ameliorative action, unless it has been done well ahead of time. In an industrial situation, earthquakes can release combustibles, break fire cutoff devices, and ignite fires or explosions. Sprinkler systems may become damaged and the fire hazards of an earthquake are often aggravated since water supplies (particularly municipal supplies) are frequently casualties of the quake. In addition to fire, the collapse of walls, buildings, and other structures is also a major hazard associated with earthquakes. Employees should clearly understand that the safest place during an earthquake is within the work area under preselected cover. Employees should know exactly where they would seek immediate cover in case of an earthquake.

In planning for an earthquake, survival of life is the main consideration. Most personnel will be anxious to get to the aid of their families after a severe earthquake occurs, yet transportation and communication systems may be out of order. Employees may be exposed to additional hazards if they attempt to leave the work area, and thus the organization should try to protect their welfare as much as possible. Employees running to get to parked cars may be exposed to such hazards as downed power lines and falling debris. If a stampede or traffic tie-up occurs, still others may be injured. Where it is impossible for employees to leave the plant area due to transportation and communication failures, their welfare becomes the organization's concern. Food and shelter, as well as first-aid treatment, may have to be provided during the continuing state of emergency.

Terrorism—The inclusion of emergency and disaster plans that address terrorism is a growing trend. As defined by the U.S. Federal Bureau of investigation, terrorisms is "the use of force or violence against persons or property in violation of the criminal laws of the United States for purposes of intimidation, coercion, or ransom." However, terrorist acts may also be the result of other emotions and psychological states rather than an effort to intimidate, coerce, or extort. Planning for such emergencies should take into consideration the general social setting and the probability of reports and rumors. Defense against persons bent on sabotage should focus on preventing them from entering the premises in the first place. If the saboteur is a member of the organization, it falls to internal security measures to reduce the risk. Whether the threat is external or internal in nature, the protection of grounds, facilities, and personnel are important. Emergency plans should be able to go into immediate and effective action. In training these personnel to handle these situations, it is essential that they understand not only what to do, but what they should not attempt to do.

Accident—Pre-planning is a major feature of effective accident prevention and investigation. Much of the organization's safety and health effort will naturally be concentrated on accident prevention measures as a matter of course, and investigative procedures will already be established. These will also play a part in emergency and disaster situations, as well as in cases of more routine or ordinary events. There are, however, some special considerations to take into account when an accident falls into the disaster or catastrophe category.

One of the urgent problems is that rumors explode in chain reaction fashion when the facts are not immediately known, especially if the accident involved many people or the loss of life. The neighborhood may be as much affected as company personnel. For example, an accident at a plant that uses hazardous chemicals may be blown out of proportion in the neighborhood causing panic in the area. In the larger community, people may believe that a company product was at fault and consumers may react strongly. Reduced sales and cancelled orders may seem to be a slow-moving reaction, but they can be as disastrous in terms of cost as an in-plant explosion. Quick, effective communications can minimize both these in-plant and community reactions.

A public information program should be an integral part of the emergency plan, and responsible management will likewise incorporate an employee communications plan, as well.

Hazardous Materials—While most hazardous materials incidents are handled prior to their becoming a disaster, a facility needs to be flexible and evolutionary in its response to a developing incident. Hazardous material incidents differ from other emergency situations because of the wide diversity of causative factors and the pervasiveness of the potential threat. Circumstances and geographic features in the vicinity of incidents vary greatly. They may occur at fixed facilities where the opportunity for development of site-specific contingency plans is great. They may also occur at any place along land, water, or air transport routes and (in the case of vessel mishaps, aircraft accidents, agricultural chemicals, and illegal dumping) may occur in unpredictable areas, relatively inaccessible by ground transportation.

The number of hazardous material shipments has increased greatly in the past ten years, and it is estimated that over 800,000 hazardous materials shipments move through the United States each day. In addition, one can only guess at the millions of in-plant handlings of hazardous materials each day. Reported accidental releases of hazardous materials numbered about 17,000 in 2014, the majority of which occurred during unloading. Only a few of the numerous regulated commodities were responsible for accidents that involved injuries and fatalities. Those most often mentioned were (1) flammable liquids, (2) corrosive materials, and (3) oxidizers (U.S. Department of Transportation, 2015).

When a chemical is released or spilled, time can be of the essence. Immediate information on the chemical involved is essential so that appropriate actions can be taken. The Chemical Transportation Center (CHEMTREC) offers a twenty-four-hour toll-free telephone emergency information service available to provide response assistance following incidents involving hazardous materials. The *Emergency Response Guidebook* is intended for provide guidance to first responders at hazardous materials incidents (PHMSA, 2012). The telephone service and *Emergency Response Guidebook* share a common objective, which is to give quick information on first aid, fire control, spill or leakage cleanup, and other pertinent facts following the release or spill of a hazardous chemical. Further information on the toxicity of chemicals can be obtained from Toxicology Literature Online (TOXLINE) bibliographical database run by the U.S. National Library of Medicine. TOXLINE provides references covering the toxicological effects of drugs and chemicals dating from the 1840s to the present.

Of further use to management, the toxicology committee of the American Industrial Hygiene Association (AIHA), National Institute for Occupational Safety and Health (NIOSH), and other organizations publish a series of Emergency Response Planning Guidelines (ERPGs) and emergency exposure limits (EELs), including Acute Exposure Guideline Level (AEGLs) and Temporary Emergency Exposure Limits (TEELs), defining how large a single brief exposure to airborne contaminants can be tolerated without permanent toxic effects. These limits do not attempt to replace accepted safe practices but can be used in emergency planning to help determine what type of emergency action to take and to assist in the proper selection of personal protective equipment.

Radiation—Emergency planning can't be too strongly stressed in the case of ionizing radiation. When radioactive materials are being handled, typical emergencies involve spills, contamination, leakage from containers or a fire involving radioactive materials or the areas in which they are used or stored. Emergency procedures for such incidents should be developed even where only small amounts of radioactive material are handled. Immediately washing the injury and the implement that caused it with a lot of water is important and requires strategically placed showers and eyewashes. For leaks or spills, the main concern is to minimize exposure by confining it to those first involved. All persons involved should be monitored for contamination before being allowed to mingle with others or go into other parts of the plant. Except for the most minor incidents, the areas should immediately be cleared of people.

Extreme Weather—Extreme weather refers to such sustained conditions that will cause normal operations to fail. This can call for the protection of personnel, customers, and even nearby residents. If a sudden blizzard closes highways near a plant, the company may suddenly be faced with buses and carloads of stranded travelers, as well as company personnel. A severe ice storm may interrupt all power in an area bringing operations to a complete standstill. A flash flood can wipe out a critical access bridge stranding employees and others at the plant. Dust or sandstorms can create similar emergency situations making travel from the plant impossible. Planning cannot cover all weather emergencies, but the most probable ones can be considered. Moreover, if a flexible attitude is kept, plans for one weather emergency can readily be adapted to another (as an unexpected dust storm can be dealt with using plans originally developed for blizzard conditions).

Energy Emergencies—Energy emergencies or shortages are becoming increasingly a part of the average person's awareness, if not of their actual experience. If anything, they are even more critical for industry than they are for the individual. Energy emergencies can include sudden fuel shortages or slowly progressing shortages caused by rising fuel prices and dwindling reserves. There may also be periodic imbalances in supplies at particular times or seasons, such as a higher demand for heating fuels during severe winter weather. Geographic imbalances can also exist. A gasoline shortage, for example, may be less critical in a community where most workers live within walking distance of the plant than it is where workers may have to travel fifty or more miles to work each day. A shortage of one type of fuel can cause subsequent shortages of substitute fuels.

Problems created by energy emergencies include quick developing safety and health problems (stoppage of elevators and life support systems), heating/cooling problems, electrical power problems, and natural gas problems. Any of these problems may involve key personnel who may have to concentrate their energies on solving plant or community problems, lessening their availability for other duties. In some energy emergency situations, the company may find itself having to work closely with local and state agencies or conform to emergency regulations. It is important to keep in mind that rarely will an energy emergency affect just a single plant or company and that the need for cooperating with larger plans may supersede parts of the corporation's formal plan.

NECESSARY STEPS IN DEVELOPING A PLAN

The following nine steps are involved in establishing an effective emergency/disaster program:

1. Assess the probability of an event
2. Assess the hazards involved
3. Designate an emergency/disaster plan coordinator
4. Develop an emergency/disaster plan
5. Approve the plan
6. Train personnel
7. Test the plan
8. Revise the plan as needed
9. Coordinate all aspects of the plan

While these nine steps are only one way to approach the development of an emergency response plan, it is essential that a systems approach be used to ensure all aspects are considered and that the developed plan is correct, complete, and comprehensive in its abilities to address an emergency event.

Assess the Probability of an Event—The first step in developing an emergency response plan is to decide how likely it is that a facility will be faced with a given emergency or disaster. What is the probability of the event occurring? One might use a simple ranking system to divide all possible events into categories, such as (1) most likely, (2) quite possible, (3) unusual but possible, (4) remotely possible, (5) has never happened but is conceivable, and (6) practically impossible. These possibilities can be weighed both subjectively and objectively for decision-making purposes. Normally, the cost and consequences of the event (should it happen) would also be considered. To a certain extent, this merges into the step of hazard assessment. The plan objectives at this stage are aimed at deciding:

1. What events should the facility prepare for?
2. What commitment of resources should the facility make?
3. What facility activities will be excluded from the plan?
4. What level of activity will plan implementation require?

Assess the Hazards Involved—Assessing the hazards merges with the probability assessment step to a certain extent. This step must consider the cost factors involved in hazards, such as the likely cost of catastrophe, the likelihood of multiple fatalities, or even a great number of minor injuries. The assessment must also take into account all company resources likely to be involved, including facilities and market share. To distinguish between these two steps, one must consider the probability of an earthquake. In step one, the planner would consider how likely it is that an earthquake would occur. For instance, the manager of a manufacturing operation in Southern California might rightly conclude that an earthquake will be very likely. In step two, he would look at his operations and assess the specific hazards and the likely impact involved. He might consider that the plant stores and uses large quantities of highly flammable material which could burst into flames following an earthquake, the fact that the plant building was constructed prior to the development of earthquake construction codes, the knowledge that the facility is the sole source of raw materials used by facilities, and the fact that it is located a great distance away from the nearest medical facility. Taken all together, this leads the plan developer to the conclusion that an earthquake is very likely to happen and will potentially be very costly to the company both in terms of property damage, injuries or fatalities, and interference with company operations elsewhere. The planner would therefore assign a very high priority to developing a thorough and effective earthquake plan.

Designate and Emergency/Disaster Plan Coordinator—The designated emergency plan coordinator should be an individual who is in a position to see the plan through development and testing, but also through actual implementation. Their duties will be discussed in greater detail later in the chapter.

Develop an Emergency/Disaster Plan—An effective safety and health plan within the SMS makes the establishment of an emergency/disaster plan immeasurably easier, as all of the organization's safety and health plans can serve as a base for an emergency response plan. For example, the company plan that covers the handling of flammable liquids can be extended to cover an emergency involving flammable liquids. The emergency response coordinator may find that the facility's safety and health program already provide for many of the trained people, much of the equipment, and many of the resources that will be needed in an emergency.

In developing the plan, it is vital to consider everything that could possibly affect the plan. The plan might be a single document, a series of manuals each covering a different emergency or function with suitable appendices, or a combination of both. Individual sections could then be broken out and organized by emergency type, department, or function. The best advice is to keep the plan as simple as possible. Unfortunately, even "as simple as possible" will have to be somewhat involved if most emergencies are to be considered. Regardless of the size and depth of the plan, it must be easy to follow and understand. The best of plans will not work if people do not understand it. Try to keep in mind the ultimate user of the entire document and individual sections, and write on a level and in a language that will facilitate a quick and correct emergency response.

Approve the Plan—An emergency response plan ultimately involves personnel and company survival. Therefore, it must have the approval of the top level of management, including the board of directors and chief executive officer. It is equally important that specific portions of the plan be signed off or approved at the operating level where the activity will take place. This should not be a mere formality indicating that the person involved has received and read the plan. It is a vital step in gaining feedback and verifying that the plan has been thoroughly read and understood.

Train Personnel—If an emergency/disaster plan is to work successfully, all participants must be trained in the responsibilities and specialized duties required by the plan. Training objectives may range from simply knowing the evacuation route from a certain section of the plant, to knowing the location of all emergency equipment, to taking charge of the welfare of all personnel for an extended period of time. For some it will involve drills and wallet-sized instruction cards. For others, it may mean a series of training courses away from home and continuous coordination with outside agencies.

Extraordinary emergencies require coordinated operations, including all domestic and governmental functions, to minimize danger and damage. Federal Emergency Management Agency (FEMA) offers a variety of online and face-to-face training courses through the Emergency Management Institute (EMI), Center for Domestic Preparedness (CDP), National Training and Education Division (NTED), and National Fire Academy (NFA). These training are intended to enhance the abilities of federal, state, local, and tribal government officials, volunteer organizations, and the public and private sectors to minimize the impact of disasters.

Test the Plan—Waiting for an actual emergency to find out if the plan works is too risky, too costly, and too late. Testing the plan is the only sensible way to find out if it will work. First, testing can point out flaws or gaps that can quickly be corrected. Second, testing can also serve as a training and refresher device. Testing exercises can be designed to try out selected elements of the plan (such as the fire or evacuation plans) or to test the entire disaster organization. Tests of individual elements should be carried out first and deficiencies corrected before the entire plan is tested. In no case should the test endanger personnel or resources in any way. This is an important point since an unpracticed or hurried response to tests is a fertile ground for accidents. Tests should be discussed and reviewed with the safety manager before they are carried out.

Revise the Plan as Needed—Any test of a plan should be critiqued immediately afterward. In this way, corrective measures can be implemented before the problems are overlooked or forgotten. The critique process should consider the comments of everyone involved in the test, since even a very minor plan failure can jeopardize successful implementation of the entire plan. This critique process should not be reserved for test runs only. Any actual activation of the plan should also be thoroughly critiqued to see if things went as planned in actual operation. It is particularly important to uncover problems that involve management deficiencies or that require management action to correct. Generally, three deficiencies are uncovered in both tests and actual emergency plan activation:

1. There are no clear guidelines on when to stop activities and get help.
2. There has been a lack of practice in actual emergency shutdowns.
3. Management has not been alert to changes in personnel, processes, or facilities that require adjustments in the plan.

The feedback and critique process, as well as the testing itself, are all aimed at allowing management to make the necessary revisions to the plan. It must further be emphasized that the plan should be tested again once any revisions are made in the plan to make sure that the revised plan now functions properly and that no new problems have been created by the changes. This last point is essential. Change is a constant in most operations. Trained personnel leave or are transferred, departments are moved from one area of the plant to another, partitions, exits and first aid cabinets are moved to accommodate new machinery or equipment. Management must revise the plan accordingly each time such a change is made and then retest the whole. This overlaps the next step to an extent, coordinating all aspects of the plan.

Coordinate all Aspects of the Plan—Every time a change is made in a plan, the change must be coordinated with all involved parties and proper adjustments made in the plan itself. Imagine the consequences of not having, for instance, the new telephone number of the medical response agency printed on wallet cards or the confusion that might result if a new type of alarm was being used in case of a flash flood. The few extra minutes needed to find the right number or ask the right question could cost lives and valuable property. Moreover, any changes in operations, facilities, personnel, or equipment that may affect the plan should be reviewed and the plan changed accordingly.

As can be seen, this requires a continual review and updating process, possibly even new tests and training from time to time, printing of new evacuation or access maps, equipment lists, wallet cards, and so on. There are so many small but critical details involved that it is essential to have one person in a position to coordinate and check them all. While effective emergency planning may involve a great commitment of time and resources, the cost is likely to be far greater if the company is unprepared to respond to an emergency.

Elements Contained in a Complete Plan

Key questions that arise during the development of an emergency response plan are "What should the plan contain?" and "Who will perform the actions it outlines?" This touches on many elements, all the way from necessary equipment and personnel to complex interfaces and backup systems. An emergency response plan should clearly define the plan components and the individuals responsible for implementing each aspect of the plan.

Clear and thorough planning for emergencies and disasters is only possible if the plans are in writing and effectively communicated. Not only must an employer with more than ten employees have a written emergency action plan whenever an OSHA standard requires one (Emergency Action Plans, 2002), but a thoroughly written and communicated plan helps to promote a culture of safety awareness among employees and management. First, the written plan can delegate specific responsibilities so that everyone knows exactly what is expected of him. Second, a plan cannot be well tested in advance if it is not precisely detailed in writing.

A thorough assessment of the hazards present at the facility is the first step in developing a complete and comprehensive emergency response plan. It is essential that the authors of the plan address those identified hazards at a facility, as doing so improves the likelihood that injuries and fatalities are prevented, property damage is reduced, the surrounding environment and communities are protected, and a quick return to normal operations is facilitated.

Guidance for the development of an emergency response plan is available from a variety of resources, including the *Principle Emergency Response and Preparedness: Requirements and Guidance* (OSHA, 2004), *Emergency Management Guide for Business and Industry* (FEMA, 1993), and *NFPA 1600: Standard on Disaster/Emergency Management and Business Continuity Programs* (NFPA, 2013). While there is no "one size fits all" format to aid in the writing of a plan, there are defined requirements for emergency response plans whenever an OSHA standard requires one:

1. Fire/emergency reporting procedures
2. Procedures for evacuations, including type and exit routes
3. Procedures for who is to operate critical operations prior to evacuation
4. Procedures to account for employees after evacuation
5. Procedures for employees performing rescue and medical duties
6. Names of those to contact for further information and explanation of the plan (Emergency Action Plans, 2002)

Additional plan requirements may exist depending on the type of facility and its operations and/or the hazards present at a facility. However, it is essential that an emergency response plan focus on solving the problems created by an emergency rather than simply responding to the chaos it creates. Existing information, procedures, and systems should be used wherever possible as the basis for emergency response procedures and systems.

The development of an emergency response plan should not be considered a "one man job." A functional committee consisting of owners, management, employees (labor, maintenance, security), human resources, community relations, and legal council should be involved in the development of the plan. The skill set of each member in such a diverse group increases the likelihood that the plan will be correct, complete, and comprehensive in its ability to identify hazards, detail specific responses, and coordinate with emergency response authorities. It is important to note that this team also fosters a sense of community and shared involvement among employees at all levels of operation and is a crucial step in developing a working culture of safety at the facility.

The specific role of the functional committee in developing the emergency response plan will be discussed in greater detail later in this chapter. Once the emergency response plan development team has been assembled, its first tasks should focus on establishing a schedule for plan development to ensure that the team effectively and efficiently works toward completion of the plan.

Executive Summary

Mission Statement—The emergency response plan should begin with a mission statement that clearly identifies the purpose of the emergency response plan, how the plan will be utilized to prepare, respond, and recover from an emergency, and who the authority is in the event of an emergency. It should be brief and general enough to give leeway for various operations, but specific enough to outline general responsibilities for the development and functioning of the plan. It should assure some flexibility, since one cannot possibly foresee every contingency and must be able to modify the plan quickly if needed. It must, however, provide for someone to be in charge at all times in all conditions. For this reason, it may be wise to specify an order of succession. The mission statement should be as lean as possible while still effectively communicating its intent, function, and authority.

Purpose of the Plan—No single emergency plan will be able to do all things for all organizations, but it is vital to spell out the purpose of a particular plan so that the wrong direction is not taken. Nothing brings disorder to a normal situation quicker than not knowing who is in charge or who is running the organization. This situation is made far worse by an emergency circumstance. One of the objectives of a disaster plan is to allow responsible individuals to concentrate on solving the major problems created by the emergency, rather than trying to bring order out of chaos due to simple lack of planning.

Authority—Effective plans do not just happen. They are shaped, built, and implemented by people who are capable of doing the complex, long-range planning required. An emergency response coordinator should be a calm, quick-thinking person, robust enough to stand the pressures and long hours of an emergency. It is equally imperative to name an alternate coordinator who can step in during emergency situations. The alternate should be kept fully informed of plan details and all aspects of managing an emergency. Because a disaster or emergency can occur at any time of the day or night, on weekends or holidays, a plan should provide that someone be in charge at all times. Top management is not always available, particularly on short notice. Naturally, a coordinator will be involved in developing the plan and perhaps implementing it because they will be the most familiar with it. However, there is still the matter of having someone at the top in overall charge. A management succession list will assure that someone is always available, regardless of the time or circumstance and that they are competent to coordinate with internal and external emergency responders.

Control Center—Having a centrally located control center from which to direct emergency operations is vitally important. Such a location should be carefully chosen, keeping several things in mind. First, it should be located in a place that is likely to survive most emergencies. Second, it should be large enough to accommodate the necessary people. Third, it should have communications and access to both the inside and outside of the plant. Some corporations have established special installations that meet all these criteria, since it is unusual to have a ready-made location that has all these qualities. Because it is always possible that a disaster condition will render a control center unusable or inaccessible, it is important to designate an alternate headquarters or control center. The disaster plan should list the alternate control center location, as well as spelling out who should report to that location and under what conditions.

Core Plan Components

Distribution of Plan—Prompt and effective action in an emergency depends on people knowing what to do, and that requires detailed and adequate information in advance. Distributing the plan to all those who need it to facilitate the sharing of information and standard operating procedures (SOPs). Plan distribution also serves as a tool for conducting employee and community responder training and provides the basis for evaluation and refinement of the plan. While plans may be revised as facility operations change or as new hazards are identified, they may also be revised based on challenges and needs identified during training. Regardless of the reason for a plan revision, it is essential that the most current version of the plan be on file and distributed to ensure that accurate information is available and that SOPs are followed in an effort to minimize hazards. It is also acceptable to distribute portions of the plan those who do not need the complete plan (i.e., community responders), but again, it is essential that the most current plan is distributed to circumvent confusion.

Emergency Equipment and Supplies—A map showing the location of all such equipment will facilitate emergency actions, regular inspections, and maintenance checks. The map can also designate temporary hospital facilities for victims who cannot be moved or taken to a local hospital immediately. Both personal protective and rescue equipment should be clearly marked on maps. Exactly what type of equipment is needed and where it should be located will be a matter of considerable discussion, since the nature of the emergency can only be hypothesized and each emergency may require somewhat different equipment.

Emergency equipment must be both readily available yet well protected. This is a tough assignment, and the solution may require considerable forethought. Not only should emergency equipment be stored where it can be easily reached, it should be reliable and able to be stored for considerable periods of time without deterioration. It should be checked, inspected, and cleaned regularly, even when it has not been used, to guard against such deterioration or malfunction due to age or disuse. In addition to locations around the facility, some equipment may be kept at the emergency operations center. Things such as sandbags, tools, generators, pumping equipment, and battens should be located at a convenient central location or at the alternate control center. Wherever possible, the plan should strive to integrate the capabilities of the facility with those of neighborhood, community, state, and federal resources.

Facility Maps and Location of Data—Maps detailing the facility layout, contents, and physical environments are essential pieces of information to aid first responders in an emergency event. In a very simple plant layout it may not be necessary to have a separate map of the entire installation, since other maps will be included as they pertain to other emergency responses. However, such a map would be useful for breaking the plant into areas of responsibility for emergency handling purposes. More detailed maps of each designated area would supplement the overall map. This would be useful, for example, in a bomb search emergency. The overall map would divide the entire installation into areas of responsibility, while the detail maps would show every nook and cranny in the area to be searched.

In the event of an emergency, it is essential that first responders have the most accurate and precise information. The location of information including chemical inventories, Safety Data Sheets (SDSs), and hazardous materials within the facility must be immediately obtainable by first responders. In addition to being readily available and current, the data must also be in a usable format that considers organization and physical mediums. For example, electronic data may not be available in the event of a power failure, and hard copies of the data may not be readable in a flash flood emergency.

Hazard Assessment—Adequate planning for emergencies facilitates adequate emergency response. A thorough list of all potential hazards and estimations of their probability of occurrence allow for a better understanding of their impacts to the business, employees, and community. Consideration of the chemical, physical, and biological hazards present must not only be given to employees and first responders, but also to off-site hazards to human populations and natural environments that may result from an emergency.

A thorough hazard assessment helps ensure that the plan is correct, complete, and comprehensive in its efforts to prepare for and respond to an emergency. Analyzing the capability of the facility to manage those hazards is also essential. A review of current policies and risk management practices and the identification of outside groups/agencies to aid in the response allows for a plan that can successfully respond to and recover from an emergency.

Auditing of General Operations and Procedures—Prevention must be the first objective of any emergency response plan. The ability of a plan to identify and correct situations that may lead to or aggravate an emergency can be determined through routine auditing. General operations including the receiving, handling, storage, and disposal of chemicals, housekeeping, employee conduct and activity, and equipment and facility maintenance should be routinely audited to identify and correct hazards, amend facility policies, and revise SOPs.

It is important to note that the individual conducting audits of general procedures must have the authority to implement immediate changes in the event that a hazardous situation is observed. General procedures including training, response to alarms, inventory control and data maintenance, operations monitoring, and evacuation procedures should also be routinely reviewed to ensure that they are consistent with the objectives of the emergency response plan.

Communication During an Emergency—The communications portion of the plan must consider two basic needs: handling internal communications and handling external communications. Internal communication is key for the proper reporting and handling of reported emergencies. The plan should have a list of key personnel to be notified in an emergency, as well as any necessary emergency phone numbers (such as fire and police and ambulance services).

Warning and alarm systems must be included in the emergency plan, giving special attention to the warning time likely to be available for a particular emergency. Except for flash floods and earthquakes, there will usually be some warning time allowing employees to get to shelters or to emergency stations. Above all, warning time should be used to undertake the most critical parts of the emergency plan.

Existing signal systems should be used to give the warning and should be connected to auxiliary power systems in the event of an electrical emergency. All alarm systems should have a specific tone to indicate a specific emergency and pre-written verbal messages used by notifying parties can help reduce confusion, fear, panic, and/or frustration. Routinely scheduled testing of each emergency notification system is absolutely necessary, however; tests should be announced in advance so employees will not panic and think it is a real emergency.

In addition to notifying community emergency responders, external communications should also perform the task of public information and public relations. The families of employees, the general public, customers, suppliers, and stockholders must also be kept informed of the emergency. Following a disaster, employees and their families will want to know several things: who is hurt and how badly; will the plant be closed down either temporarily or permanently; how long will it take to rebuild or repair the plant; will I have my job when it reopens; what do I do in the meantime; what should I do about hospitalization; and so on.

Customers will want to know about deliveries they were expecting, probable delays involved, and whether they should seek other sources of supply in the meantime. Likewise, suppliers will want to know about existing and future needs. Should certain services or supplies be discontinued for a period, or should they simply continue at another location. Arrangements may be necessary to restore and repair utility services, as well as those provided by private firms. Over all of this hangs the need of the news media, who have a legitimate interest in the emergency or disaster. If they are not fully informed of the situation, they will draw their own conclusion. Having such partial or distorted reports published or broadcast can have far worse results than a complete and honest account of the event.

Evacuation Procedures—In an emergency, it is essential to guide employees and visitors to safety, direct employees away from hazardous areas, minimize hazards associated with the evacuation procedure itself, and to avert panic. Wherever possible, every effort should be made to get and keep people away from the immediate scene of the disaster or emergency. To achieve these aims, both evacuation plans and evacuation maps are an essential element of any emergency disaster plan. An evacuation plan covering most emergencies can be most clearly shown via a map. An overall facility map supplemented by area maps may be useful; small area maps can simplify the evacuation and eliminate the possibility of error.

First, when developing exit and evacuation plans, the National Fire Protection Association standards and local codes should be consulted. Each building should have enough exits to permit the prompt escape of all personnel in an emergency. Indicating secondary evacuation routes on maps can also facilitate prompt evacuation in the event that an exit may be blocked due to fire or debris. Exits should be kept unobstructed and accessible at all times. Exit doors should swing in the direction of travel, particularly if a large number of persons will evacuate via that exit. Moreover, exits should be well marked in highly visible lettering that is plainly legible.

Every office or section should have designated monitors who will see that persons are safely guided to safety. These monitors should be clearly designated in the emergency plan and should participate in all drills and tests so they will be prepared in the event of an evacuation emergency. While the monitor may *guide* an evacuation, it is up to the person in charge to *authorize* the evacuation, as well as the safe return to the building. The evacuation maps should also clearly indicate post-emergency assembly points so, that the designated monitors can account for personnel in their section once they have evacuated. Visitors' safety may be more difficult to assure, since the visitor probably does not know anything of the corporate emergency plans, the evacuation routes, or the location of the shelters. It is the duty of the monitor to take care of the visitors in their area. Equally important, there should be a firm policy and procedure to keep any other visitors from entering a facility or area after an emergency situation has developed.

Shelter Procedures—Some emergency and disaster situations require sheltering in place rather than evacuation. Shelter areas are necessary for emergencies such as floods, hurricanes, and tornadoes. There is some psychological value in having shelters in places that are normally used for other purposes, such as a well-designed and protected cafeteria. However, it may be necessary to designate several different shelter areas since the nature of the emergency itself may render one location more vulnerable than another. A well-designed and protected basement, suitable for tornado protection, may be the first place to be flooded. Likewise, a glass-walled cafeteria located on high ground may be ideal as a shelter during a flood emergency, but particularly vulnerable during a windstorm.

In choosing the shelter, both location and construction should be carefully considered. Height above flood lines, presence or absence of breakable glass, flexibility in withstanding wind or earthquake shock should all be considered. Following a widespread disaster, employees may be in the care of management for quite a while. Food, sleep, shelter, dry clothing, and medical care may be needed. Cold weather may be a particular problem if sources of heat are damaged or unavailable. Most employees are anxious to leave company shelters as they want to find out how their families are or how their homes survived the disaster. Communications and transportation may be their first concern, even before food, shelter, or first aid. The company should be aware of this and seek to provide safe means of transport as soon as it is possible to do so.

Transportation—Like communications, transportation is an essential function in handling an emergency promptly and efficiently. It serves such vital roles as getting people out of endangered areas, moving necessary emergency equipment to the right locations, and transporting injured or disabled persons to treatment facilities. A thorough emergency transportation plan would include maps, diagrams and instructions for drivers, including locations of emergency facilities, routes to nearby hospitals, and location of fuel supplies. A special high-priority aspect of emergency transportation is the transportation of victims to hospitals or treatment centers. Besides an inventory of area ambulance facilities, company vehicles should be inventoried for the task, and designated for that purpose in an emergency.

Employee Welfare—While it is desirable to have employees involved in the disaster plan, keep in mind that their welfare is the responsibility of company management. They must be protected from additional hazards and are not likely to understand those hazards unless management provides that information. Employees should know the do's and don'ts of what to do after a disaster. They should be warned and periodically reminded of the hazards that might be encountered, such as live wires, broken gas mains, or collapsing buildings. They should be given specific instructions, and reminders should be posted in prominent locations.

Special Procedures—Special procedures that detail the response to those specific hazards identified at a facility should be included in the plan. First, the plan should clarify what constitutes an emergency. For example, it would describe what is meant by a radiation emergency or spell out the difference between a "flash flood" and a "normal flood" emergency. Clearly defining what constitutes an emergency and subsequent emergency response helps to ensure that those exposed to the hazard are able to recognize unsafe conditions.

Thoroughly written directions for responding to hazards such as fires/explosions, floods, and hazardous chemical releases are essential in the emergency response plan but should also be communicated during training at a level that speaks to the audience based on job activity and physical location within the facility. While detailed special procedures are included in the written emergency response plan, the full plan may not be readily available in the event of an emergency. Quick, "go-to" references including signs, placards, and posters can be made available in those areas where an emergency requiring those procedures is likely to occur.

Equipment Shutdown—In the event of some emergencies, an orderly and controlled shutdown may be necessary for the protection of employee lives, community welfare, and company resources. This is particularly true in certain industries, engineering and research laboratories, storage facilities, or high-technology operations. Explosives, fumes, high voltage lines, or other hazards may constitute an imminent danger following a disaster or emergency. Failure to execute an orderly shutdown could result in the destruction of the entire facility, or worse, cause widespread injury and death both among workers and those in the neighborhood. An adverse condition may develop either with or without attendant disasters. Management should be able to identify such a condition as soon as it becomes abnormal and dangerous, and before it has had time to generate injuries or damage.

With suitable emergency plans, such injuries and damage can be avoided. This interim period is the most critical time for emergency action. The most desirable approach would be to continue normal operations while steps are taken to remedy the abnormal condition. However, this is not always possible or safe, so an emergency plan must also contain provisions for partial or complete shutdown. Keep in mind that some equipment may take several hours to shut down fully and that flood, fire, explosion, or earthquake will probably not give that much warning. In many cases, the shutdown crew may be the last to leave the plant. For that reason, it is vital to provide maximum protection for these people. Crews should be kept as small as possible and should be provided with personal protective equipment and other necessary equipment for actual shutdown. Careful training and good simulation drills are essential. First, they will greatly speed the actual operation, and second, they will assure that the crews will not take any unnecessary risks.

Generally, the responsibility for executing an orderly and efficient shutdown rests on one person. This is usually someone from the plant engineering office who is familiar with the facilities to be shut down. Sometimes referred to as the "utilities officer," this person should help develop designs and checklists to spell out the steps to be taken for orderly shutdown. A simple matrix can be used to show the shutdown task, the time required for the task, the name of the person or persons assigned to the task, and the date or time assigned and checked. The person in charge of the shutdown would also identify all processes that would prove dangerous in an emergency and determine what should be done about each, know who to contact in the locality regarding protective action in case of failures involving lines, valves, and switches, and identify the employees best qualified to evaluate and report conditions following a disaster, and those best able to take corrective actions. These steps are all fairly time-consuming. However, keep in mind that emergency plans can be developed from existing information and the person in charge of the shutdown might already have such information available, ready to be put into a formal emergency plan format.

Restoration, Repair, and Salvage—Getting things back to normal, or at least into shape to resume operations, calls for special skills and knowledge. This, too, should be provided for in the emergency plan. The facility engineering crew should survey the building and grounds thoroughly, paying particular attention to the structural integrity of buildings and installations before restoration, repair, or salvage can be undertaken. The survey would look for unsupported walls, precarious equipment, live or exposed wiring, escaping gases or liquids, dangerous chemicals, and damaged containers of volatile materials. Any part of the shutdown process that was not completed before the emergency should be taken care of at this time. Only after a thorough investigation should the utilities or processes that were shut down be turned on again.

The survey also provides a basis for the repair of facilities and equipment. It can determine what can and cannot be repaired, and how much effort will be required to do so. Whatever cannot be economically restored to working order should at least be salvaged. That is, it should be retrieved and disposed of as profitably as possible. Repair, restoration, and salvaging operations can be very hazardous in the abnormal or disrupted conditions following an emergency. Only personnel who are properly qualified, trained, and equipped to do so should perform these activities, and they should be adequately supervised throughout the operations. Only these qualified work crews and supervisors should be allowed in the area, and all others should be kept out of the hazardous area until work is completed inspected by engineering and safety personnel. A facility should only be considered safe to resume normal operations when the facility poses no further risk to employees and alarms/notification systems have been repaired, replaced, and are functioning.

Personnel Assignment—The fact that people behave in an amazingly responsible manner during emergencies is good reason to have an emergency task for many of them. Moreover, when an employee knows they have a certain task to perform, they will not be as likely to succumb to panic or confusion. Be certain that every employee with an assigned emergency task is included in tests and critiques of the plan. Their involvement will be a key to the amelioration of the emergency. Every person who has a specific task to perform during an emergency should know and understand it well in advance of need. Wherever possible, the person in charge of the normal function should remain in charge of it during an emergency.

If the person has no specific assignment, they should still know exactly what to do whether it be to proceed to the nearest exit or shelter, report to a particular person, or simply wait for instructions. Faced with disaster, especially with fear for their lives, most people can be counted on to respond energetically, participating actively in rescue and recovery operations. How well they do this, however, is a matter of preplanning, particularly advance training for emergencies and disasters.

Training—Personnel assigned to emergency duties must know what they are to do and how to do it. Only training provides this information. Emergency plans should specify how and when training occurs for all personnel and for those with specific assignments. Training activities including walk-throughs, table top exercises, and functional drills (i.e., evacuation drills) not only provide an opportunity to ensure that each individual responds appropriately during an emergency, but they also promote a culture of safety among personnel.

Provisions should be made for realistic training, or at least the realities of an emergency should be stressed during each training session. However, tests should neither be so realistic as to endanger anyone nor should the plan be tested so often that it becomes a routine. Tests should be conducted frequently enough to ensure that personnel don't forget the procedures, anytime there have been changes in personnel or in the plan itself since the last test. Topics covered in personnel training should include individual roles and responsibilities, hazards that exist, notification procedures, special emergency response procedures, evacuation/shelter/accountability procedures, location and use of common and emergency management equipment, and emergency shutdown procedures.

Training schedules should also strive to coordinate with emergency response personnel and community emergency response agencies so that all components of the plan can be evaluated. The outcomes of a training activity should be communicated following its completion, and the procedures should be critically evaluated. This will allow for the identification of those procedures that are effective, those procedures that may require revision, and the assessment of employees tasked with specific assignments and job duties.

Documentation—The destruction of records essential to conducting business can be as disastrous as the damage to personnel or facilities. So vital are these records that an emergency response plan must fully provide for their protection. Such records may pertain to corporate existence, financial and legal documents, and operating records. Plans for protecting these vital records might include duplicate records stored at an off-site location and/or storage of records in different formats (hard copy vs electronic records). This helps to avoid the loss of records considered essential to tracing past operations and continuing present ones.

The protection and continuance of vital documents including proprietary or confidential data, such as formulas, patented processes, and specifications of a proprietary nature, should be given special consideration, both in preserving them physically and protecting their integrity. This applies likewise to company confidential material and to other security documents. If the company is involved in operations that fall under government security regulations, they will already be aware of the requirements and guidelines established for protecting documentation from breach of security during an emergency or disaster.

Appendices—An emergency response plan should be correct, complete, and comprehensive in its writing, yet it must also be "user friendly" and immediately available. In order to keep the emergency plan "lean," only essential information should be located in the plan itself. Appendices can facilitate such leanness and are often included to allow for quick and easy access to information regarding specific procedures. Items frequently included in appendices pertain to utilities and shut-off points, the location of hazardous materials, floor plans, alarm systems, and chemical inventories. In addition to considering what information should be included in appendices, determining what specific procedure or informational appendices should be distributed to community responders also warrants consideration. Community responders should be given those special procedures that describe their role in an emergency response so that they may act accordingly.

Functionaries, Agencies, and Individuals Involved

In discussing the elements of an emergency response plan, the need to designate who is responsible for certain functions and actions has been emphasized. One of the objectives of a disaster plan is to allow responsible individuals to concentrate on solving the major problems created by the emergency, rather than trying to bring order out of chaos due to simple lack of planning or preparedness. If plans provide for the predictable and routine, those responsible can then concentrate on the unusual or unpredictable.

Some managers have difficulty grasping the requirements and rigors of emergency/disaster planning. They do not fully realize that the enterprise could face a disaster crisis of massive proportions at any time, so they give little or no thought to developing a thorough plan. The damage from this short-sighted management attitude could be irreparable. It is essential that top management recognizes the magnitude of the problem and concentrates its best efforts on developing a workable, effective plan. They should also recognize that the best-made plan will not function on its own, as it takes the proper people and agencies to carry it out. This section deals with those main functionaries and their roles in the emergency plan.

Executive Board—An emergency situation calls for a tremendous effort from the entire organization. Emergency planning should, therefore, have early and complete backing from the executive board. In fact, the plan may well originate with the board. They should play a key role in setting the policy and purpose of the plan and may well be the prime movers in some areas, such as the line of succession or emergency finances.

Chief Operating Officer—The chief operating officer has overall responsibility for carrying out the directions of the board. This is particularly true in plan development. He or she will also, if he or she is available, activate the plan in the event of an emergency. In some corporations (particularly smaller ones) he or she may personally direct the execution of the plan, as well. In larger corporations, however, he or she may simply direct the emergency response plan coordinator to activate and execute the plan. The chief operating officer would normally deliver the succession list directly to those involved and occupy the company control center or alternate headquarters. Once the emergency is over, he or she will put into operation plans for repair and restoration in order to get back into production or operation.

Emergency Response Plan Coordinator—Ordinarily, the coordinator has a key role both in development and execution of the plan, and generally is responsible for the overall direction of the emergency plan. Upon direction of the chief operating officer, or by some prior arrangement, the coordinator authorizes the giving of the alarm and sets the emergency plan in operation, either by phone or delegation. If on-site, the coordinator will then proceed to the scene unless the disaster prevents them from doing so.

Functional Committee—The functional committee is made up of representatives of various key departments. Their duty is to assist and advise the coordinator in developing and organizing the plan. Forming a committee of existing department heads is essential as existing organizations, departments, and functions can serve as the best base for handling an emergency.

Under the guidance of the coordinator, the functional committee works to

- Make continuing studies of the facility's vulnerability preparedness and recommend actions based on this vulnerability and preparedness.
- Establish close liaison with local governments on the plan.
- Prepare and publish the emergency response plan.

- Assign emergency duties as provided in the plan.
- Develop emergency financial procedures.
- Designate post-emergency assembly points.
- Test and revise the emergency response plan.
- Arrange for training or briefings by appropriate local agencies.
- Review and update the plan with particular attention to internal company changes and outside influences.

In addition to these development-phase functions, the committee may also serve special functions during an actual emergency or disaster, such as sending advance parties to activate shelters, alternate headquarters, or evacuation points, instructing monitors to begin movement of personnel to shelters or evacuation points, or commencing direction of emergency operations from the control center or alternate headquarters.

Fire Agencies—From 2006–2010, approximately 42,800 fires at or in industrial or manufacturing facilities were reported to U.S. fire departments (Evarts, 2012). Every possible effective use of in-house fire capabilities must be considered, and the use of public fire departments and other facilities promptly secured and coordinated. This is normally that task of the fire chief or the person who has the fire responsibility at the facility. However, special instructions and plans may be needed where the fire emergency is attended by any other emergency or disaster situation, including widespread disaster.

Security—An emergency or disaster always throws an extra burden on the existing security system or department. They may have additional duties, such as directing ambulances, guiding medical assistance teams to the proper locations, guarding access and exit gates against stampede or intruders, patrolling facilities to protect them against vandals or looting, and taking care of additional traffic or parking problems. They may also be needed to provide extra guard or security protection in plant areas where confidential or secret papers, formulas, or items are kept to prevent a breach of security or compromise of integrity.

All of this means that during any serious emergency, the normal ranks of security personnel will need to be augmented. Prior arrangements should be made for this, either by contracting with an outside agency or providing extra backup from the ranks of existing employees. Certainly some employees can be readily assigned to assist in the less sensitive or hazardous aspects of security work, at least until outside reinforcement can be obtained. Whatever arrangements are made, they should be fully documented in the emergency plan.

Medical Services—The organization is naturally concerned about medical care in any emergency, including first aid and the handling and transport of an injured person from the time they are found until treatment has begun. It goes beyond the temporary care of the injured to include the prevention of further injuries, secondary trauma, or death. Planners almost always have a tendency to overestimate the load that can be handled by a hospital unless unusual foresight, planning, and training had taken place beforehand. This places a greater burden on plant medical facilities, particularly in cases of serious or widespread emergencies or disasters.

A member of the functional committee familiar with emergency medical care should lead the planning effort regarding first aid and medical services. They should advise on the plan, and the emergency equipment and facilities needed to handle an emergency. The emergency response plan should also consider those cases that demand fast action, such as breathing stoppages, hemorrhage or severe bleeding, poisoning, burns, and snake and insect bites. For crucial cases such as these where time is of the essence, all personnel should be encouraged to take first aid training, and the coordinator should take steps to learn who among exiting employees has completed basic first aid training, so that they may be incorporated into emergency plans.

The medical department or unit may also be responsible for coordinating the activities of search and rescue units. This multi-talented crew would not ordinarily include medical personnel, though all rescue unit members should have some basic first aid training as they may be the first to find or contact injured employees. Their main job is to reach and free personnel who are trapped rather than render extensive medical aid. Their tools should include such things as shovels, block and tackle, ropes, cutting torches, portable telephones and radios, safety goggles, hard hats, linesman's gloves, and so on. None of this is precisely "medical equipment," but all are necessary to the emergency medical rescue effort.

City, County, and State Governments—City, county, and state governments have a variety of emergency disaster plans, often backed by federal funding and direction. Many cities have developed their own emergency response plans. The coordinator and functional committee should evaluate available assistance well in advance to ensure meshing of plans and efficiency of effort. In some cases, the local government can furnish guidance in every aspect of the corporate emergency plan development. A facility may likewise be obligated to furnish local and state governments with their emergency response plans, particularly if hazardous materials are on-site. Facilities that manufacture, process, or store designated hazardous chemicals must submit a list of those chemicals or copies of their SDSs to local fire departments, Local Emergency Planning Committees (LEPCs), and the State Emergency Response Commission (SERC), as required by the Emergency Planning and Community Right-to-Know Act (EPCRA, 1986).

Federal Government—The National Response System (NRS) is a system consisting of individuals from local, state, and federal agencies, industry, and

Table 18.2 Roles and Responsibilities of Federal Agencies in Emergency Response (EPA, 2015)

Federal Agency	Roles and Responsibilities
The Environmental Protection Agency (EPA)	<ul><li>Chair the NRT and cochair all regional response teams</li><li>Provide scientific support coordinators, and support for site assessments, health and safety issues, action plan development, and contamination monitoring</li></ul>
U.S. Coast Guard	<ul><li>Vice chair for the NRT</li><li>Manages the NRC</li><li>Maintains a National Strike Force trained to respond to major marine pollution incidents</li><li>Commands, controls surveys releases in coastal waters</li></ul>
Federal Emergency Management Agency (FEMA)	<ul><li>Advises and aids lead agencies in coordinating relocation assistance</li><li>Provides guidance, policy, and technical assistance in emergency preparedness, planning, training, and exercising activities for state and local governments</li></ul>
Department of Defense (DOD)	<ul><li>Acts when oil or hazardous substances are released from a facility or vessel under its jurisdiction</li><li>Allocates U.S. Army Corps of Engineer equipment and expertise for removing navigational obstructions and performing ship structural repairs</li><li>Provide U.S. Navy oil spill containment and recovery equipment and manpower</li></ul>
Department of Energy (DOE)	<ul><li>Provides on-scene coordinators when hazardous substances are released from DOE facilities or when materials are being transported under DOE's control are spilled</li><li>Aids in the control of immediate radiological hazards</li></ul>
Department of Agriculture (USDA)	<ul><li>Measures, evaluates, and monitors situations where natural resources have been affected by hazardous substances</li><li>Contributes expertise from the U.S. Forest Service, Agricultural Research Service, National Resources Conservation Service, Food Safety and Inspection Service, and Animal and Plant Health Inspection Service</li></ul>
Department of Commerce (DOC)	<ul><li>Through the National Oceanic and Atmospheric Administration (NOAA), provides scientific support for resources and contingency planning in coastal and marine areas to predict movement and dispersion of oil and other hazardous substances</li><li>Works to remediate damage to coastline and marine resources caused by oils and hazardous substance releases</li></ul>

Department of Health and Human Services (DHHS)	• Assess health hazards at an emergency response • Maintain and provide information on health effects, including the Agency for Toxic Substances Disease Registry (ATSDR) and the National Institutes for Environmental Health Sciences (NIEHS) • Provides technical guidance regarding worker health and safety through the National Institute for Occupational Safety and Health (NIOSH)
Department of Interior (DOI)	• Contributes expertise on natural resources, endangered species, and federal lands and waters • Responsible for native Americans and U.S. territories • Contributes expertise from Fish and Wildlife Service, Geological Survey Bureau of Indian Affairs, Bureau of Land Management, Minerals Management Service, Bureau of Mines, National Park Service, Bureau of Reclamation, Office of Surface Mining and Reclamation Enforcement, and Office of Territorial Affairs
Department of Justice (DOJ)	• Provides expert advice on legal questions arising from discharges or releases, and federal agency responses • Represents the federal government in litigation relating to discharges or releases
Department of Labor (DOL)	• Through the Occupational Safety and Health Administration (OSHA), conducts safety and health inspections of hazardous waste sites to ensure on-site employee safety and protection from hazards • Determines if a site is in compliance with safety and health standards and regulations
Department of Transportation (DOT)	• Provides response expertise through the Pipeline and Hazardous Materials Safety Administration (PHMSA), which offers specialized advice on requirements of packaging, handling, and transporting regulated hazardous materials • PHMSA promulgates and enforces hazardous materials regulations, provides emergency response guidebooks, and supports protective action decision strategies and exercise scenarios
Nuclear Regulatory Commission (NRC)	• Responds in accordance with its incident response plan when radioactive materials are released by its licensees
Department of State	• Leads in developing international contingency plans • Coordinates international response efforts when discharges or releases cross international boundaries or involve foreign flag vessels • Coordinates requests for aid from foreign governments
U.S. Navy	• U.S. Navy Superintendent of Salvage and Diving provides equipment and expertise for supporting responses to open-sea pollution incidents

other organizations with the intent of minimizing threats to human health and the environment following emergencies. The National Contingency Plan (NCP) is the core of the NRS and describes the federal government's role in ensuring that resources and expertise are immediately available for emergencies beyond the capabilities of local and state responders and establishes a National Response Team (National Oil and Hazardous Substances Pollution Contingency Plan, 2015).

If emergencies are serious enough to be considered "nationally significant incidents," the National Response Framework (NRF) is activated. Working in conjunction with the NCP and NRS, the NRF provides for the coordination of federal assistance to state and local governments. Designed to be adaptable to a variety of emergencies, the NRF describes the roles and responsibilities, and coordinating structures for responding to an emergency. For disasters (either natural or man-made) where Federal assistance is provided under the Stafford Act, FEMA coordinates emergency assistance (U.S. Department of Homeland Security, 2013). Following recent fires, floods, hurricanes, and earthquakes, FEMA has functioned successfully in helping the area to recover and return to normal life. To allow for a true all-hazards approach to emergency response, several federal agencies are available for emergency planning and response assistance (table 18.2).

Mutual Aid—A mutual aid association is a cooperative organization of firms within a given industrial and business community. They have a voluntary agreement to assist each other by providing materials, equipment, and personnel during a disaster or emergency. Any company has a better chance of emergency survival if its knowledge and capabilities are combined and coordinated with those of neighboring concerns. It is nearly impossible to have adequate emergency supplies, equipment, and personnel available for every possible disaster, so the best defense is the combined resources of several concerns, each providing some of the manpower and equipment. Another arrangement is to contract for emergency services. This standby aid is usually prearranged through an annual retainer. It might include a wrecking company, a truck rental firm, or an electrical contractor.

CONCLUSION

While effective emergency/disaster planning may involve a great commitment of time and resources, the manager should not attempt to cut corners or stint in the effort. The consequences of even a small emergency, if the company is unprepared to meet it, can be costly at best and catastrophic at worst. Emergency response planning is a critical element of the SMS.

This examination of emergency and disaster planning is far from complete. It is merely intended to give the reader an overview of what is involved when planning for the unexpected. Keep in mind that nearly all of the things to be covered in an emergency disaster planning effort are a part, or extension of, normal activities. The plan need only draw together the information already available, in a manner suitable to its particular needs. Very little has to be researched or developed from scratch.

Just as good planning is essential to any business operation, it is vital to effective emergency and disaster handling. Company response will be no more adequate than the plan that directs it. The importance of sound planning cannot be too strongly stressed, but it can perhaps be summarized in one simple statement: it is literally a matter of corporate life or death.

REFERENCES

Aaron, James E. et al. (1979). *First Aid and Emergency Care*. Macmillan, New York.

Bird, F.E. and R.G. Loftus. (1976). *Loss Control Management*. Institute Press, Loganville, GA.

CSB. (2014). *Preliminary Findings of the U.S. Chemical Safety Board from its Investigation of the West Fertilizer Explosion and Fire*. U.S. Chemical Safety Board, Washington, D.C.

EPA. (2015). *National Response Team (NRT) Member Roles and Responsibilities*. U.S. Environmental Protection Agency, Retrieved from: http://www2.epa.gov/emergency-response/national-response-team-nrt-member-roles-and-responsibilities on 11/18/15.

Emergency Action Plans, 29 CFR § 1910.38 (2002).

Emergency Planning and Community Right-to-Know Act, 40 CRF § 311–12.

Erickson, P.A. (2006). *Emergency Response Planning for Corporate and Municipal Managers*, 2nd edition. Elsevier, Burlington, MA.

Evarts, B. (2012). *Fires in U.S. Industrial and Manufacturing Facilities*. National Fire Protection Association, Fire Analysis and Research, Quincy, MA.

FEMA. (1993). *Emergency Management Guide for Business and Industry*. Federal Emergency Management Agency, U.S. Department of Homeland Security, FEMA 141.

FEMA. (2005). *Emergency Management Guide for Business and Industry: A Step-by-Step Approach to Emergency Planning, Response and Recovery for Companies of All Sizes*. University Press of the Pacific, Honolulu, HI.

Fine, W. (1971). Mathematical Evaluations for Controlling Hazards. *Journal of Safety Research*, 3(4): 157–66.

Grimaldi, J.V. and R.H. Simonds. (1975). *Safety Management*, 5th edition. Richard D. Irwin, Homewood, IL.

Hammer, W. (1993). *Product Safety Management and Engineering*, 2nd edition. Prentice-Hall, Englewood Cliffs, NJ.

Handley, W. (1977). *Industrial Safety Handbook*, 2nd edition. McGraw-Hill, New York.

Heinrich, H.W. (1980). *Industrial Accident Prevention: A Safety Management Approach*, 5th edition. McGraw-Hill, New York.

Johnson, W.G. (1980). *MORT Safety Assurance System*. Marcel Dekker, New York.

National Oil and Hazardous Substances Pollution Contingency Plan, 40 CFR § 300.

National Safety Council, (1974). *Accident Prevention Manual for Industrial Operations*, 7th edition. National Safety Council, Chicago, IL.

National Safety Council. (2012). *Fundamentals of Industrial Hygiene*. National Safety Council, Chicago, IL.

NFPA. (2013). *NFPA 1600: Standards on Disaster/Emergency Management and Business Continuity Programs*. National Fire Protection Association, Quincy, MA.

OSHA. (2004). *Principal Emergency Response and Preparedness Requirements and Guidance*. Occupational Safety and Health Administration, U.S. Department of Labor, OSHA 3122-06R.

Petersen, D. *Analyzing Safety Performance*. Garland Press, New York.

PHMSA. (2012). *Emergency Response Guidebook: A Guidebook for First Responders During the Initial Phase of a Dangerous Goods/Hazardous Materials Transportation Incident*. Pipeline and Hazardous Materials Safety Administration, U.S. Department of Transportation.

PHMSA. (2015). *Hazmat Intelligence Portal*. Pipeline and Hazardous Materials Safety Administration, U.S. Department of Transportation.

Planer, R.G. (1980). *Fire Loss Control*. Marcel Dekker, New York.

Thygerson, A.L. (1977). *Accidents and Disaster: Causes and Countermeasures*. Prentice-Hall, Inc, Englewood Cliffs, NJ.

United States Chemical Safety Board. (2014). *Preliminary Findings of the U.S. Chemical Safety Board from its Investigation of the West Fertilizer Explosion and fire*. U.S. Chemical Safety Board, Washington, D.C.

Walsh, T.J. and R.J. Healy. (1987). *Protection of Assets Manual*. Silver Lake Publishing, Aberdeen, WA.

Whitman, L.E. (1979). *Fire Prevention*. Nelson-Hall, Chicago, IL.

Worick, W. (1975). *Safety Education: Man, His Machines, and His Environment*. Prentice-Hall, Inc. Englewood Cliffs, NJ.

Chapter 19

Hazardous Materials and Waste Management

Tracy L. Zontek and Kimberlee K. Hall

CASE STUDY

In Houston, Texas, a plant manufactures and refurbishes residential air handling equipment. There are over 500 line workers and 25 managerial and supervisory staff. The refurbishment of heat pump motors entails multiple solvent applications of various toxicities and physical properties. These materials include cutting oils, solvents, metal working fluids, metals, paints, and peroxides with properties such as corrosivity (low and high pH), ability to be absorbed though intact skin, toxicity, ignitability, reactivity, and volatility under normal working conditions. Many aromatic hydrocarbons cause narcosis. The chemicals are used in small quantities throughout various departments. Among the service/maintenance staff, twenty-eight are non-English-speaking employees responsible for collecting chemical waste from processes throughout the plant and depositing them into 55-gallon drums, known as *bulking waste*. The practice of bulk waste collection is commonly done to save on the costs of disposal of hazardous materials. This bulked material is often used as fuel for cement kilns. Five of the twenty-eight employees who bulk this waste experienced dizziness, nausea, and vomiting but do not want to report it to their supervisor or medical department out of fear of losing their jobs. No training was provided for personal protective equipment; although, personal protective equipment (aprons, gloves) were provided. On occasion, employees would report that the drums were fuming and smoked from the bung openings when materials were added and were somewhat hot to the touch during this process. One day, two employees collapsed on the job, were unconscious after several hours, and were taken to the hospital only to be declared dead on arrival. OSHA was called into investigate the fatality and found a number of violations related to *Hazard Communication* and employee exposure.

Although the company had designated Satellite Accumulation Areas (SAAs), no two-way communication, no intrinsically safe equipment, no standard safety construction including containment, or areas to segregate incompatible chemicals existed in the bulking areas. There was no standard process to identify what substances were allowed to be bulked into drums and no compatibility testing. The company is adjacent to hundred single family homes, an elementary school, daycare, and playgrounds. The neighbors commonly complained of a sweet chemical smell emanating from the plant and storm drain. The OSHA compliance officer noted some of the drums were found to be leaking, potentially into the storm drain.

Questions for consideration:

- Should the Occupational Safety and Health Administration (OSHA) compliance officer address Environmental Protection Agency (EPA) violations?
- What are potential violations that should be addressed by EPA?
- What EPA regulations may have been violated?
- Do members of the community have any rights regarding the use and release of chemical waste at the facility?
- Can an Safety Management System (SMS) effectively address these problems in the future?

Hazardous Materials and Waste Management

Hazardous materials and waste management are complex, multifaceted, global problems that continue to affect more people, more industries, and more processes every day. These problems touch every business manager, every plant manager, and every government official. Indeed they can, and probably do, affect all.

In this chapter, the conditions, history, and legislative attempts and subsequent evaluation of both environmental regulation and law will be examined. The problems inherent in hazardous materials, waste handling, and management are real, and corporate managers have to come to grips with the need to fully understand the multitude of sometimes overlapping regulations and local, state, and federal jurisdictions. To recognize the importance of this need to know, managers must take note of the increasing numbers of indictments against corporate officials for failure to comply with hazardous waste storage, transportation, and disposal regulations. In addition, violations of hazardous materials and waste management create negative reports in the media and with consumers.

The introduction and sections to follow will alert the reader to the regulations that may apply to their operation and present recommendations for a safety management systems approach to developing a program for their own plant or operation. An important element of any hazardous materials and waste management program is the need for upper management involvement and accountability. The evidence from recent legal actions, brought by local and federal enforcement agencies against high ranking officials of major corporations, helps drive home the point of where the ultimate responsibility lies.

INTRODUCTION

What is meant exactly by the term *hazardous waste*? It is a complex and confusing issue, at best, since some waste materials are dubbed "hazardous" simply because the Environmental Protection Agency (EPA) has listed them on its hazardous substances list. In addition, some states and local governments have added their own agents to the EPA list. To make it even more confusing, the general public, through the impact of generally exaggerated press and television coverage, tends to assume that every chemical represents a toxic hazard. In every case, the dose makes the poison, and many factors may contribute to adverse effects to human health and the environment.

Hazardous waste materials range from synthetic organic chemicals to heavy or toxic metals, to inorganic sludges, to solvents, and to aqueous waste streams. They may be solid, liquid, or gaseous. They may be pure materials, complex mixtures, residues and effluents from operations, discarded products, contaminated containers, or tainted soil. Most hazardous waste materials are managed on the site where they are generated, more so in some states and industries than others, and more likely by larger plants and generators than by small ones.

As of 2009, EPA had on record approximately 460 treatment, storage, and disposal facilities, 14,700 large quantity generators, and 18,000 transporters (EPA, 2015). Approximately 16,000 hazardous waste (HW) generators contributed over thirty-three million tons of Resource Conservation and Recovery Act (RCRA) HW in 2011 (EPA, 2013).

Early Seeds of a Problem

The HW problem in the United States began to manifest itself as early as the 1950s at Love Canal in New York State. At that time, however, people had no idea of just how uncontrolled the situation could or would become. As a result of the amount of public concern and outrage over such incidents as Love Canal and Woburn, Massachusetts, Congress developed and enacted the Resource Conservation and Recovery Act of 1976 (RCRA) and the Comprehensive Environmental Response, Compensation, and Liability Act of 1980 (CERCLA). Until these laws came about, if the problem was handled at all, it was handled in a very disorganized fashion.

The identification of priority uncontrolled waste disposal sites has been a slow and tedious process, and remedial action at such sites is finally taking place. The complexity of the overall situation has significantly contributed to the amount of disorganization involved in carrying out remedial action. Literally every site is unique in itself. This fact has made progress difficult, because it has not been possible to come up with an easy "cookbook" approach to corrective action at any given site.

In light of the overall uncertainty and ambiguity in this situation, it is imperative that a formal plan of action be developed. The plan must be flexible enough to apply to the unique characteristics of a given site, yet structured enough to apply certain practices to every situation. What is needed is a framework which is universal, but with space to adapt to subtle differences in its application. The answer, in this instance, is a systems approach to remedial action. By its inherent nature, this approach will provide the structural framework, which applies to very site, while allowing for the flexibility needed to perform the remedial action in a variety of situations using a variety of available technologies.

Legal Aspects for Compliance

In addressing the legal issues associated with remedial action, much of the legislative and regulatory concern was started at the federal level. However, because federal law did not set minimum national standards that could be administered by state programs, all fifty states, the District of Columbia, and Commonwealth of Puerto Rico have developed Superfund-type laws and regulations. The absence of a requirement to submit their programs for federal review enables states to develop innovative and effective clean-up programs, yet the major area of concern is compliance with federal regulations. In addition to federal and state laws, an organization must be cognizant of the many local ordinances and referendums being passed which affect its given locations.

For the most part, federal and state efforts follow along similar lines. There are some problems though, which will be addressed in the discussion at the end of this section. The local level, however, is a separate consideration. Many municipalities and other subdivisions have taken this whole subject to heart and devised sets of rules deemed to be in the best interest of their communities. Unfortunately, a common result of such efforts is one of confusion and often conflict. Facility managers should have a thorough understanding of the all the local, state, and federal requirements that may affect their operations.

A complete discussion of all of the federal and state regulations pertaining to HW would require a lengthy text, numbering in the hundreds of pages, along with the requirement of an experienced attorney to assist in their interpretation. A summary of the federal legislation follows.

Federal Legislation for Hazardous Materials and Waste Management

Hazardous materials and waste management falls under the following major pieces of federal legislation.

1. The Resource Conservation and Recovery Act of 1976 (RCRA)
2. The Comprehensive Environmental Response, Compensation, and Liability Act of 1980 (CERCLA)
3. The Occupational Safety and Health Act of 1970 (OSHA)
4. Superfund Amendments and Reauthorization Act (1986) (SARA)
5. Oil Pollution Act (OPA) (1990)
6. Underground Storage Tanks (UST) (2015)

While RCRA and CERCLA specifically address HW management, OSHA provides for protection of the workers carrying out work with hazardous materials and remedial actions at HW sites. SARA Title III closed the loop by providing community involvement in the planning and response of hazardous materials. The OPA and UST regulations serve to require better materials, controls, and emergency response planning in the event of an incident. Each of these laws is complex in its own right. In general, the laws themselves are independent of each other. They are somewhat interrelated by reference, but for the most part each addresses separate parts of the overall situation. Collectively, these laws are the most comprehensive regulatory effort thus far in attempting to control the hazardous materials and waste problem in the United States.

Resource Conservation and Recovery Act

The primary objective of RCRA is to provide for a comprehensive means of regulating HW. The concept of "cradle to grave" was derived from RCRA because of the intended scope of the law and respective regulations. It was the intent of Congress to protect the environment through every stage of the material's life. To this end, RCRA provided for development of regulations, which would cover HW generators, transporters, and treatment, storage, and disposal facility operators.

RCRA defines *HW* as

> a solid waste, or combination of solid wastes, which because of its quantity, concentration, or physical, chemical, or infectious characteristics may (1) cause, or significantly contribute to an increase in serious irreversible, or incapacitating reversible illness; or (2) pose a substantial present or potential hazard for human health or the environment when improperly treated, stored, transported, or disposed of, or otherwise managed.

Identifying the criteria for and listing of HW is described in Section 3001 of RCRA. This section requires the development of specific criteria for identification of HW and the creation of a list of materials considered as HW. The criteria developed for determining if a solid waste is hazardous include:

1. Ignitability
2. Corrosivity
3. Reactivity
4. Toxicity

In addition to these criteria, defined in detail in the regulations, a material may be specifically listed as a hazardous material as a result of the source from which it originated. Examples include wastewater treatment sludges, plating materials, specific industrial processes, and others.

A major portion of the burden of the RCRA law and subsequent regulations is placed upon the HW generators. Determination of whether a waste is hazardous, manifest documentation, and the responsibility of assuring that a waste shipment is sent to an EPA-permitted disposal facility are among the specific obligations placed solely on the generators.

Comprehensive Environmental Response, Compensation, and Liability Act

The Comprehensive Environmental Response, Compensation, and Liability Act (CERCLA) of 1980, better known as *Superfund*, deals with a number of problems related to the uncontrolled release of hazardous substances into the environment, a subset of which is the category of uncontrolled HW sites which may include abandoned dumps, inactive sites, or still operating facilities. In addition, this law designates those parties liable for the costs related to cleanup activities. Superfund came about, for the most part, because of widespread recognition of the many uncontrolled HW sites throughout the nation and the need to provide a mechanism to accelerate their control and cleanup.

Implementation of CERCLA has proven to be a complex and multifaceted process. Essentially, Congress provided government with the authority to address HW emergencies, established a trust fund to provide for response and cleanup activities, and created a taxing mechanism to generate revenue for the trust fund. The act established broad liability for all who generate, transport, store, treat, and dispose of HW. The EPA implements most aspects of these statutes, although a number of other federal agencies also have defined responsibilities. Moreover, the states have a major role in implementing these programs under the guidance and supervision of EPA.

Under Section 211 of CERCLA, the establishment of the Hazardous Substances Response Trust Fund was developed with the purpose of creating a taxing structure for the generation of the Superfund. Initially, the $1.6 billion fund acquired money from taxes on oil compounds and feedstock chemicals and environmental tax assessed on corporations to pay for clean-up activities at abandoned waste sites. Authority for taxes expired in 1995, but revenue is received from appropriations from the general fund; fines, penalties, and recoveries from potentially responsible parties; and interest accrued on the balance of the fund (GOA, 2008). This fund is used (1) if potentially responsible parties (PRPs) have not been identified or are not financially liable, (2) if litigation against a PRP is pending, (3) if there is insufficient evidence linking PRP to waste, or (4) if a health threat is imminent enough to warrant immediate action. It is funded primarily by general tax revenues and recovery costs from potentially responsible parties who polluted the site.

As of 2015, 1323 sites have been added to the National Priorities List (NPL) to address hazardous waste removal and remediation needs, and an additional fifty-three sites have been proposed for addition to the NPL. While the number of sites added to the NPL list continues to grow, there has been success in deleting sites from the NPL list through effective removal and remediation projects. Since 1983, 390 sites have been removed from the NPL list, six of those deletions occurring in 2015 (EPA, 2015). Successes in deleting sites from the NPL are not without cost, however, as EPA was posed to spend nearly $187 million on 275 removal actions and approximately $543 million on 105 remedial programs to initiate cleanup and construction projects in 2015 (EPA, 2014a).

RCRA versus CERCLA

Having provided a detailed breakdown of these laws, an overview comparison is necessary to explain the relationship between these legislative acts. One distinct difference between the RCRA and CERCLA is the noticeable absence of regulatory activity associated with CERCLA. For the most part, only the National Contingency Plan (NCP) and certain requirements for liability required additions/revisions to the Code of Federal Regulations. RCRA, however, required extensive efforts on the part of EPA for the development of regulations.

The second major difference lies in the concepts of proactive and reactive. RCRA functions in a proactive manner to address present and future activities in HW management. CERCLA functions to remedy problems now and in the future for improper disposal methods in the past. RCRA addresses people or businesses while CERCLA addresses places. Finally, both laws provide for establishing trust funds for post closure activities for disposal sites. The one item both specifically preclude, though, is the compensation of victims who suffer deleterious health effects subsequent to exposure to released hazardous substances.

Occupational Safety and Health Act

The Occupational Safety and Health Act of 1970 was enacted with the objective of providing a safe and healthful workplace for all working Americans. To this end, specific standards were promulgated into 29 CFR 1910 and under specific state programs, as applicable.

Protecting employees requires a working knowledge of the types of hazards to which the worker is exposed, the protective measures applicable to the specific hazards, and applicable OSHA standards. Apart from specific OSHA standards which are intended to provide at least the minimum of protection, each facility with hazardous materials or waste should have on staff, or within reach through consultation, a professional health and safety specialist who is familiar with good industrial hygiene and safety practices. In addition, it would be advisable that this person be experienced in the unique problems associated with the industry and locality.

For the purpose of this writing, site hazards and protection measures are addressed generically. This approach facilitates making generalities applicable to activities at any site. There is no doubt, however, that within time constraints that may arise, the safety program should be reviewed and updated for each site. This recommendation is made in light of the fact that, although the types of hazards may be very similar from site to site, the specific materials present and specific physical characteristics of a given site will vary.

Superfund Amendments and Reauthorization Act

The Superfund Amendments and Reauthorization Act (SARA) of 1986 focused on experience in administering CERCLA to identify more efficient ways to eliminate or minimize future cleanup costs. Furthermore, SARA closes the loop by providing the community input into emergency planning and response. SARA builds on other regulatory acts and attempts to set uniform requirements with that of the Hazard Communication Standard of the Occupational Safety and Health Act (OSHA). SARA provided the following updates to previous legislation:

- Stressed the importance of permanent remedies and innovative treatment technologies in cleaning up HW sites

- Required Superfund actions to consider the standards and requirements found in other state and federal environmental laws and regulations
- Provided new enforcement authorities and settlement tools
- Increased state involvement in every phase of the Superfund program
- Increased the focus on human health problems posed by HW sites
- Encouraged greater citizen participation in making decisions on how sites should be cleaned up
- Increased the size of the trust fund to $8.5 billion

The SARA legislation is quite complex with finely detailed information and reporting requirements and overlying responsibilities by public agencies at all levels. The purpose of this discussion is to provide an overview of the act sufficient to understand its implications and the steps needed to be taken to meet the requirements. The most impactful change is Section 313 of the act that sets federal, state, and local government requirements and guidelines for businesses and industry for the gathering and submission of information relating to the release of toxic chemicals into the environment. The intent of this legislation is to inform the general public and communities surrounding covered facilities about releases of toxic chemicals, aid in research, develop regulations and standards, and coordinate emergency response planning.

Background on CERCLA and SARA

The Comprehensive Environmental Response, Compensation, and Liability Act of 1980 provides authority for federal cleanup of HW sites. CERCLA also requires reporting to the National Response Center (NRC) the release of a hazardous substance into the environment greater than a certain quantity.

CERCLA did not address emergency response to accidental chemical releases. The incidents in Bhopal, India, and Institute, West Virginia, showed that highly toxic chemicals are a special problem for emergency response. Often, by the time emergency personnel arrive on the scene, the spill or released cloud of toxic gas has already caused damage to public health or the environment and may already have dissipated. For such chemicals early warning and emergency response planning are needed for effective public and environmental protection.

To meet this need, the SARA was enacted in 1986. SARA reauthorizes the 1980 law, originally funded at $1.6 billion for five years. The new amendments fund Superfund at $8.5 billion. SARA authorizes the EPAEPA to order responsible parties to clean up hazardous substances from sites or vessels where there is a release or threatened release into the environment. CERCLA funds are used where parties responsible for hazardous releases do not voluntarily clean up the problem or refuse to participate in cleanup procedures, and the government proceeds with the task. The government then attempts to collect the funds it spent. SARA does not change the basic structure of CERCLA but greatly expands and defines it. It also establishes new settlement procedures with responsible parties.

CERCLA Natural Resource Claims procedures became effective in 1986. The claims are for the cost of restoring, rehabilitating, replacing, or acquiring the equivalent of natural resources injured as a result of the release of a hazardous substance and the costs for assessing injury to such natural resources. This creates a new defense for "innocent landowners" who buy property without knowledge that it was used for disposal of hazardous substances. The new owner, however, must have adequately evaluated the site. Civil penalties for violations up to $25,000 per day may be imposed and can increase to $75,000 for later violations. Criminal penalties also increase under the amendments.

Part of SARA is Title III—The Emergency Planning and Community Right-to-Know Act (EPCRA) of 1986. Title III sets requirements for industry and federal, state, and local governments for emergency planning and "community right-to-know" of the presence and releases of certain hazardous and toxic chemicals in use at and by facilities subject to the act. It focuses upon chemical releases which could immediately affect the public's health. SARA establishes a new category of 406 extremely hazardous substances (EHS) that can cause acute effects upon release. This legislation builds upon the EPA's Chemical Emergency Preparedness Program (CEPP) and similar state and local government programs.

Title III aims are separate from the original HW cleanup provisions. Incorporated in the new law are requirements for coordinated emergency response planning and a chemical release inventory reporting and notification system for making information available to the public about the chemical substances in use at and released from a facility.

Title III has four subtitles of interest:

a. Framework for emergency planning—Sections 301–303, developing state and local government emergency response and preparedness capabilities and emphasize better coordination and planning, especially within the local community.
b. Emergency notification—Section 304.
c. Community right-to-know reporting requirements—Sections 311–312, and toxic chemical release reporting—emissions inventory—Section 313.
d. Trade secret protection, citizen suits, and the availability of information to the public. Sections 321–330.

The right-to-know provisions are a separate title to the Superfund legislation, not merely an amendment. The new law does not preempt state or local community right-to-know laws.

Safety Data Sheet

SARA defines hazardous materials as those requiring safety data sheets (SDSs). The idea behind an SDS is that if a company knows the hazards associated with a chemical, it will safeguard those using it. Also, if a worker knows what the hazards are, he or she will more likely use the safeguards provided by the employer. An organization can reduce the complexity in meeting reporting requirements and avoid some work by knowing which materials are not required to have an SDS:

1. Pesticides that are already labeled as required by the EPA under the Fungicide, Insecticide, and Rodenticide Act
2. Food, drugs, and cosmetics labeled as required by the FDA
3. Any distilled spirits not intended for nonindustrial uses when labeled as required by the Federal Alcohol Administration Act
4. Consumer products or hazardous substances when under the regulations of the Consumer Product Safety Commission
5. HW covered by the Solid Waste Disposal Act
6. Tobacco or tobacco products
7. Wood or wood products
8. Manufactured items formed to a specific shape so that its use depends on that shape and, when used, does not release or cause exposure to a hazardous chemical
9. Food, drugs, or cosmetics intended for personal consumption by the employees while in the workplace

In 2003, the United Nations (UN) adopted the Globally Harmonized System of Classification and Labeling of Chemicals (GHS) in order to standardize hazardous materials labeling worldwide. The GHS includes criteria for the classification of health, physical, and environmental hazards and specifies what information should be included on labels of hazardous chemicals and SDSs. As commerce and trade continue to have global reach, this system simplifies the recognition and safe handling of hazardous materials. OSHA updated the Hazard Communication standard in 2013 (with implementation through 2016) to reflect the changes in GHS. These changes, in turn, affect parts of Title III that refer to the use of Material Safety Data Sheets (MSDS), now simply called SDS. The new system is also being adopted by the U.S.

Department of Transportation and Consumer Product Safety Commission. The main changes include the following:

- Hazard classification of chemicals that are produced or imported based on specific criteria to address health and physical hazards as well as classification of chemical mixtures
- Labels that includes a signal word, pictogram, hazard statement, and precautionary statement for each hazard class and category
- SDS with a new sixteen-section format
- Information and training on the new label elements and safety data sheet format, in addition to the current training requirements.

Emergency Planning

The purpose of the emergency planning section is to develop community chemical emergency response plans and capabilities. The plans will list available resources for preparing for and responding to a chemical emergency.

Companies that produce, use, or store extremely hazardous substances (EHS) over a specified quantity (threshold planning quantity—TPQ) are subject to the requirements of Sections 301–303. The company must advise the State Emergency Response Commission (SERC) and the Local Emergency Planning Committee (LEPC) of its facilities that are subject to the requirement and provide information on their chemical inventory. Title III requires designated state commissions to have a scheme to receive and process public requests for information collected under the other sections of Title III. States must also coordinate activities of local emergency planning committees and review local chemical emergency plans. Community involvement is ensured by the requirement that the LEPC include representatives from following: elected officials, local environmental groups, media, community groups, law enforcement, transportation, health care facilities, emergency responders, and local regulated facilities.

Emergency planning should include the following elements: description of local emergency equipment and facilities and the persons responsible for them; an outline of evacuation plans; a training program for emergency responders (including schedules); and methods and schedules for exercising emergency response plans. These should be coordinated between all affected parties, most significantly the facility and local emergency responders.

Emergency Reporting

Covered facilities must immediately notify the LEPC and SERC if there is a release of certain EHSs (Section 304). Notification is also required for any of the EHSs or any substance subject to emergency notification requirements under CERCLA (Section 103[a]), if the release is over a specified threshold quantity. The EPA has developed a consolidated list of chemicals subject to reporting requirements under the EPCRA, CERCLA and Section 112(r) of the Clean Air Act titled the "List of Lists." The List of Lists provides easy access to evaluate quantities and chemicals that may fall under multiple regulatory requirements.

The requirements of the Emergency Release Reporting section apply to facilities where hazardous substance (HS) and/or extremely hazardous substances (EHS) are produced, used, or stored. If either an HS or an EHS is released above a threshold amount, a report is required to be made immediately to the LEPC, the SERC, and the National Response Center (NRC). The initial report can be by telephone, radio, or in person. A follow-up, written report is required to be made to the SERC and LEPC.

Special note: Section 304—emergency notification—is not subject to trade secret protection. All other sections of Title III are subject to trade secret protection if it is requested with initial documents.

COMMUNITY RIGHT-TO-KNOW— HAZARDOUS CHEMICAL INVENTORY

It is important to understand what the EPA considers to be a hazardous chemical. For the purpose of these regulations, a *hazardous chemical* is any chemical which has an SDS requirement. SDSs can be extremely important as a central source of information and data needed to comply with the reporting and training requirements of the EPA, OSHA, state, and local government agencies. The sections of SARA most dependent upon SDSs are Sections 311 and 312—the community right-to-know reporting requirements.

Section 311 calls for preparing or maintaining SDSs. In practice, a correctly prepared SDS that is in compliance with GHS can form the basis for and provide the information needed for much of the reporting, labeling, and hazard communication right-to-know communications and training programs required by EPA and OSHA regulations.

When the complexities and far-reaching effects of EPA regulations are realized, many companies seek quick ways to meet the requirements of the standard. As a result, there has been an increased demand for off-the-shelf chemical labels, SDSs, and training programs. There are several commercial producers of ready-made SDSs who offer a wide collection of data sheets covering the more well-known chemicals and chemical substances that are available online and on demand. This eliminates the need for hard copies that quickly become outdated. Furthermore, an electronic database that is linked with a company procurement department can close the loop when ensuring all hazardous materials on-site have an SDS available. This represents the types of systems approach that is needed for hazardous materials management. Some companies use the hazard communication standard as a survey vehicle to find out exactly what chemicals, and in what containers and quantities, are actually on hand, what is actually being used, and what can be disposed of and removed from the chemical inventory. The survey approach helps identify chemicals no longer needed that can be disposed of and removed from the chemical inventory. A careful review of existing inventories has led to the removal of as many chemicals as possible from use, a reduction in permissible inventory, and an improved chemical inventory control.

Senior managers, in particular safety managers, need to keep in mind that when a project is set up to carry out a certain concern, such as a new compliance program, the project is often thought finished once the initial tasks are completed. Activities are then directed toward a new project. Within a year or two, the lack of a programmatic approach is felt, and often within three to five years, few traces of the original effort remain—the project is a shambles or forgotten. SMS is one way to assure that projects become programs and needed programs are ongoing. SMS requires reviewing the contents of a program for accuracy, upgrading whenever possible, and having methods to measure its effectiveness. Building the hazard communication program into the SMS will trigger timely and consistent actions and also provide evidence of a program's success. It helps to start from the ground up. That is to make sure the basics are in place and then set up a plan whereby one continues to upgrade the program as needed.

Once correctly assembled and maintained, companies must submit a list of hazardous chemicals, by location, to the SERC, LEPC, and local fire department. Section 312 requires annual submission of an emergency and hazardous chemical inventory in a two tier approach.

Tier I has information about average quantities and locations of chemicals, depicted by hazard class. Tier I reporting is more general. Tier II includes a chemical name as it appears on an SDS, amount present and location, storage and use process, and trade secret claims. Both Tier I and II information must be available to the public on request from local and state governments during normal working hours, subject to trade secret protection.

Section 313 established the Toxic Release Inventory (TRI), making information about industrial use of chemicals available to the public. This creates a strong incentive to generate positive public relations for companies to maintain and improve environmental performance.

The emissions reporting requirements apply to facilities that meet three criteria.

- The facility is included in a TRI-covered North American Industry Classification System (NAICS) code.
- The facility has ten or more full-time employee equivalents (i.e., a total of 20,000 hours or greater).
- The facility manufactures (defined to include importing), processes, or otherwise uses any EPCRA Section 313 chemical in quantities greater than the established threshold in the course of a calendar year.

The TRI-covered North American Industry Classification System (NAICS) replaces the four-digit Standard Industrial Classification (SIC) codes identified initially with the EPCRA legislation. The NAICS codes covered by section 313 include: 212 (Mining), 221 (Utilities), 31-33 (Manufacturing), all other miscellaneous manufacturing (includes 1119, 1131, 2111, 4883, 5417, 8114), 424 (Merchant Wholesalers, Non-durable Goods), 425 (Wholesale Electronic Markets and Agents Brokers), 511, 512, 519 (Publishing), 562 (Hazardous Waste), and all federal facilities.

The TRI list includes chemicals that cause cancer or other chronic human health effects, significant adverse acute human health effects, or significant adverse environmental effects. The current list contains 594 individually-listed chemicals and thirty-one chemical categories (including four categories containing sixty-eight specifically listed chemicals). Reporting under EPCRA Section 313 is triggered by the quantity of a chemical; for most TRI chemicals, the thresholds are 25,000 pounds manufactured or processed or 10,000 pounds otherwise used. Sixteen TRI chemicals and four TRI chemical categories that meet the criteria for persistence and bioaccumulation have lower thresholds, such as 10 or 100 pounds and 0.1 grams. Since a 55-gallon drum of liquid weighs about 500 pounds, the requirement is triggered if a firm uses about twenty drums a year.

A full range of information is required. A covered firm should provide information on each chemical according to these criteria:

—How is the chemical used?
—What quantity of the chemical is present at the facility at any time during a calendar year?
—For each waste stream, what are the waste treatment or disposal methods employed?
—What quantity of the toxic chemical is entering the environment by release to the air, ground, and water?

Under Section 313, anyone can petition the EPA to add or delete chemicals from the mandated toxic chemicals inventory list. Furthermore, the Pollution Prevention Act of 1990 required companies to report additional data on waste management and source reduction activities to TRI. A total of 3362 facilities (16% of all TRI facilities) in 2013 reported initiating 10,623 source reduction activities, such as operating practices, process modifications, raw material modifications, and spill and leak prevention.

Besides releases that are made on-site, when there is an in-transit release and no notification is made, both owner and operator of transport can be held liable. The owner may be held liable even if he or she didn't know of the incident. The EPA holds the owner responsible through the entire process of transportation, handling, and disposal known.

Trade Secrets

EPCRA allows firms to withhold certain proprietary data on products which they claim as trade secrets or vital to their stake in the market for their product.

Companies who wish to maintain proprietary chemical information must submit this intention to EPA at the same time that an EPCRA report is submitted. For example, the Tier II form is required to be submitted to the SERC, the LEPC, and the local fire department on or before March 1, annually. Therefore, the EPCRA Trade Secret Substantiation Form must be submitted to EPA at the same time that the Tier II is submitted to the SERC, the LEPC, and the fire department. When a Tier II form is submitted, the local fire department also has the power the request an inspection of the facility to review. When submitting a Trade Secret Substantiation Form, only specific chemical identities can be claimed as proprietary; a generic class must be identified. Health care professionals can request specific trade secret information for diagnostic/treatment purposes in emergency situations. Local health officials can also request in writing access to trade secrets for non-emergency purposes in order to plan for prevention or treatment in the event of a catastrophic event. Furthermore, anyone can challenge trade secrets by petitioning EPA.

Any citizen can enter a civil lawsuit if the owner/operator of a covered facility does not follow the law's data reporting requirements. The same holds true for local and state governments who feel that a certain industry is not supplying the required SDSs. The law specifies the federal district courts as the place for such actions to originate.

Enforcement and Liability

Common violations of SARA Title III include failure to notify to appropriate agencies immediately of reportable release (regardless of whether information was not complete at the time); not notifying all potentially affected

LEPC/SERCs of releases; and not providing EPCRA Section 311 reports within 90 days of a previously unreported substance being brought on-site. Civil and administrative penalties can be enforced, typically $10K, $75K per violation or per day per violation for failing to comply with reporting requirements. In addition, criminal penalties of up to $50K or five years in prison may be imposed for knowingly and willfully failing to provide emergency release notification. Lastly, up to $20K and/or up to one year in prison can be enforced for knowingly and willfully disclosing trade secrets. Companies must be vigilant in their review of hazard materials inventory and management to ensure all regulatory requirements are met. Again, an SMS approach to this process will save time, money, and potential liability.

Underground Storage Tanks

In 2015, EPA published the final rule on Underground Storage Tanks (UST), updated from the 1988 regulations of the Solid Waste Disposal Act. A *UST system* UST is defined as a tank and any underground piping such that >10 percent of combined volume is contained underground. Excluded from the standard are farm and residential tanks <1100 gallons; on premise heating oil tanks; tanks in basements or tunnels; septic and storm water tanks; emergency spill and overflow tanks. Specific requirements vary by new or existing tanks but contain some provisions for leak detection, spill and overflow protection, corrosion protection, and financial responsibility. Leak detection options include monthly monitoring; inventory control plus tank tightness testing; pressurized piping; automatic shut-off device, flow restrictor, or continuous alarm; and annual line tightness test. Spill and overflow protection options include catchment basin, auto shut-off devices, overfill alarms, or ball float valves. Corrosion protection choices are coated and cathodically protected steel or fiberglass reinforced plastic (FRP) or steel tank clad with FRP. Financial responsibility must be maintained to include the cost of cleaning up a leak and compensating other people for bodily injury and property damage. Coverage amounts are based on group (industry, government) and require $½–1 million per occurrence coverage and aggregate coverage of $1–2 million. EPA can take corrective action and sue the tank owner/operator to recover cleanup costs or, require the owner to do the cleanup. There is a $500 million leaking UST trust fund, financed by taxes on motor fuels. The updated standard also included more rigorous operator training requirements and codes of practice. Underground storage tanks are a leading cause of groundwater contamination due to previously used steel tanks, so these regulations should prevent future contaminated sites that require CERCLA intervention.

Oil Pollution Act

Promulgated in 1990 as an amendment to the Clean Water Act, the Oil Pollution Prevention Act (OPA) sought to address severe voids in the pollution prevention and response to oil spills in navigable U.S. waters. Stimulated by the Exxon Valdez spill of greater than eleven million gallons of crude oil into Prince William Sound, Alaska, the United States determined that there were inadequate resources to respond to spills and that the scope of damages compensable under federal law to those impacted by spill was small in comparison to the damage. The OPA developed a comprehensive prevention, response, liability, and compensation plan to handle oil spills and greatly increased oversight of maritime oil transportation. Specifically, oil storage facilities and vessels must submit plans detailing how they will respond to large discharges. In addition, geographic regions developed Area Contingency plans to respond to oil spill events. The OPA authorized the Oil Spill Liability Trust Fund from a tax on oil. This fund provides up to $1 billion for oil spill removal and uncompensated damages, per incident. The fund can be used for claims for removal or damage; appropriations for prevention, research and development; removal costs incurred by Coast Guard and EPA; state access for removal; payments for natural resource damage assessments and restorations; and claims for uncompensated removal costs and damages. In addition, there is an emergency fund, $50 million that may be used at the discretion of the president annually for removal activities and natural resource damage assessment. Unfortunately in 2010, another catastrophic spill occurred in the Gulf of Mexico–Deepwater Horizon. Oil discharged for eighty-seven days from a subsea drilling system owned and operated by British Petroleum (BP). It is estimated that 206 million gallons were discharged into U.S. waters; of that, it is estimated that 50 percent has evaporated, dissolved, or been effectively removed via human activities, but hundred million gallons still remain (consumed by bacteria, settled on ocean floor). EPA estimates that BP has paid approximately $15 billon to the federal government, state governments, individuals, and businesses. Federal and state reimbursements were used to compensate for oil spill response costs, behavioral studies, tourism promotion, seafood testing and marketing, and others. Individuals and businesses were compensated for economic or medical costs. The effectiveness of this legislation will be evaluated as the impact of this spill continues to evolve.

Other Environmental Law Issues

Apart from the sheer complexity of these laws and regulations, there are several other problems associated with the legal issues surrounding HW. The primary area of concern lies in addressing the needs of those people who suffer health problems as a result of exposure to released materials near HW sites. Neither RCRA nor CERCLA specifically address a means of compensating these victims for the illness and suffering associated with exposure. For the present time, such areas are handled in court under tort laws. Toxic torts are associated with exposure to toxic substances (industrial chemicals, pesticides, lead-based paint, pharmaceutical drugs, and other environmental toxins). Tort claims can be brought by individuals or groups who have been exposed to a toxic substance and have suffered injuries, illnesses, or damages. The plaintiff must prove the substance was harmful, dangerous, or toxic; they were exposed to this substance; and it caused harm or damage. A variety of legislative acts have attempted to address this problem, and some of the specific state laws do make provisions for toxic waste victims.

Some of the other legal issues involved in HW include general liability; bankruptcy, as a means of avoiding liability; state program versus federal programs; and local referendums with the attitude of "not in my backyard." Environmental law and the concept of toxic torts are still relatively new to the legal profession. In many cases the case law is sparse, and the complexity of individual waste streams and confounding factors is immense. It will be many years before the evolutionary process of law provides answers to the complex legal issues associated with HW.

SOCIAL AND LEGAL INFLUENCES

As can be seen from the brief overview of the major federal regulations, managing hazardous materials and waste issues can be a major undertaking requiring significant resources and personnel. Hazardous materials use is complex. When considered in combination with other social and legal influences, this problem becomes compounded. In light of this complexity and the unique factors involved in addressing Superfund sites, any attempts to remedy the situation must take an integrated approach. It is imperative that all of the factors influencing hazardous materials and remedial activities be identified and accounted for. In addition, it must be recognized from the beginning that any solution attempted usually involves compromises and trade-offs. It can be nearly impossible to satisfy the individual needs of each of the parties or system parameters that must be addressed. Integration of the process into a strong SMS can help assure its operational effectiveness and continuous improvement.

Remedial Actions Quantified

A systems approach to remedial action brings together all of the influencing factors and accounts for the system parameters and any outside constraints placed upon the system. To understand this approach, several concepts need to be explained. Malasky defines a *system* as:

> A composite, at any level of complexity of personnel, materials, tools, equipment, facilities, environment, and software. The elements of this composite entity are used together in the intended operational or support environment to perform a given task, to achieve a specific production, or to support a mission requirement.

Based upon this definition, a hazardous materials use may, as a whole, be considered a system. In addition, any of its individual components (employees, process, soil, groundwater, air, etc.) may independently be considered a system. For the purposes of this project, the entire process will be considered one system. Expanding upon the concept of a system, Malasky defines *system safety* as:

> An optimum degree of safety, established within the constraints of operational effectiveness, time, and cost, and other applicable interfaces to safety that is achievable throughout all phases of the system lifecycle.

This definition provides the starting point for the systems approach. Because of the stipulation of optimal safety within certain constraints, this approach falls in line with the realism of understanding that there must be trade-offs and compromises in coming to a workable solution. Furthermore, the systems approach presupposes an understanding of, and accounting for, all of the interfacing parties.

Identification of Stakeholders and Interfacing Considerations

The first step in the systems approach is to identify the interfacing concerns, formally called stakeholders. Major stakeholders include

1. Site owner/operator
2. HW transporters using the site
3. HW generators using the site
4. Surrounding area populations
5. Future generations to be located at or near the site
6. Shareholders in all companies concerned
7. Regulatory agencies: DOT, EPA, OSHA, state, and local
8. Employees of companies and their families
9. Employees of the regulatory agencies (inspectors)
10. Consumers of products manufactured by generators
11. Lawmakers and other professionals

With the exception of shareholders and consumers, the specific interests and concerns of the stakeholders have been addressed in previous sections. For the most part, shareholders' and consumers' concerns are of a financial nature unless they also happen to fall into one of the other groups. To a certain extent, the financial interests are covered under the consideration of cost in evaluating the feasibility of a given measure.

The final concern from a systems standpoint is the process of bringing everything together in a workable framework that covers the situation as thoroughly as possible. The mechanism for accomplishing this task is a safety management system that provides for a formal effort of systematically dealing with the problems. This program is specifically designed for a given site, tailored to the unique needs of the site, and structured in such a way as to include the entire life cycle of the system.

CONCLUSIONS

The regulations discussed only begin to address the hazardous materials dilemma. Unless the wastes now produced are better managed, new threats will simply replace the old ones. New standards favor chemical waste treatment and reduction and discourage untreated land disposal. SARA is one more step in an ongoing effort to face up to the difficulties and complexities in hazardous materials and HW management.

The experience of several European countries offers valuable lessons in improving industrial waste management, though few have adequately tackled controlling the total volume of waste generated. Without concerted efforts to reduce, recycle, and reuse more industrial waste, even the best treatment, disposal, and information systems will be overwhelmed.

The hazardous materials management and waste problem is perhaps one of the most complex problems facing society today. There are any number of conflicts and interfaces occurring simultaneously at a given site. When one considers the thousands of facilities and waste sites located throughout the United States and their individual characteristics, the depth and breadth of the problem is vast. In an attempt to gain some level of control over the situation, or at least begin to gain control, society must deal with all of the influencing factors for this problem. Satisfaction of all the needs of all of the individual stakeholders may never be possible. However, to at least account for those needs and recognize the need to prioritize them, society begins to attain some semblance of balance and control. The systems approach provides the mechanism necessary to identify and account for all of the stakeholders and their individual needs. Furthermore, this approach provides the flexibility needed to adapt to the unique characteristics of each site, as well as changes occurring outside of the site, while maintaining a structural framework that allows for accomplishment of the overall objective of controlling an uncontrolled hazardous waste site.

REFERENCES

Birns, C. (November/December, 1983). Grappling with the Problem of Uncontrolled Hazardous Waste Sites. *Hazardous Materials and Waste Management*, pp. 18–25.

Ehrenfeld, J. and J. Bass. (1982). *Handbook for Evaluating Remedial Action Technology Plans*, USEPA 625/6-82-006.

EPA. (1992). *CERCLA/Superfund Orientation Manual*. United States Environmental Protection Agency, Office of Solid Waste and Emergency Response, Washington, D.C. EPA/542/R-92/005.

EPA. (2013). *The National Biennial RCRA Hazardous Waste Report (Based on 2011 Data)*. United States Environmental Protection Agency, Washington, D.C.

EPA. (2014a). *FY2015 EPA Budget in Brief*. United States Environmental Protection Agency, Office of the Chief Financial Officer (2710A), Washington, D.C. EPA-190-S-14-001.

EPA. (2014b). *RCRA Orientation Manual*. United States Environmental Protection Agency, Office of Resource Conservation and Recovery Program Management, Communications, and Analysis Office, Washington, D.C. EPA530-F-11-003.

EPA. (2015a). *NPL Site Status Information*. United States Environmental Protection Agency, Washington, D.C. Retrieved from: http://www.epa.gov/superfund/npl-site-status-information.

EPA. (2015b). *Toxic Release Inventory*. United States Environmental Protection Agency, Washington, D.C. Retrieved from: http://www.epa.gov/toxics-release-inventory-tri-program.

EPA. (2015c). *Emergency Planning and Community Right-to-Know Act*. United States Environmental Protection Agency, Washington, D.C. Retrieved from: http://www.epa.gov/epcra.

EPA. (2015d). *Underground Storage Tanks*. United States Environmental Protection Agency, Washington, D.C. Retrieved from: http://www.epa.gov/ust/revising-underground-storage-tank-regulations-revisions-existing-requirements-and-new#rule-summary.

GAO. (2008). *Superfund: Funding and Reported Costs of Enforcement and Administration Activities*. United States Government Accountability Office, Washington, D.C. GAO-08-841R.

Malasky, S.W. (1982). *System Safety: Technology and Application*. Garland STPM Press, New York, NY.

Mallow, A. (1981). *Hazardous Waste Regulations—An Interpretive Guide*. Van Nostrand and Company, New York, NY.

Ness, S.A. (1988). Quality Control in an Employer's Hazard Communication Program. *Applied Industrial Hygiene*, (3)1: p. F29–33.

OSHA. (n.d.). *Hazard Communication*. Retrieved from: https://www.osha.gov/dsg/hazcom/index.html.

Petak, W. and A. Atkisson. (1982). *Natural Hazard Risk Assessment and Public Policy* Springer-Verlag, New York, NY.

Sullivan T.F.P. (Ed). (2013). *Environmental Law Handbook*, 22nd ed. Berman Press; Plymouth, UK.

Index

ABIH. *See* American Board of Industrial Hygiene (ABIH)
Accident Causation. *See* Swiss-Cheese Model
accidents, 352–53;
   case study, 209;
   and hazards, 28–29;
   impact of, 210–13;
      and cost estimation, 211;
      management inefficiencies, 212;
      and morale, 212–13;
      and public concern, 213;
   investigations:
      committee, 216–18;
      priorities for, 218–20;
      quality of, 220–29;
   management:
      overview, 229–30;
      stake in, 209–10;
   and stakeholders, 213–15;
      line manager, 215;
      safety professional, 215–16;
      staff managers, 216;
      supervisor/foreman, 214–15
accountable executive (AE), 73
ACGHI. *See* American Conference of Governmental Industrial Hygienists (ACGIH)
active and passive surveillance, 135
active errors, 24
Acute Exposure Guideline Level (AEGLs), 354
ADA. *See* Americans with Disabilities Act (ADA)
administrative controlling, 141
administrative program, 310–12
AE. *See* accountable executive (AE)
AED. *See* Automated External Defibrillators (AED)
AEGLs. *See* Acute Exposure Guideline Level (AEGLs)
AIHA. *See* American Industrial Hygiene Association (AIHA)
ameliorative function, 349
American Board of Industrial Hygiene (ABIH), 297
American Conference of Governmental Industrial Hygienists (ACGIH), 14, 84
American Industrial Hygiene Association (AIHA), 14, 353
American National Standards Institute (ANSI), 13, 84;
   and SMS, 52–54
American Society of Safety Engineers (ASSE), 13
Americans with Disabilities Act (ADA), 328

American Trucking Association, 286
Annual Summary of Occupational Injuries and Illnesses, 275
ANSI. See American National Standards Institute (ANSI)
anthropogenic ionizing radiation sources, 306
a posteriori probability, 198
a priori probability, 198
APUs. See auxiliary power units (APUs)
Area Contingency plans, 396
"as low as reasonably achievable (ALARA)," 308, 312, 314
ASSE. See American Society of Safety Engineers (ASSE)
assessment tool controlling, 143
audits:
   and PSM, 172;
   and safety program controls, 85–89;
   authority, 87;
   management, 86–87;
   process, 86;
   steps in making, 88–89;
   types of, 87 88
Automated External Defibrillators (AED), 335
auxiliary power units (APUs), 202–3

back injuries, 127
BCPE. See Board of Certification in Ergonomics (BCPE)
BCSP. See Board of Certified Safety Professionals (BCSP)
BLS. See Bureau of Labor Statistics (BLS)
Board of Certification in Ergonomics (BCPE), 138
Board of Certified Safety Professionals (BCSP), 138, 297
Boolean Algebra, 204–5
brainstorming, 56
Breast Cancer Awareness, 339
budgeting, 256–57;
   evelopment of, 259–62;
   optimum, 257–59;
      methods of determining, 258;
      task definition, 258–59;
      practical reasons for, 257;
      time factors in, 257;
      total effort, 261–62
bulking waste, 379
Bureau of Labor Statistics (BLS), 330

CAAA. See Clean Air Act Amendment (CAAA)
cardiopulmonary resuscitation (CPR), 264, 283
"Cargo Security Advisory Standards," 286
carpal tunnel syndrome, 124–25
car seal program, 176
cash flow analysis, 255–56
causation chain, 24
CBT. See computer-based training (CBT)
CDP. See Center for Domestic Preparedness (CDP)
Center for Domestic Preparedness (CDP), 357
CEPP. See Chemical Emergency Preparedness Program (CEPP)
CERCLA. See Comprehensive Environmental Response, Compensation, and Liability Act (CERCLA)
Certified Health Physicist (CHP), 307
Certified Industrial Hygienist (CIH), 297
Certified Professional Ergonomist (CPE), 138
Certified Safety Professional (CSP), 138, 297, 326
Chemical Emergency Preparedness Program (CEPP), 388
Chemical Safety Board (CSB), 346
Chemical Transportation Center (CHEMTREC), 353
CHEMTREC. See Chemical Transportation Center (CHEMTREC)
child labor, and powered machinery, 4
chlorine dioxide (CLO2), 295

Index   405

CHP. *See* Certified Health Physicist (CHP)
CIH. *See* Certified Industrial Hygienist (CIH)
Clean Air Act, 309
Clean Air Act Amendment (CAAA), 166, 168
Clean Water Act, 396
CLO2. *See* chlorine dioxide (CLO2)
collaboration, 26
common law doctrines, 7–8
communications:
   and emergency response planning, 364–65;
   and information systems, 80–84;
      and feedback, 82–84;
      roles and responsibilities, 80–81;
      and security, 81–82
Complimentary Law, 202
Comprehensive Environmental Response, Compensation, and Liability Act (CERCLA):
   and hazardous waste, 381, 384–85;
   versus RCRA, 385–86;
   and SARA, 387–89
computer-based training (CBT), 129
confidence intervals, 200
construction safety, 264–65
continuous audits, 88
contractors, of PSM, 172
corporate safety strategies, 32
cost-benefit analysis, 250–54
cost estimation, and accidents, 211
covered entities, 281
CPE. *See* Certified Professional Ergonomist (CPE)
CPR. *See* cardiopulmonary resuscitation (CPR)
crane safety, 265
CSB. *See* Chemical Safety Board (CSB)
CSP. *See* Certified Safety Professional (CSP)
cut sets, 203–4

data location, and facility maps, 363
DCS. *See* demand-control-support (DCS) model

deferred future earnings (DFE), 250
demand-control-support (DCS) model, 323
Department of Transportation (DOT), 272, 286–92
depreciation, 245–46
De Quervain's tenosynovitis, 124
DFE. *See* deferred future earnings (DFE)
discomfort surveys, 159–61
documentation, and emergency response planning, 369–70
DuPont Sustainable Solutions, 340

earthquakes, 351–52
education. *See* training and education
EELs. *See* emergency exposure limits (EELs)
EIS. *See* environmental impact statement (EIS)
"electronic protected health information" (e-PHI), 281
emergency action plan, 110
emergency/disaster response planning:
   appendices, 370;
   approval of, 357;
   assessing hazards, 356;
   assessing probability of event, 355–56;
   authority, 361;
   chief operating officer of, 371;
   and communication, 364–65;
   control center, 360–62;
   coordinating aspects of, 358–59;
   core plan components, 362–70;
   development of, 356–57;
   distribution of, 362;
   and documentation, 369–70;
   elements of, 359–60;
   emergency equipment and supplies, 362;
   and emergency plan coordinator, 356, 371;
   and employee welfare, 366;
   and equipment shutdown, 367–68;
   evacuation procedures, 365;

executive board of, 371;
facility maps and location of data, 363;
and federal government, 373–76;
and fire agencies, 372;
and functional committee, 371–72;
functionaries and agencies involved, 370–76;
and hazard assessment, 363;
and hazardous waste, 390;
and medical services, 372–73;
mission statement, 360–61;
and mutual aid association, 376;
occupational medical team role in, 334–41;
off-the-job safety, 335–41;
and personnel assignment, 368–69;
personnel training, 357, 369;
and PSM, 172;
purpose of, 361;
restoration, repair, and salvage, 368;
revision of, 358;
and security, 372;
shelter procedures, 365–66;
and SOPs, 363–64;
and special procedures, 366–67;
steps in developing, 355–56;
testing, 357–58;
training, 110;
and transportation, 366;
types of, 349–55;
and West Fertilizer Plant fire and explosion, 346–47
emergency equipment and supplies, 362
emergency exposure limits (EELs), 353–54
Emergency Management Institute (EMI), 357
emergency plan coordinator, 356, 371
Emergency Planning and Community Right-to-Know Act (EPCRA), 347, 388
emergency reporting, and hazardous waste, 390–91
Emergency Response Guidebook, 353

Emergency Response Planning Guidelines (ERPGs), 353
EMI. *See* Emergency Management Institute (EMI)
employee benefit association, 7
employee participation, 172
employee welfare, and emergency response planning, 366
employee wellness programs, 322–23
employer responsibilities, 142
energy emergencies, 354–55
enforcement trends, and PSM, 174–76;
car seal program, 176;
work system permit, 175–76
environmental discharges, 314
environmental factors and stressors, 299–300
environmental impact statement (EIS), 317
Environmental Protection Agency (EPA), 309, 381
environmental safety:
hazardous materials, 265–66;
hazardous waste, 266;
oil spill control, 266
EPA. *See* Environmental Protection Agency (EPA)
EPCRA. *See* Emergency Planning and Community Right-to-Know Act (EPCRA)
e-PHI. *See* "electronic protected health information" (e-PHI)
equipment shutdown, and emergency response planning, 367–68
equivalent uniform annualized cost (EUAC), 248
ergonomics:
case study, 123;
committee establishment, 133–39;
active and passive surveillance, 135;
awareness of, 136;
composition, 133–34;
and ergonomics expert, 138–39;
hazard prevention and control, 135–36;

Index 407

management techniques and
    interpretation, 139;
medical management program,
    136–38;
training, 134–35, 138–39;
work/job design and redesign,
    139;
and discomfort surveys, 159–61;
management commitment, 133;
manual handling checklist, 143–46;
musculoskeletal disorders (MSDs),
    124–27;
    back injuries, 127;
    carpal tunnel syndrome, 124–25;
    De Quervain's tenosynovitis, 124;
    rotator cuff tendonitis, 126;
    tendonitis and tenosynovitis, 125;
    thoracic outlet syndrome, 126;
    trigger finger, 127;
    wrist ganglion, 126–27;
overview, 123–24;
process, 130–33;
program management, 128;
quantitative methods, 146–59;
    data collection, 148–50;
    Liberty Mutual Lifting Formula,
        147–48;
    NIOSH Lifting Equation, 150–59;
    preparation, 148;
    RULA assessments tools, 146–47;
training:
    and education, 128–29;
    general, 129;
    job-specific, 129;
    management briefing, 130;
    for Supervisors, 130;
    support, 130;
and workplace hazards, 139–40;
    controlling, 140–43;
    risk factors, 140
ergonomics expert, 138–39
*Ergonomics Program Management
    Guidelines for Meatpacking
    Plants,* 132
EUAC. *See* equivalent uniform
    annualized cost (EUAC)

evacuation procedures, and emergency
    response planning, 365
event systems, 200–3
explosion disasters, 350
extremely hazardous substances (EHS),
    388, 390, 391
extreme weather, 354
Exxon Valdez spill, 396
eye protection, 262–63

FAA. *See* Federal Aviation
    Administration (FAA)
FAA Safety Management Systems Final
    Rule, 345
facility management, and PSM, 169–70
facility maps, and data location, 363
failure modes effect analysis (FMEA),
    196–97
fault hazard analyses (FHAs), 198
fault tree analysis (FTA), 196, 204–7
Federal Aviation Administration (FAA),
    330
Federal Coal Mine Health and Safety
    Act of 1969, 284
Federal Emergency Management
    Agency (FEMA), 357
FEMA. *See* Federal Emergency
    Management Agency (FEMA)
FHAs. *See* fault hazard analyses (FHAs)
fiberglass reinforced plastic (FRP), 395
FILI. *See* Frequency-Independent
    Lifting Index (FILI)
financial statements, 254–56;
    cash flow analysis, 255–56;
    income statement, 255
fire agencies, 372
fire disasters, 350
FIRWL. *See* Frequency-Independent
    Recommended Weight Limit
    (FIRWL)
fishbone diagram, 226
flood disasters, 350
FMEA. *See* failure modes effect
    analysis (FMEA)
Frequency-Independent Lifting Index
    (FILI), 151

Frequency-Independent Recommended Weight Limit (FIRWL), 151
front loading, 261
FRP. *See* fiberglass reinforced plastic (FRP)
FTA. *See* fault tree analysis (FTA)
FTE. *See* full-time equivalent (FTE)
full-time equivalent (FTE), 15
functional committee, and emergency response planning, 371–72
future value, 245

GCWR. *See* gross combined weight rating (GCWR)
GHS. *See* Globally Harmonized System of Classification and Labeling of Chemicals (GHS)
Globally Harmonized System of Classification and Labeling of Chemicals (GHS), 302, 389
gross combined weight rating (GCWR), 287
gross vehicle weight rating (GVWR), 287
Gulf of Mexico–Deepwater Horizon, 396
GVWR. *See* gross vehicle weight rating (GVWR)

hardware requirements, and job processes, 100–2;
 fabrication, 101;
 installation, 101;
 life cycle study, 101;
 occupancy-use, 101–2;
 operation, 102
hazard assessment, and emergency response planning, 363
hazard communication program, 110
hazard control, 103, 194–95
hazard identification, 103
hazardous chemical inventory, 391–97
hazardous materials, 288, 353–54;
 and environmental safety, 265–66
Hazardous Materials Table (HMT), 288
hazardous substance (HS), 288, 391

hazardous waste (HW), 288;
 and CERCLA, 381, 384–85;
 emergency planning, 390;
 emergency reporting, 390–91;
 enforcement and liability, 394–95;
 environmental law issues, 396–97;
 and environmental safety, 266;
 federal legislation for, 383;
 hazardous chemical inventory, 391–97;
 legal aspects for compliance, 382;
 and OPA, 396;
 and OSHA, 386;
 overview, 379–82;
 and RCRA, 381, 383–85;
 and remedial actions, 397–398;
 and SARA, 386–87;
 and SDS, 389–90;
 social and legal influences, 397–399;
 and trade secrets, 394;
 and underground storage tanks (USTs), 395
hazard prevention and control, 135–36
hazard risk index (HRI), 194
hazard severity, 192–94
Health and Human Services (HHS), 281
health care provider responsibilities, and ergonomics, 142
Health Insurance Portability and Accountability Act (HIPAA), 281, 327
health physicists, 307
hearing conservation, 263–64
Herzberg, Frederick, 69
HIPAA. *See* Health Insurance Portability and Accountability Act (HIPAA)
HMT. *See* Hazardous Materials Table (HMT)
HRI. *See* hazard risk index (HRI)
HS. *See* hazardous substance (HS)
human behavior modification, 112–20;
 internal safety and health training, 117;
 program development, 113;
 requirements, 112;

responsibilities by operating function, 114–17;
safety and health staff as resources, 117–18;
special safety and health training, 119–20;
strategy, 118–19;
training program content, 113–14
hurricanes, 350
HW. *See* hazardous waste (HW)

ICRP. *See* International Commission on Radiological Protection (ICRP)
ICRU. *See* International Commission on Radiation Units and Measurements (ICRU)
ICS. *See* Incident Command System (ICS)
improvised scaffold, 30–31
Incident Command System (ICS), 295
incident investigation, 172
income statement, 255
industrial hygiene:
art and science of, 300–2;
case study, 295–96;
environmental factors and stressors, 299–300;
overview, 296–298;
safety and health team responsibilities, 298–99;
and workplace safety, 263
industrial safety, and PSM, 169
injury and accident statistics, 278–79
innovation, 26
integration, 27
intermittent audits, 87
internal safety and health training, 117
International Atomic Energy Agency (IAEA), 308–9
International Commission on Radiation Units and Measurements (ICRU), 308–9
International Commission on Radiological Protection (ICRP), 308–9
international standards, for radiation safety, 308–9

job breakdown, 103
job safety analysis (JSA), 56, 102–4
job selection, 103
job-specific training, 129
JOEH. See Journal of Occupational and Environmental Hygiene (JOEH)
Journal of Occupational and Environmental Hygiene (JOEH), 14
JSA. *See* job safety analysis (JSA)

labor legislation:
common law doctrine and worker injury suits, 7–8;
and employee benefit association, 7;
mine safety inspections, 5;
and on and off job safety, 15;
powered machinery and child labor, 4;
and safety organizations, 13–14;
and United States, 14–15;
and U.S. Movement, 6–7;
women labor force, 4–5;
and workers compensation, 8–10;
workshop safety in 1800s, 5–6
latent errors, 24
leading indicators, 279
LEPC. *See* Local Emergency Planning Committee (LEPC)
Liberty Mutual Lifting Formula, 147–48
life cycles, and systems safety, 189–207;
concept phase, 189–90;
confidence intervals, 200;
construction phase, 190;
cut sets, 203–4;
definition phase, 190;
event systems, 200–3;
failure modes effect analysis (FMEA), 196–97;
fault hazard analyses (FHAs), 198;

fault tree analysis (FTA), 196, 204–7;
hazard control, 194–95;
hazard severity, 192–94;
operating and support hazard analysis (OaSHA), 197;
operation phase, 191;
parallel systems, 202–3;
preliminary hazard analysis (PHA), 195;
risk analysis, 194;
risk reduction, 197;
series system, 200–1;
statistical tools, 198–200;
subsystem hazard analysis (SSHA), 196;
system hazard analysis (SHA), 196;
termination phase, 192
line managers, 215
Local Emergency Planning Committee (LEPC), 347, 373, 390
lockout/tagout, 111
Log and Summary of Occupational Injuries and Illnesses, 275
loss control concept, 24
Lost-Time Accidents (LTAs), 252
LTAs. *See* Lost-Time Accidents (LTAs)

management commitment, of ergonomics, 133
management inefficiencies, and accidents, 212
management of change (MOC), 171, 176–78
Manhattan Project, 306
marine pollutant, 288–89
market standing, 25
Maslow, Abraham, 68–69
Material Safety Data Sheets (MSDS), 389
McClelland, David, 69
McGregor, Douglas, 72
mechanical integrity, 171
Medical Health Physicist (MHP), 308
medical management:
and ergonomics, 142;
program, 136–38
medical records, 281–82
medical services, and emergency response planning, 372–73
medical treatment, 278
MHP. *See* Medical Health Physicist (MHP)
Mine Safety and Health Act (MSHA), 272, 282–85
mine safety inspections, 5
mishaps. *See* accidents
mission statement, 360–61
MOC. *See* management of change (MOC)
morale, and accidents, 212–13
motivation-hygiene theory, 69–70
MSDs. *See* musculoskeletal disorders (MSDs)
MSDS. *See* Material Safety Data Sheets (MSDS)
MSHA. *See* Mine Safety and Health Act (MSHA)
musculoskeletal disorders (MSDs), 124–27;
back injuries, 127;
carpal tunnel syndrome, 124–25;
De Quervain's tenosynovitis, 124;
rotator cuff tendonitis, 126;
tendonitis and tenosynovitis, 125;
thoracic outlet syndrome, 126;
trigger finger, 127;
wrist ganglion, 126–27
mutual aid association, and emergency response planning, 376

NAICS. *See* North American Industry Classification System (NAICS)
National Contingency Plan (NCP), 376, 385
National Council on Radiation Protection and Measurements (NCRP), 309, 312
National Fire Academy (NFA), 357
National Fire Protection Association (NFPA), 84, 346
National Fire Safety Month, 339

National Highway Transportation Administration, 336
National Institute for Occupational Safety and Health (NIOSH), 340, 353
National Priorities List (NPL), 385
National Response Center (NRC), 290, 387, 388, 391
National Response Framework (NRF), 376
National Response System (NRS), 373
National Response Team, 376
National Safety Council (NSC), 13, 335
National Training and Education Division (NTED), 357
NCP. See National Contingency Plan (NCP)
NCRP. See National Council on Radiation Protection and Measurements (NCRP)
Network of Employers for Transportation Safety, 336
NFA. See National Fire Academy (NFA)
NFPA. See National Fire Protection Association (NFPA)
9/11 World Trade Center disaster, 324
NIOSH. See National Institute of Occupational Safety and Health (NIOSH)
NIOSH Lifting Equation, 150–59;
 for single-task lifting analysis, 155–59;
 using, 152–55
North American Industry Classification System (NAICS), 393
NPL. See National Priorities List (NPL)
NRC. See National Response Center (NRC); Nuclear Regulatory Agency (NRC)
NRF. See National Response Framework (NRF)
NRS. See National Response System (NRS)
NSC. See National Safety Council (NSC)

NTED. See National Training and Education Division (NTED)
Nuclear Regulatory Agency (NRC), 308

OaSHA. See operating and support hazard analysis (OaSHA)
occupational illness, 278
occupational injury, 278
occupational medical team:
 case study, 319–20;
 elements of, 321–22;
 employee wellness programs, 322–23;
 and stress, 322–23, 332;
 evaluations, 330–34;
 screenings, 332–33;
 surveillance, 333–34;
 overview, 320–21;
 role in emergency response, 334–41;
 off-the-job safety, 335–41;
 surveillance team, 323–30;
 functions of, 328–30;
 management of, 326–28;
 overview, 323–25
Occupational Safety and Health Act (OSHA), 165, 298, 324;
 and hazardous waste, 386;
 and records/recordkeeping, 273–74;
 and SMS, 47–49;
 coordination and communication on multiemployer worksites, 51–52;
 hazard prevention and control, 50–51;
 management commitment and employee involvement, 49–50;
 program evaluation and improvement, 51;
 safety and health education and training, 51;
 worker participation, 50;
 worksite analysis, 50
occupational safety systems, 187–89
off-the-job safety, 15, 335–41
Oil Pollution Act (OPA), 396
oil spill control, 266

Oil Spill Liability Trust Fund, 396
operating and support hazard analysis (OaSHA), 197
operating procedures, 171
operational program:
    external exposure control, 312–13;
    internal exposure control, 313–14
order-of-magnitude methodology, 237–38
OSHA. *See* Occupational Safety and Health Act (OSHA)

parallel systems, 202–3
PBS. *See* performance-based safety (PBS)
PELs. *See* Permissible Exposure Limits (PELs)
people requirements, and job processes, 97–100;
    evaluation, 100;
    personnel training, 99;
    staffing, 98–99;
    testing and qualification, 99–100
performance-based safety (PBS), 191
periodic audits, 87
Permissible Exposure Limits (PELs), 49
permit-required confined space plan, 111
personal protection equipment (PPE), 110–11
personnel assignment, and emergency response planning, 368–69
personnel needs, 68–71
PHA. *See* preliminary hazard analysis (PHA)
pilot-error syndrome, 222
potentially responsible parties (PRPs), 385
powered machinery, and child labor, 4
PPE. *See* personal protection equipment (PPE)
pregnancy, and radiation safety, 315
preliminary hazard analysis (PHA), 56–57, 195
present worth, 243–44

process hazard analysis (PHA), 171, 172
process safety information, 171
process safety management (PSM):
    case study, 165;
    definition of, 168;
    elements of, 170–74;
    enforcement trends, 174–76;
        car seal program, 176;
        work system permit, 175–76;
    and facility management, 169–70;
    historical background, 166–68;
    overview, 165–66;
    system and industrial safety, 169;
    systems approach to, 178–81;
    trade secrets, 174
process safety reviews, 172
productivity, 25
product safety, 266–67
profitability, 25
PRPs. *See* potentially responsible parties (PRPs)
psychological wellness, 322
public concern, and accidents, 213

quantitative methods, for ergonomics, 146–59;
    data collection, 148–50;
    Liberty Mutual Lifting Formula, 147–48;
    NIOSH Lifting Equation, 150–59;
    preparation, 148;
    RULA assessments tools, 146–47

radiation disasters, 354
radiation protection technologist (RPT), 308
radiation safety:
    education and training programs, 316–17;
    elements of, 310–15;
        administrative program, 310–12;
        environmental discharges, 314;
        monitoring, 314–15;
        operational program, 312–14;
    international standards, 308–9;

overview, 305–6;
pregnancy and, 315;
professionals, 306–8;
  health physicists, 307;
  medical physicist, 308;
  radiation protection technologist (RPT), 308;
recordkeeping, 315–16;
U.S. Standards, 309;
and workplace safety, 263
radioactive contamination, 313
rate of return, 246–48
RCRA. *See* Resource Conservation and Recovery Act (RCRA)
Reason, James, 23, 219
Recommended Weight Limit (RWL), 150
records/recordkeeping:
  case study, 271–72;
  and Department of Transportation (DOT), 286–92;
  description, 272–74;
  injury and accident statistics, 278–79;
  location, retention, and access to, 279–80;
  medical records, 281–82;
  and OSHA, 273–74;
  overview, 272;
  quick reference, 279;
  and radiation safety, 315–16;
  and SMS, 292;
  types and forms of, 274–82
reinforcement, 71
replacement analysis, 248–50
Research and Special Programs Administration (RSPA), 287
Resource Conservation and Recovery Act (RCRA):
  versus CERCLA, 385–86;
  and hazardous waste, 381, 383–85
respiratory protection, 265
revitalization, 26–27
risk analysis, and systems safety, 194
Risk Management Program (RMP), 347
risk mapping:

application example, 239–43;
order-of-magnitude methodology, 237–38;
overview, 234–36;
risk determination, 238–39;
scope of, 236–37
risk reduction, 197
RMP. *See* Risk Management Program (RMP)
rotator cuff tendonitis, 126
RPT. *See* radiation protection technologist (RPT)
RSPA. *See* Research and Special Programs Administration (RSPA)
RULA assessments tools, 146–47
RWL. *See* Recommended Weight Limit (RWL)

SA. *See* safety assurance (SA)
SAAs. *See* Satellite Accumulation Areas (SAAs)
Safe Drinking Water Act, 309
safe performance, and SMS, 71–73;
  discipline, 72–73;
  trust, 73
safety and health:
  alignment of objectives, 25–32;
    accidents and hazards, 28–29;
    and board of directors, 31–32;
    hypothesis, 29–31;
  as avoidance of loss, 23–24;
  case study, 1, 17;
  consultants, 43–44, 120–21;
  evolution of, 2–3;
  importance of, 17–18;
  as interdisciplinary functions, 18;
  and loss control concept, 24;
  as management function, 19–24;
    evaluation, 21–23;
  management program, 267–69;
  During the Middle Ages, 3;
  on and off job, 15;
  overview, 1–2;
  planning function, 35–37;

414   *Index*

policy and objectives, 37–41;
responsibility, authority, and accountability, 41–43;
symptoms of management failure, 18–19;
uncertain times of, 10–13;
and United States, 14–15
safety and health expenditures:
administration, 267;
and budgeting, 256–57;
development of, 259–62;
optimum, 257–59;
practical reasons for, 257;
time factors in, 257;
total effort, 261–62;
environmental safety:
hazardous materials, 265–66;
hazardous waste, 266;
oil spill control, 266;
financial statements, 254–56;
cash flow analysis, 255–56;
income statement, 255;
product safety, 266–67;
risk mapping:
application example, 239–43;
order-of-magnitude methodology, 237–38;
overview, 234–36;
risk determination, 238–39;
scope of, 236–37;
time value of money:
cost-benefit analysis, 250–54;
depreciation, 245–46;
future value, 245;
overview, 243;
present worth, 243–44;
rate of return, 246–48;
replacement analysis, 248–50;
training:
construction safety, 264–65;
crane safety, 265;
respiratory protection, 265;
workplace safety, 262–64;
eye protection, 262–63;
hearing conservation, 263–64;
industrial hygiene programs, 263;

radiation safety, 263
safety and health program controls:
and audits, 85–89;
authority, 87;
management, 86–87;
process, 86;
steps in making, 88–89;
types of, 87–88;
case study, 79;
committees, 89–93;
communications and information systems, 80–84;
and feedback, 82–84;
roles and responsibilities, 80–81;
and security, 81–82;
measurement standards, 84–85;
overview, 79–80
safety and health team:
responsibilities of:
employee, 299;
management, 298–99;
supervisor, 299;
responsibility, authority, and accountability, 41–43;
as training resources, 117–18
safety assurance (SA), 57–61
safety culture, and SMS, 54
safety data sheets (SDSs), 363;
and and hazardous waste, 389–90
safety errors, 19
safety management system (SMS), 298, 320;
ANSI Z10, 52–54;
case study, 47, 67;
and OSHA regulations, 47–49;
coordination and communication on multiemployer worksites, 51–52;
hazard prevention and control, 50–51;
management commitment and employee involvement, 49–50;
program evaluation and improvement, 51;
safety and health education and training, 51;

Index 415

worker participation, 50;
worksite analysis, 50;
overview, 67–68;
personnel needs, 68–71;
and records/recordkeeping, 292;
responsibilities, 73–76;
and safe performance, 71–73;
   discipline, 72–73;
   trust, 73;
and safety assurance (SA), 57–61;
and safety culture, 54;
and safety policy, 54–55;
safety promotion, 61–64;
and safety risk management (SRM):
   brainstorming, 56;
   checklists, 55–56;
   job safety analyses, 56;
   observation and experience, 56;
   preliminary hazard analysis (PHA), 56–57;
worker involvement, 76
safety organizations, 13–14. *See also specific organizations*
safety policy:
   and objectives, 37–41;
   and SMS, 54–55
safety professionals:
   and accidents, 215–16;
   role in planning, 43–44
safety program, 44–46
safety promotion, 61–64
safety risk management (SRM):
   brainstorming, 56;
   checklists, 55–56;
   job safety analyses, 56;
   observation and experience, 56;
   preliminary hazard analysis (PHA), 56–57
safe work practices/hot work permits, 171–72
SARA. *See* Superfund Amendments and Reauthorization Act (SARA)
Satellite Accumulation Areas (SAAs), 380
security, and emergency response planning, 372

SERC. *See* State Emergency Response Commission (SERC)
series system, and systems safety, 200–1
service, 25
SHA. *See* system hazard analysis (SHA)
shelter procedures, and emergency response planning, 365–66
SIC. *See* Standard Industrial Classification (SIC)
single-task lifting analysis, 155–59
SMS. *See* safety management system (SMS)
social contribution, 26
social influence, 26
Solid Waste Disposal Act, 395
SOPs. *See* standard operating procedures (SOPs)
SPCC. *See* spill prevention, containment, and countermeasure (SPCC)
spill prevention, containment, and countermeasure (SPCC), 266
SSHA. *See* subsystem hazard analysis (SSHA)
SSPP. *See* System-Safety Program Plan (SSPP)
staffing and job processes:
   case study, 95;
   decommission and disposal, 104;
   hardware requirements, 100–2;
      fabrication, 101;
      installation, 101;
      life cycle study, 101;
      occupancy-use, 101–2;
      operation, 102;
   interfacing elements, 106–7;
   job safety analysis (JSA), 102–4;
   modification, 104;
   overview, 95–96;
   people requirements, 97–100;
      evaluation, 100;
      personnel training, 99;
      staffing, 98–99;
      testing and qualification, 99–100;
   procedural requirements, 104–6
staff managers, and accidents, 216

stakeholders, and accidents, 213–16;
  line manager, 215;
  safety professional, 215–16;
  staff managers, 216;
  supervisor/foreman, 214–15
Standard Industrial Classification (SIC), 393
standard operating procedures (SOPs), 362
State Emergency Response Commission (SERC), 347, 373, 390
state of resources, 25
statistical tools, and systems safety, 198–200
stress, and employee wellness programs, 322–23, 332
subsystem hazard analysis (SSHA), 196
Superfund. *See* Comprehensive Environmental Response, Compensation, and Liability Act (CERCLA)
Superfund Amendments and Reauthorization Act (SARA):
  and CERCLA, 387–89;
  and hazardous waste, 386–87
Superfund and Atomic Energy Act, 309
supervisor/foreman, and accidents, 214–15
Supplementary Record, 275
Swiss-Cheese Model, 23
system and industrial safety, 169
system hazard analysis (SHA), 196
System-Safety Program Plan (SSPP), 186–87
system-safety tools, 186–87
systems safety management and engineering:
  case study, 185;
  and life cycles, 189–207;
    concept phase, 189–90;
    confidence intervals, 200;
    construction phase, 190;
    cut sets, 203–4;
    definition phase, 190;
    event systems, 200–3;

failure modes effect analysis (FMEA), 196–97;
fault hazard analyses (FHAs), 198;
fault tree analysis (FTA), 196, 204–7;
hazard control, 194–95;
hazard severity, 192–94;
operating and support hazard analysis (OaSHA), 197;
operation phase, 191;
parallel systems, 202–3;
preliminary hazard analysis (PHA), 195;
risk analysis, 194;
risk reduction, 197;
series system, 200–1;
statistical tools, 198–200;
subsystem hazard analysis (SSHA), 196;
system hazard analysis (SHA), 196;
termination phase, 192
occupational safety systems, 187–89
overview, 185–86
system-safety tools, 186–87

Take Safety Home, 340
task definition budgeting, 258–59
TEELs. *See* Temporary Emergency Exposure Limits (TEELs)
Temporary Emergency Exposure Limits (TEELs), 354
tendonitis, and tenosynovitis, 125
tenosynovitis:
  De Quervain's, 124;
  and tendonitis, 125
terrorism, 352
thermoluminescent dosimeter (TLD), 314
thoracic outlet syndrome, 126
threshold planning quantity (TPQ), 390
time value of money:
  cost-benefit analysis, 250–54;
  depreciation, 245–46;
  future value, 245;

overview, 243;
present worth, 243–44;
rate of return, 246–48;
replacement analysis, 248–50
TLD. *See* thermoluminescent dosimeter (TLD)
tornadoes, 350
Total Worker Health, 340
Toxicology Literature Online (TOXLINE), 353
Toxic Release Inventory (TRI), 392
TPQ. *See* threshold planning quantity (TPQ)
trade secrets:
  and hazardous waste, 394;
  and PSM, 174
Trade Secret Substantiation Form, 394
training and education:
  case study, 109;
  ergonomics, 128–29;
  general, 129;
  governmental requirements, 110–11;
  and human behavior modification, 112–20;
    internal safety and health training, 117;
    program development, 113;
    requirements, 112;
    responsibilities by operating function, 114–17;
    safety and health staff as resources, 117–18;
    special safety and health training, 119–20;
    strategy, 118–19;
    training program content, 113–14;
  job-specific, 129;
  management briefing, 130;
  overview, 109;
  of PSM, 172;
  purpose of, 110;
  safety and health consultants, 120–21;
  for Supervisors, 130;
  support, 130

Training Requirements in OSHA Standards, 279
transportation:
  and emergency response planning, 366;
  safety, 339
TRI. *See* Toxic Release Inventory (TRI)
trigger finger, 127

ndeclared hazardous material, 289
underground storage tanks (USTs), 395
United States:
  and labor legislation, 6–7;
  and safety and health, 14–15;
  workers compensation in, 8–10
U.S. Public Health Service (USPHS), 295
U.S. Standards, for radiation safety, 309
USPHS. *See* U.S. Public Health Service (USPHS)
USTs. *See* underground storage tanks (USTs)

West Fertilizer Plant fire and explosion, 346–47
willingness to pay (WTP), 250
window of opportunity, 24
women labor force, 4–5
work environment, 278
worker injury suits, 7–8
workers compensation, and labor legislation, 8–10
work/job design and redesign, 139
workplace hazards, and ergonomics, 139–40;
  controlling, 140–43;
    administrative, 141;
    assessment tools, 143;
    employer responsibilities, 142;
    health care provider responsibilities, 142;
    medical management, 142;
    treatment, 143;
    work practice, 141–42;
  risk factors, 140
workplace safety, 262–64:

eye protection, 262–63;
hearing conservation, 263–64;
industrial hygiene programs, 263;
radiation safety, 263
work practice controls, 141–42

work system permit, 175–76
wrist ganglion, 126–27
WTP. *See* willingness to pay (WTP)

X-Ray and Radium Protection, 309

About the Authors

Ted S. Ferry, Ed.D.
Ted S. Ferry was a distinguished author, educator, and safety specialist whose professional safety career spanned more than forty years. He was Emeritus Professor at the University of California, Los Angeles Institute of Safety and Systems Management, where he was instrumental in developing safety curricula for the USC Bachelor of Science in Safety degree. Dr. Ferry was a Registered Professional Engineer (California), a Board-Certified Safety Professional, and a Certified Hazard Control Manager.

Mark A. Friend, Ed.D., CSP
Dr. Mark A. Friend is Professor of Doctoral Studies at Embry-Riddle Aeronautical University in Daytona Beach, Florida, with more than forty years in higher education. He is a Professional Member of the American Society of Safety Engineers (ASSE) where he helped establish the Academic Practice Specialty. He currently serves on the Judicial Commission of the Board of Certified Safety Professionals (BCSP), and he has worked as a safety consultant, trainer, expert witness, and author. His top-selling book in the field, *Fundamentals of Occupational Safety and Health*, is currently in its sixth edition. He has received the *PPG Safety Educator of the Year*, the National Safety Management Society/West Virginia University *Alumni Achievement Award*, the ASSE *Culbertson Award for Volunteer Service*, and the *ASSE Academic Practice Specialty Safety Professional of the Year Award*. He and his wife Kathy live in New Smyrna Beach, Florida.

Kathy S. Friend, MSOSM
Ms. Kathy Friend is Director of Enrollment Operations for the Worldwide Campus of Embry-Riddle Aeronautical University. She is a member of the

ASSE and has a background in operations, safety, and personnel management from both a civilian and a military perspective. As a U.S. Air Force veteran, and later DOD civil service employee, Ms. Friend worked in Intelligence, and Military Personnel. She's currently employed in as Director, Enrollment Operations, for a university and has many years of management-related experience. She serves as a Course Facilitator for Eastern Kentucky University in the Master of Science Safety, Security, and Emergency Management Program in the College of Justice and Safety. She has presented at the national conferences of the National Safety Council and American Society of Safety Engineers. Kathy Friend holds a B.S. degree in Technical Management and an M.S. degree in Occupational Safety Management from Embry-Riddle Aeronautical University.

Kimberlee K. Hall, Ph.D.
Dr. Kimberlee Hall is Assistant Professor in the Environmental Health Program in the School of Health Sciences at Western Carolina University. Her responsibilities include teaching undergraduate courses in water quality control, food sanitation and protection, and institutional and residential environment health and safety. Additionally, her research interests include the evaluation, development, and modeling of fecal pollution and source indicators in surface water for the development of microbial risk assessment models. She holds a B.S. degree in Ecology and Environmental Biology from Appalachian State University and a Ph.D. in Environmental Health Sciences from East Tennessee State University.

Mark D. Hansen, CSP, PE, CPEA, CPSA, CPE
Mark Hansen is the Vice President, Risk Management and HSE, for Osburn Contractors, Inc., in Garland, Texas. With over thirty years of experience in the field, he is a licensed professional engineer (Texas) and a Certified Safety Professional (CSP). Mr. Hansen has a B.S. degree in Psychology and an M.S. degree in Industrial Engineering, specializing in Safety, both from Texas A&M University. He is currently the Past-President and fellow of the ASSE. He has authored over 150 technical publications, including his third book on career development entitled *Climbing the Corporate Ladder-Safely!* (May 2011), where some of the materials presented are discussed in detail. His other books are entitled *Out of the Box: Skills for Developing Your Own Career Path* (April 2002); *Out of the Box: More Skills for Developing Your Own Career Path* (May 2010); and most recently, *Climbing the Corporate Ladder-Safely* (March 2011).

Burton R. Ogle, Ph.D., CIH, CSP
Dr. Burton Ogle is a tenured Professor of Environmental Health at Western Carolina University. He is a Certified Industrial Hygienist and Certified

Safety Professional. He served on the North Carolina Board of Registered Environmental Health Specialist for the past twelve years. In addition to his sixteen years of academic experience, Dr. Ogle has been awarded a faculty fellowship at Oak Ridge National Laboratory—Center for Nanophase Materials Sciences each summer since 2004 to focus on the environment, safety, and health management of nanoscale material research and its subsequent applications. This research has resulted in guidance documents for the Department of Energy as well as scientific papers. Dr. Ogle is an active member of the American Industrial Hygiene Association, American Society of Safety Engineers, and the National Environmental Health Association.

Michael F. O'Toole, Ph.D.

Dr. Michael F. O'Toole is Associate Professor and Program Coordinator of the Aerospace and Occupational Safety Program at Embry-Riddle Aeronautical University where he teaches both graduate and undergraduate courses in safety and health. He is also the owner of O'Toole & Associates and works with a variety of businesses, industries, trade associations, and governmental agencies around the world. Prior to his current position, he was the Director, Safety and Health for Aggregate Industries located in Rockville, Maryland. From 1997 until 2006, he taught at Purdue University Calumet where he was Associate Professor of Organizational Leadership and Supervision. Prior to his academic career, he held a variety of Human Resource and Corporate Safety positions over a twenty-year career with USG Corporation. Dr. O'Toole received a B.A. degree in Behavioral Psychology and an M.A. degree in Industrial Psychology from Western Michigan University, an M.S. degree from Northern Illinois University in Safety Engineering Technology and a Ph.D. from the University of Illinois at Chicago in Public Health Policy Administration.

Henry A. Walters, Ph.D., M.B.A.

Dr. Henry Walters retired after working over forty years in a myriad of fields which always included aspects of safety. He left the U.S. Army to work in the aerospace industry for Link Flight Simulation where his duties included incorporating safety and ergonomics into the development of new simulator systems for military applications. After serving as Assistant Professor in Occupational Safety and Health at both Murray State and Indiana State Universities where he taught primarily System Safety Engineering and Ergonomics at both the undergraduate and graduate levels, Dr. Walters concluded his professional career as a Senior Consultant with Behavioral Science Technology. During his career, he authored numerous professional articles, chapters for two books, and a book entitled *Statistical Tools of Safety Management*. Throughout his career he was active in the American Society of Engineers and the National Safety Council. He holds a B.S. degree in Engineering from the U.S. Military Academy, an M.B.A. from Rollins College, and a Ph.D. from the Union Institute.

Tracy L. Zontek, Ph.D., CIH, CSP
Dr.Tracy Zontek is a tenured, Associate Professor of Environmental Health at Western Carolina University. She is also an Adjunct Professor at Embry-Riddle University in the Master of Science Occupational Safety Management program. She is a Certified Industrial Hygienist and a Certified Safety Professional. In addition to her academic experience, Dr. Zontek has been selected as Faculty Fellow at Oak Ridge National Laboratory—Materials Science and Technology Division each summer since 2006 to focus on the environment, safety, and health management of nanoscale material research and its subsequent applications. This research has resulted in guidance documents for the Department of Energy as well as scientific papers. Dr. Zontek is an active member of the ASSE and the National Environmental Health Association.

Special thanks go to **Marisa Aguiar**, a Ph.D., a residential student within the Department of Doctoral Studies at Embry-Riddle Aeronautical University. She received her bachelors in Psychology from the University of Central Florida in 2014. Ms. Aguiar helped edit the final text.

Authors from previous text, originally authored and edited by Ted S. Ferry, include the following, along with their affiliations at the time of the publication:

    Dr. Charles I. Barron, University of Southern California
    Lewis Bass, attorney and president of Kairos Company
    Judith Anne Erickson, University of Southern California
    Dr. Leo Greenberg, writer and consultant in Ra'anana, Israel
    Dr. Earl D. Heath, P.E., C.S.P. formerly with New York University and Johns Hopkins University
    Raymond Kuhlman, C.S.P., OSHA Training Institute
    Jerome F. Lederer, formerly with U.S. Airmail Service
    Charles B. Mitchell, Citation Insurance Company
    Dr. Julius Morris, C.S.P., International Welfare Associates
    Dr. James Pierce, University of Southern California
    Dr. Jack ReVelle, Hughes Aircraft Company
    Dr. Walter F. Wegst, University of Southern California
    Eric Wigglesworth, Menzies Foundation